Frontiers of Astrophysics

Frontiers of Astrophysics

Eugene H. Avrett, editor

Harvard University Press
Cambridge, Massachusetts
and London, England
1976

Library of Congress Cataloging in Publication Data
Main entry under title:

Frontiers of astrophysics.

Includes bibliographical references and index.
1.   Astrophysics.   I.   Avrett, E. H.
QB461.F79   523.01   76-10135
ISBN 0-674-32659-8
ISBN 0-674-32660-1 pbk.

Preface

Astrophysics is a science that uses physics to interpret astronomical events. This field is now rapidly developing; each year brings an increased number of significant and exciting discoveries based on data from spacecraft, aircraft, rockets, balloons, and ground-based observatories. Thus there is need for a current, intermediate-level text that will introduce the interested reader to the frontiers of astrophysical research. This book has been written to meet such a need.

The manuscript was drafted between October 1975 and January 1976 and used during that period as part of a tutorial course for undergraduates in astronomy at Harvard. Each of the twelve chapters was prepared by a specialist, actively engaged in a particular branch of astrophysical research, who describes and evaluates the most recent work in his field. Although the chapters are uniform in purpose, they reflect the different requirements of each subject and the individual style of each author.

The book was written on the assumption that the reader has a good knowledge of basic physics. It can be used as a supplement for the advanced student in an introductory-level astronomy course or as the textbook in an upper-level course. It is also intended as a readable introduction to astrophysical research for those studying physics and astronomy at the graduate and professional level.

The text starts with our solar system and its central star, proceeds to other stellar and interstellar phenomena, and ends with extragalactic astronomy and cosmology. This is an arbitrary sequence; the individual chapters are units in themselves and may be read in any order.

For their contributions to the final draft of this book I thank John Mariska and Stephen Perrenod, teaching fellows with the Harvard Astronomy Department, and a number of undergraduates: Michael Brown, David Erb, Arthur Goldberg, William Kraus, Matthew Malkan, Bruce Rosen, Gary Taubes, and Andrew Zachary. I also thank Donald Cooke for his help with the manuscript and the many illustrations.

We are extremely grateful to Professor William A. Fowler of the California Institute of Technology for reading the manuscript; his comments and suggestions have led to many improvements.

Eugene H. Avrett
Lecturer on Astronomy,
Harvard University
Physicist, Smithsonian
Astrophysical Observatory

Contents

Frontiers of Astrophysics

1

The Formation of
the Solar System

William R. Ward

Of the topics discussed in this volume, the problem of the formation of the solar system is probably the oldest. Indeed, current theories owe much to the seventeenth-century writings of René Descartes, who proposed what we now refer to as the nebular hypothesis, i.e., that the planets and satellites formed out of a rotating disk of gas and dust surrounding the primordial sun. Most contemporary work concerns some variation of this theme. And yet, like many of the other subjects to be discussed in this text, unusually rapid progress has been made in recent years. This is due, in a large part, to the explosive growth of solar-system information gained by the chemical analysis of meteorites and lunar samples, and by the modern use of planetary probes.

Several review articles treating various aspects of solar-system formation have been published in past years (e.g., Williams and Cremin, 1968; Woolfson, 1969; Anders, 1971; Grossman and Larimer, 1974). Recent detailed models have been proposed by Cameron (1973), Cameron and Pine (1973), Alfvén and Arrhenius (1970a, 1970b, 1973, 1974), and Öpik (1973). An excellent book on this subject has been written by the Russian astronomer V. S. Safronov (1969) and a symposium on the origin of the solar system was held in Nice in 1972, the proceedings of which have been published (Reeves, 1972). A fairly comprehensive list of research papers covering the last four years has been prepared by Ward (1975).

Rather than attempt the monumental task of providing a complete picture of the solar-system formation process, I have chosen the less ambitious but more tractable approach of presenting an introductory look at three important areas of current research: (1) chemical condensation in the solar nebula, (2) the accumulation of solid material by gravitational instabilities, and (3) the dynamics of planetary accretion. The calculations of chemical condensation involve determining what minerals condense from a cooling gas of solar composition. These calculations can be compared with the chemical analyses of meteorites in an attempt to ascertain the conditions in the nebula at the time of their formation. However, direct chemical condensation can produce objects no larger than centimeters in size. The growth of larger objects from such small debris has long been one of the most puzzling questions regarding the accumulation of solid material. A mechanism which can apparently accomplish this involves the development of gravitational instabilities in a thin disk of condensed debris. On the other hand, it seems unlikely that this process can produce objects much in excess of a few kilometers in radius—still far short of planetary dimensions. The final stage of accumulation seems to require mutual collisions between such planetesimals at low enough velocities to favor net accretion over erosion, a situation that is thought to be generated by the gravitational relaxation of the planetesimal disk.

We turn now to a discussion of these events in more detail.

Sequence of Chemical Condensation

One of the most fruitful areas of research on the origin of the solar system in recent years has been that of cosmochemistry, particularly in the establishment of a mineral condensation sequence for the cooling solar nebula. Table 1-1 lists the twenty most abundant elements relative to silicon in the nebula, which is assumed to be of solar composition. (Helium is ignored because it is chemically inert.) From such a gas, solids will begin to condense when the temperature drops below about ~ 1700 K. Which minerals condense and exactly at what temperatures must be determined from the details of the chemical kinetics. We shall briefly discuss the calculation methods employed by various researchers (e.g., Lord, 1965; Larimer, 1967; Lewis, 1972a; Grossman, 1972) and then describe some of the results.

Since the principal constituent of a gas of solar composition is molecular hydrogen, the partial pressure, p_{H_2}, of hydrogen roughly determines the total pressure, p. The number density, N_{H_2}, of hydrogen molecules is, by virtue of the ideal gas law,

$$N_{H_2} \simeq p/kT, \tag{1-1}$$

k being Boltzman's constant and T the temperature. Because molecular hydrogen is diatomic, the number density of hydrogen atoms is

Table 1-1 Gaseous species contributing more than 10^{-7} of the total moles of their common constituent element between 2000 K and 1200 K.

Element	Abundance (Si = 10^6)	Gaseous species
Hydrogen	2.6×10^{10}	H_2, H, H_2O, HF, HCl, MgH, HS, H_2S, MgOH
Oxygen	2.36×10^7	CO, SiO, H_2O, TiO, OH, HCO, CO_2, PO, CaO, COS, MgO, SiO_2, AlOH, SO, NaOH, MgOH, PO_2, $Mg(OH)_2$, AlO_2H
Carbon	1.35×10^7	CO, HCN, CS, HCO, CO_2, COS, HCP
Nitrogen	2.44×10^6	N_2, HCN, PN, NH_3, NH_2
Magnesium	1.05×10^6	Mg, MgH, MgS, MgF, MgCl, MgO, MgOH, $Mg(OH)_2$
Silicon	1.00×10^6	Si, SiS, SiO, SiO_2
Iron	8.90×10^5	Fe
Sulfur	5.06×10^5	SiS, CS, S, HS, H_2S, PS, AlS, MgS, NS, S_2, COS, SO, CS_2, SO_2
Aluminum	8.51×10^4	Al, AlH, AlF, AlCl, AlS, AlO, Al_2O, AlOH, AlOF AlO_2H
Calcium	7.36×10^4	Ca, CaF, CaO, $CaCl_2$
Sodium	6.32×10^4	Na, NaH, NaCl, NaF, NaOH
Nickel	4.57×10^4	Ni
Phosphorus	1.27×10^4	P, PN, PH, P_2, PH_2, PS, PO, PH_3, PO_2, HCP
Chromium	1.24×10^4	Cr
Manganese	8800	Mn
Fluorine	3630	HF, AlF, CaF, F, MgF, NaF, NF, KF, PF, CaF_2, AlOF, TiF_2, MgF_2, MgClF, TiF
Potassium	3240	K, KH, KCl, KF, KOH
Titanium	2300	Ti, TiO, TiF_2, TiO_2, TiF
Cobalt	2300	Co
Chlorine	1970	HCl, Cl, AlCl, NaCl, KCl, MgCl, $CaCl_2$, $MgCl_2$, AlOCl, MgClF

From Grossman (1972).

$N_{\text{H, TOTAL}} \approx 2 N_{\text{H}_2}$. If α_i is the abundance of element i relative to hydrogen, then

$$N_{i, \text{TOTAL}} \simeq 2\alpha_i p/kT. \qquad (1\text{-}2)$$

Element i will, in general, reside in many gaseous molecular species j. Table 1-1 shows the most important species according to Grossman (1972). For each element there is a mass balance equation which equates the expression (1-2) to the sum of the contributions of each molecular species. In terms of the partial pressures, $p_{ij} = N_{ij}kT$; the mass balance equations can be written as

$$2\alpha_i p = \sum_j \gamma_{ij} p_{ij}, \qquad (1\text{-}3)$$

where γ_{ij} is the number of atoms of element i in molecule j. In the case of nitrogen, for example,

$$2\alpha_N p = 2p_{N_2} + p_{HCN} + p_{PN} + p_{NH_3} + p_{NH_2} + p_N + \left(\begin{array}{c}\text{small}\\\text{contributions}\end{array}\right). \quad (1\text{-}4)$$

The partial pressures of the various molecular species are not independent but are related through various chemical equilibria. For instance, ammonia exists in equilibrium with its monatomic components,

$$N + 3H \rightleftharpoons NH_3, \quad (1\text{-}5)$$

and their relative partial pressures in atmospheres are related through an equilibrium constant K,

$$p_{NH_3}/p_N p_H^3 = K_{NH_3}. \quad (1\text{-}6)$$

The equilibrium constant is calculated from the change in the Gibbs free energy ΔG^0,

$$K_j = \exp(-\Delta G_j^0/RT), \quad (1\text{-}7)$$

where ΔG_j^0 is the difference in the free energies of the right side (products) and of the left side (reactants) at the standard pressure of one atmosphere and R is the universal gas constant equal to 1.987 cal K^{-1} mole^{-1}. These values can be determined experimentally in the laboratory. An important source of this information is the JANAF tables.

   All the molecular species exist in equilibrium with each other, this situation being maintained by the various possible chemical reactions. Of course, not all of these reactions furnish independent information since often one equilibrium constant can be expressed in terms of other equilibrium constants. An independent set of reactions chosen by Grossman are the molecular dissociation reactions like equation (1-5), which break down each species into its monatomic components. For each corresponding equilibrium equation the partial pressure of the molecule is expressed in terms of the partial pressures of its constituents, i.e., from equation (1-6), $p_{NH_3} = K_{NH_3}p_N p_H^3$. In this way all molecular species in the mass balance equations (e.g., equation 1-3) are eliminated, leaving twenty (one for each element in Table 1-1) simultaneous, nonlinear equations in the twenty unknown partial pressures of the monatomic component elements. These are then solved numerically. The most ambitious programs have considered nearly three hundred gaseous molecular species.

To determine when condensation will ensue from such a gas, the equilibrium constants for the decomposition reactions of various crystalline phases are used. For corundum, Al_2O_3,

$$Al_2O_3(s) \rightleftharpoons 2Al(g) + 3O(g) \qquad (1\text{-}8)$$

and

$$p_{Al}{}^2 p_O{}^3 / A(Al_2O_3) = K_{Al_2O_3}, \qquad (1\text{-}9)$$

where $A(Al_2O_3)$, the activity of corundum, is unity for pure crystalline phase. Whenever the quantity on the left, computed by substituting the appropriate calculated monatomic partial pressures, exceeds the equilibrium constant on the right, the mineral will condense.

Of course, once condensation ensues, the partial pressures of the various gaseous constituents are also altered. These must be solved for repeatedly as the temperature is lowered and more material condenses. In addition, the equilibrium constants are temperature-dependent, i.e., typically $\log K_j \propto 1/T$. Finally, solid material can continue to react with the vapor phase as in the case of Mg_2SiO_4, which converts at lower temperature to $MgSiO_3$ consuming gaseous Si in the process.

As an example, consider a gas at 500 K and a pressure of 10^{-4} atm composed of the three most abundant elements—H, C, and O—which are bound up in the molecules CO, CO_2, CH_4, H_2, and H_2O (see Kaula, 1968). We take as the important chemical reactions:

$$CO_2 + H_2 \rightleftharpoons CO + H_2O \quad (\Delta G_1{}^0 = 4.90 \text{ kilocal/mole}) \qquad (1\text{-}10)$$

$$CO_2 + 4H_2 \rightleftharpoons CH_4 + 2H_2O \quad (\Delta G_2{}^0 = -18.16 \text{ kilocal/mole}). \qquad (1\text{-}11)$$

Hence the partial pressures (in atmospheres) at equilibrium satisfy the relations

$$p_{CO} p_{H_2O} / p_{CO_2} p_{H_2} = e^{-\Delta G_1{}^0 / RT} = K_1, \qquad (1\text{-}12)$$

$$p_{CH_4} p_{H_2O}^2 / p_{CO_2} p_{H_2}^4 = e^{-\Delta G_2{}^0 / RT} = K_2, \qquad (1\text{-}13)$$

while the mass balance equations are

$$2p = 2p_{H_2} + 2p_{H_2O} + 4p_{CH_4}, \qquad (1\text{-}14)$$

$$2\alpha_0 p = p_{CO} + 2p_{CO_2} + p_{H_2O}, \qquad (1\text{-}15)$$

$$2\alpha_c p = p_{CO} + p_{CO_2} + p_{CH_4}. \qquad (1\text{-}16)$$

Treating p_{H_2} and p_{H_2O} as independent variables, equations (1-14)–(1-16) can be solved for $p_{CO_2}, p_{CO}, p_{CH_4}$ and substituted into equation (1-12), which becomes

$$p_{H_2O}^2 + \tfrac{1}{2}[pA + p_{H_2}(1 + \tfrac{3}{2}K_1)]p_{H_2O} + \tfrac{1}{4}K_1p_{H_2}[p_{H_2} - 2pB] = 0, \quad (1\text{-}17)$$

and into equation (1-13) divided by equation (1-12), which becomes

$$p_{H_2O}^2 - [p - p_{H_2} - 4K_3p_{H_2}{}^3] + 2K_3p_{H_2}{}^3[pA + p_{H_2}] = 0, \quad (1\text{-}18)$$

where $A = 4\alpha_c - 2\alpha_0 - 1$, $B = 2\alpha_0 - 2\alpha_c + 1/2$, and $K_3 = K_2/K_1$. Noting the relative abundances in Table 1-1, we expect $p_{H_2} \approx p$ and anticipate that $p - p_{H_2} = \Delta p$ will be a small quantity of the same order as the other partial pressures. Hence, we can rewrite equations (1-17) and (1-18) as

$$p_{H_2O}^2 + \tfrac{1}{2}[(4\alpha_c - 2\alpha_0 + \tfrac{3}{2}K_1)p - \Delta p]p_{H_2O}$$
$$- \tfrac{1}{4}K_1p[\Delta p + 4(\alpha_0 - \alpha_c)p] = 0, \quad (1\text{-}19)$$

$$p_{H_2O}^2 - (\Delta p - 4K_3p^3)p_{H_2O} + 2K_3p^3[(4\alpha_c - 2\alpha_0)p - \Delta p] = 0. \quad (1\text{-}20)$$

Numerically, $\alpha_c = 5.19 \times 10^{-4}$, $\alpha_0 = 9.08 \times 10^{-4}$, $K_1 = 7.21 \times 10^{-3}$, $K_3 = 1.20 \times 10^{10}$, and $p = 10^{-4}$ atm. Equation (1-20) can be written as

$$\Delta p = 2p_{H_2O} + 2.62 \times 10^{-8} + p_{H_2O}(p_{H_2O} - \Delta p)/2.40 \times 10^{-2}. \quad (1\text{-}21)$$

Since p_{H_2O} and Δp should be of the order $2\alpha_0 p \sim \mathcal{O}(10^{-7})$, the last term of equation (1-21) is negligible compared to the first. Combining equations (1-19) and (1-21) enables one to solve for p_{H_2O} and Δp. Equation (1-14) can then be used to find p_{CH_4}, while equations (1-12) and (1-13) yield p_{CO} and p_{CO_2}, respectively. The completion of this calculation has been left as an exercise for the reader.

Figure 1-1 shows the condensation sequence of a gas of full solar composition at 10^{-4} atm. The first condensates consist of a group of refractory trace elements such as osmium (Os), rhenium (Re), and zirconium (Zr) which condense well above the ~ 1680 K condensation temperature of Al_2O_3. Most of the Ca and all of the Ti have condensed as $CaTiO_3$ and $Ca_2Al_2SiO_7$ by 1500 K (common mineral names are listed in Table 1-2). Uranium (U), protactinium (Pa), thorium (Th), tantalum (Ta), and niobium (Nb) condense in solid solution in $CaTiO_3$. At 1387 K, $CaMgSi_2O_6$ appears, followed by metallic iron at 1375 K. Nickel and cobalt are carried with the iron. At 1370 K, Mg_2SiO_4 appears, condensing most of the Mg. However, as mentioned above, it later reacts with the vapor to form $MgSiO_3$, consuming the

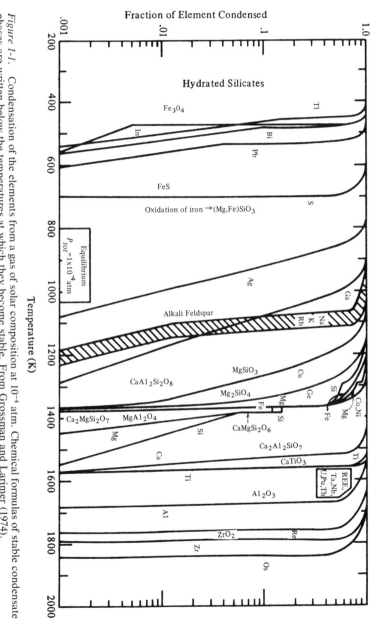

Figure 1-1. Condensation of the elements from a gas of solar composition at 10^{-4} atm. Chemical formulas of stable condensate phases are written below the temperatures at which they become stable. From Grossman and Larimer (1974).

Table 1-2   Chemical formulas of common minerals.

Solid solution series	Mineral	Formula
Olivine	Forsterite	Mg_2SiO_4
	Fayalite	Fe_2SiO_4
Pyroxene		
Orthorhombic	Enstatite	$MgSiO_3$
	Ferrosilite	$FeSiO_3$
Monoclinic	Diopside	$CaMgSi_2O_6$
	Tschermak's molecule	$CaAl_2SiO_6$
	Ti-Al-pyroxene	$CaTiAl_2O_6$
Plagioclase	Anorthite	$CaAl_2Si_2O_8$
feldspar	Albite	$NaAlSi_3O_8$
Melilite	Gehlenite	$Ca_2Al_2SiO_7$
	Akermanite	$Ca_2MgSi_2O_7$
	Soda-melilite	$CaNaAlSi_2O_7$
Nickel-iron		(Fe, Ni, Co, Cr)
	Corundum	Al_2O_3
	Perovskite	$CaTiO_3$
	Hibonite	$CaO \cdot 6Al_2O_3 \pm MgO, TiO_2$
	Spinel	$MgAl_2O_4$
	Magnetite	Fe_3O_4
	Schreibersite	$(Fe, Ni)_3P$
	Troilite	FeS
	Feldspathoids	
	Nepheline	$NaAlSiO_4$
	Sodalite	$3NaAlSiO_4 \cdot NaCl$
	Wollastonite	$CaSiO_3$
	Grossularite	$Ca_3Al_2Si_3O_{12}$
	Cordierite	$(Mg, Fe)_2Al_4Si_5O_{18}$

From Grossman and Larimer (1974).

remaining gaseous Si. At about 1200 K, sodium (Na), potassium (K) and rubidium (Rb) form solid solutions with previously condensed $CaAl_2Si_2O_8$. Metallic iron begins to oxidize below 750 K and troilite (FeS) becomes stable at 700 K. Lead (Pb), bismuth (Bi), indium (In), thallium (Tl) condense between 600 K and 400 K. Magnetite appears at 450 K and magnesium silicates become hydrated below 350 K. Many of these results are summarized in Table 1-3. (Also evident, in a comparison of Table 1-3 with Figure 1-1, is a slight dependence of condensation temperature on nebular pressure.)

Figure 1-2 (from Lewis, 1972a) shows the cumulative mass and density of the condensate. The situation has been continued below 200 K where condensation of water ice begins at ~ 180 K and is followed by the conversion of NH_3 to solid $NH_3 \cdot H_2O$ at ~ 110 K. Conversion of ice to solid $CH_4 \cdot 8H_2O$ does not occur until ~ 60 K.

The chemical properties of the condensate at any point in the nebula depend upon the local physical conditions, i.e., temperature, pressure, etc. These, in turn, are related to the structure of the solar nebula. Ideas on the solar nebula vary widely. For example, there is a minimum nebula mass necessary to account for the observed mass of the planetary system, which is on the order of a few percent of a solar mass. However, solar nebula models range from this minimum value up to several solar masses. The most detailed models yet constructed are those of Cameron and Pine (1973), who consider a two-solar-mass nebula. Figures 1-3 and 1-4 show the surface den-

Table 1-3 Stability fields of equilibrium condensates at 10^{-3} atm total pressure.

Phase		Condensation temperature (K)	Temperature of disappearance (K)
Corundum	Al_2O_3	1758	1513
Perovskite	$CaTiO_3$	1647	1393
Melilite	$Ca_2Al_2SiO_7–Ca_2MgSi_2O_7$	1625	1450
Spinel	$MgAl_2O_4$	1513	1362
Metallic iron	(Fe, Ni)	1473	
Diopside	$CaMgSi_2O_6$	1450	
Forsterite	Mg_2SiO_4	1444	
	Ti_3O_5	1393	1125
Anorthite	$CaAl_2Si_2O_8$	1362	
Enstatite	$MgSiO_3$	1349	
Eskolaite	Cr_2O_3	1294	
Metallic cobalt	Co	1274	
Alabandite	MnS	1139	
Rutile	TiO_2	1125[a]	
Alkali feldspar	$(Na, K)AlSi_3O_8$	~1000	
Troilite	FeS	700	
Magnetite	Fe_3O_4	405	
Ice	H_2O	≤200	

From Grossman (1972).

[a] Below this temperature, calculations were performed manually using extrapolated high-temperature vapor composition data. In some cases, gaseous species which had been very rare assumed major importance at low temperature (CH_4).

Figure 1-2. Cumulative mass and density for equilibrium condensation sequences at 10^{-6}, 10^{-4}, and 10^{-2} bar total pressure. Temperature scales are given for (A) 10^{-6} bar, (B) 10^{-4} bar, and (C) 10^{-2} bar pressure. The 15 steps indicated are (1) condensation of Ca, Al, and Ti silicates; (2) $MgSiO_3$ and Fe condensation; (3) conversion of H_2S to FeS; (4) conversion of CO to CH_4; (5) conversion of Fe to FeO; (6) conversion of enstatite to talc plus brucite; (7) conversion of N_2 gas to NH_3; (8) condensation of water ice; (9) conversion of NH_3 to solid $NH_3 \cdot H_2O$; (10) conversion of ice to solid $CH_4 \cdot 8H_2O$; (11) condensation of $CH_4(s)$; (12) condensation of Ar(s); (13) condensation of Ne(s); (14) condensation of $H_2(s)$; (15) condensation of He(s): Steps (3) and (5) always occur at 670 and 510 K, respectively, and should be read only on scale B. Steps (4) and (7) are not condensations, and are indicated on the temperature scales only. From Lewis (1972a).

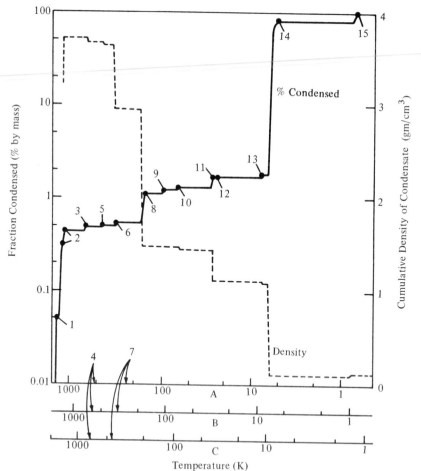

sity, temperature, and pressure at mid-plane for two such models. These models are the end product of the gravitational collapse of a spherical distribution of gas with an angular momentum per unit mass estimated from the assumption of fully developed turbulent motion. (Uniform and linear refer to two types of initial density distribution.)

Figures 1-5 and 1-6 (from Lewis, 1972b) compare uncompressed planetary densities with that expected of the total condensate at the various appropriate orbital distances in the solar nebula modeled by Cameron and Pine. Figure 1-5 shows an adiabat for the nebula superposed on the mineral stability field. The locations of planetary formation are indicated. Figure 1-6 shows the density of condensed material at 10^{-3} atm and compares this with the actual planetary densities. Good agreement is obtained from the adiabat of the Cameron-Pine nebula.

However, even as this is being written, these models are being drastically

Figure 1-3. The surface densities for the two cases of the primitive solar nebula models. From Cameron and Pine (1973).

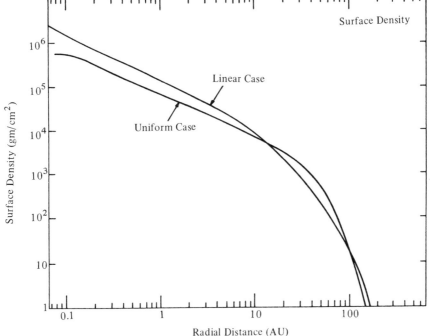

revised by Cameron (personal communication), for two reasons. First, the
nebula evolves dynamically, owing to such mechanisms as Eddington-Sweet
circulation, turbulent transport of angular momentum, etc., on a time scale
shorter than the gravitational collapse time of the gas. Hence, the structure
is forming and evolving simultaneously. Second, theoretical arguments relating to global stability can be made that suggest the nebula may be turbulent throughout. This issue is at present unsettled but will clearly bear heavily on other ideas concerning the formation of the planetary system.

Lewis predicts the following chemical characteristics of the planets implied by Figures 1-5 and 1-6. Mercury has a massive Fe-Ni alloy core surmounted by a small mantle of Fe^{+2}-free magnesium silicates. Refractory

Figure 1-4. The temperatures and pressures at the mid-plane in the nebular models computed
with the Larson adiabat. From Cameron and Pine (1973).

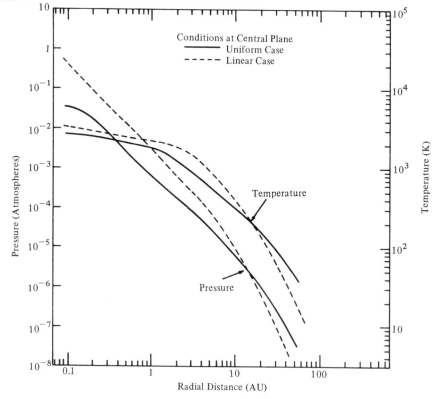

Figure 1-5. Some major features of the chemistry of solar material, 0 to 2000 K and 10^{-7} to 10^1 bar total pressure. The condensation curves of $CaTiO_3$ (a representative refractory mineral), Fe, $MgSiO_3$, and ice; the appearance temperatures of FeS, tremolite (a hydrous calcium silicate) and talc (a hydrous magnesium silicate); and the line at which Fe metal is wholly oxidized to FeO (as $FeSiO_3$ and Fe_2SiO_4) are indicated. An adiabat for the nebula is drawn in, and nebular pressures corresponding to the formation conditions of the planets in Cameron's models are marked on the adiabat. The symbols, reading from the high-temperature end downward, are for Mercury, Venus, earth, Mars, the asteroids, Jupiter, Saturn, Uranus, and Neptune. Temperatures of formation deduced from correlating the density data in Figure 1-6 with the observed densities of the terrestrial planets are in excellent agreement with this adiabat. From Lewis (1972b).

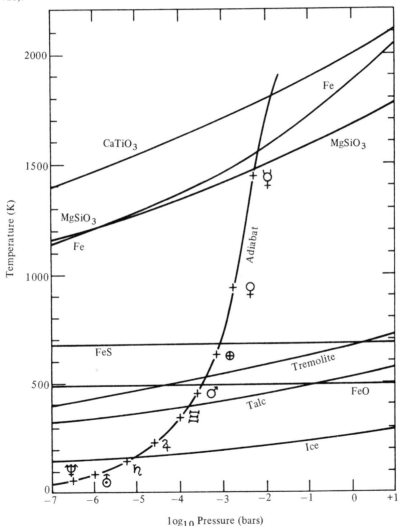

Figure 1-6. Density of condensed material in equilibrium with a solar-composition gas, 400–1600 K at 10^{-3} bars. A simplified chemical system (the 20 most abundant elements) is employed for three different values of the Fe:Si ratio. The densities of the planets are excellently consistent with an Fe:Si ratio of 1.08, but the omission of rare elements and uncertainties in the abundances of major elements could displace the entire manifold of curves slightly. A true Fe:Si ratio below 1.0 is still possible. From Lewis (1972b).

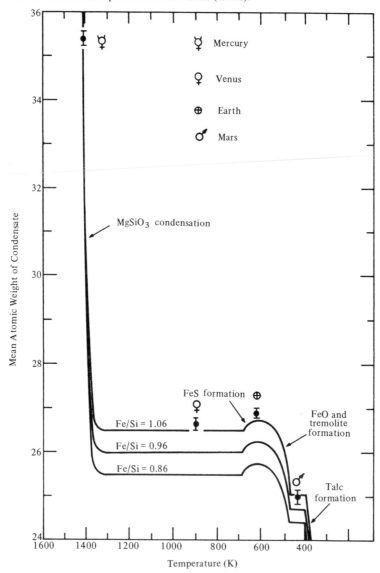

oxides are present but only traces of alkali metals, sulfur, FeO, etc. Venus has a Fe-Ni core, a massive mantle of Fe^{+2}-free magnesium silicates, and a silica-rich crust similar to earth's. Sulfur is probably absent. The earth has an inner core of Fe-Ni and an outer core of Fe-FeS melt. Certain chalcophile elements are deficient in the mantle and crust but enriched in the outer core. The mantle contains 10% FeO. Deficiencies of S, K, Rb, and Cs in the crust and upper mantle are due to their extraction into an FeS-rich melt. Mars is almost devoid of free iron, may contain a core of FeS, and has a mantle rich in FeO. Hydrous minerals were probably retained. The crust should be more iron-rich than the earth's.

The most important application of the condensation sequence calculations concerns the chemistry of meteorites. By comparison of the experimentally determined chemical composition of various classes of meteorites with that of solar composition material, one can, with the aid of the condensation sequence, deduce certain events and conditions at the time of their formation. An excellent review of this subject has been given by Grossman and Larimer (1974). Based on the relative abundances of refractory materials, volatile materials, metals, and silicates, the following conclusions have appeared in the literature and are summarized by those authors.

A relatively small number of fractionation processes (at least four) are required to explain the abundances of nearly all the elements. The processes presented in outline form are:

1. Highly refractory material rich in such elements as Ca, Al, and Ti moved around in the inner solar system, becoming preferentially depleted in some types of chondrites (enstatite and ordinary) and possibly enriched in the outer regions of the earth and moon. Some fraction of metal, also rich in refractory siderophiles, may have been involved.

2. A major fractionation of metal from silicate grains took place between 1000° and 700°K. This event certainly affected the parent material of chondrites and may in part account for the density differences among the planets. This upper temperature limit is close to the ferromagnetic Curie point of the metal (950° to 900°K), suggesting that the onset of magnetism may have triggered the event.

3. Some time just prior to accretion a fraction of the dustlike condensate was flash-heated, possibly by electric discharges or by collisions between dust grains, to produce volatile-deficient chondrules.

4. The unaltered dust continued to equilibrate with the gas, becoming gradually enriched in volatiles (e.g., In, Tl, and Xe). This dust (matrix) accreted together with varying amounts of chondrules at $T = 450 \pm 50°K$ and a pressure of about $10^{-5\pm1}$ atm for ordinary chon-

drites and $T = 350° \pm 50°$ K and a pressure of about 10^{-6} atm for C chondrites. Each group of ordinary chondrites accreted over a fairly narrow temperature range ($\pm 10°$ to $20°$K), the H group forming near $470°$K and the L and LL groups near $450°$K.

Although we apparently have some idea of the chemical composition of condensed debris as a result of nebula cooling, the size distribution of this debris is relatively obscure. Many meteorites contain *both* centimeter-sized material called chondrules and a finely ($\sim$ micron) textured matrix. A growth rate for condensing grains can be crudely estimated as

$$dR/dt \sim v_j(\rho_j/\rho_p), \qquad (1\text{-}22)$$

where ρ_j is the spatial density of the condensing constituent, v_j is the thermal velocity of the gaseous molecules, ρ_p is the solid density of the grain, and R is the grain radius. For the more abundant species, this is typically of the order of centimeters per year. However, the final particle size will depend on the number of nucleation sites formed. The situation is further complicated by two other considerations: the temperature of the nebula in some places may never have risen high enough to evaporate pre-existing micron-sized interstellar grains, and clumping of debris during the collapse of the interstellar gas may have produced some centimeter-sized particles even before the nebula formed (Cameron, 1975a).

If a substantial amount of material was present in the centimeter size range, this would be helpful to further accumulation since these particles would quickly settle out of the (nonturbulent?) solar nebula and form a thin disk of debris. This sets the stage for the development of gravitational instabilities, a mechanism that is very efficient at forming kilometer-sized objects quickly. This mechanism is the topic of the next section.

Gravitational Instabilities

In this section we shall look at the possible role of gravitational instabilities in the early accretion of solid material. The chemically condensed grains are envisaged as having settled to the mid-plane of the solar nebula under the influence of solar gravity to form a thin disk of debris. The finite thickness of the gaseous solar nebula is maintained by a vertical pressure gradient which balances the vertical component of the solar gravity,

$$-\frac{1}{\rho_g}\frac{dp}{dz} \sim \frac{GM_\odot}{r^2}\left(\frac{z}{r}\right) = z\Omega^2, \qquad (1\text{-}23)$$

where ρ_g is the gas density, p is the pressure, $M_\odot$ is the solar mass, r is the solar distance, and Ω is the local orbital angular velocity. Newly condensed particles lose the support of the pressure gradient (because of their high body density, $\rho_p \gg \rho_g$) and begin to settle toward the mid-plane. This motion is retarded by a gaseous drag force

$$F_D \sim \tfrac{4}{3}\pi R^2 \rho_g c_g v_z, \tag{1-24}$$

where c_g is the gas sound speed, v_z the particle's descent velocity, and R the particle's radius. A characteristic settling time can be found by equating equation (1-23) to the particle's deceleration implied from equation (1-24) to find v_z and computing

$$\tau_{\text{settling}} \sim z/v_z \sim \sigma_g/\rho_p \Omega R, \tag{1-25}$$

where σ_g is the surface density of the nebula. Of course, if this time is long, the particle will continue to grow by equation (1-22). The maximum particle size and, hence, minimum settling time, can be found by equating $\tau_{\text{settling}} \sim R(dR/dt)^{-1}$, the right-hand side being evaluated with the aid of equation (1-22).

For numerical work in this section we shall adopt conditions appropriate for a minimum-mass solar nebula at a distance of ~ 1 AU. The surface density is estimated by spreading out the masses of the terrestrial planets over the inner solar system and augmenting up to solar composition. The surface densities of solid and gaseous materials are $\sigma \sim 7.5$ gm/cm^2 and $\sigma_g \sim 1.5 \times 10^3$ gm/cm^2. The half-thickness of the nebula is $z \sim c_g/\Omega$ which, for a gas sound speed of $c_g \sim 2 \times 10^5$ cm/sec, is about $z \sim 10^{12}$ cm. The resulting gas density is $\rho_g \sim 7.5 \times 10^{-10}$ gm/cm^3. With these values the largest grains to form by chemical condensation are of the order of a few centimeters and settle to the mid-plane in a few years. Smaller grains settle in correspondingly longer times according to equation (1-25).

The debris collecting in the mid-plane is regarded as obeying the hydrodynamic equation of motion,

$$\partial v/\partial t + (v \cdot \nabla)v = -(1/\rho)\, \nabla P + \nabla\Phi, \tag{1-26}$$

where v is the local velocity, P is the equivalent pressure related to the *particle's* dispersion velocity, ρ is the spatial density of particles, and Φ is the gravitational potential. In addition, we have Poisson's equation

$$\nabla^2\Phi = -4\pi G\rho, \tag{1-27}$$

the conservation of mass

$$\partial\rho/\partial t + \nabla \cdot (\rho\mathbf{v}) = 0, \tag{1-28}$$

and, for adiabatic processes,

$$\partial P/\partial t + (\mathbf{v} \cdot \nabla)P = \gamma P/\rho[\partial\rho/\partial t + (\mathbf{v} \cdot \nabla)\rho], \tag{1-29}$$

where γ is the adiabatic index, i.e., the ratio of the specific heats.

Presumably there is some basic equilibrium state satisfying equations (1-26)–(1-29) in which the debris orbits the primary as a flattened, differentially rotating disk. (The particles will execute nearly circular Keplerian orbits.) The stability of this state against self-gravitation can be investigated by first-order perturbation analysis of equations (1-26)–(1-29) (Goldreich and Ward, 1973). Such a procedure is performed in the appendix of this chapter. However, the basic physics underlying the gravitational instabilities can be illustrated by several order-of-magnitude arguments presented below.

For the sake of simplicity, let us consider a uniformly rotating sheet of material and seek the conditions that permit a region to contract under self-

Figure 1-7. Schematic diagram of a uniformly rotating disk undergoing self-gravitationally induced fragmentation.

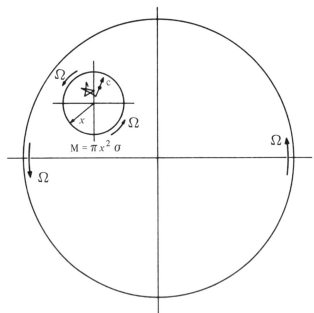

gravitation (Figure 1-7). Two aspects of the particle motions are relevant: First, there is a rotation of the region about its center due to the rotation of the disc as a whole and, of course, there is angular momentum associated with this. Second, the random motions of the particles superimposed on the orbital motion act somewhat like a gas pressure in supporting the region. Let us start by considering each effect separately.

Although a proper analysis must examine the incremental forces arising from perturbations of the equilibrium configuration (see appendix), the correct dependence of stable and unstable lengths on Ω, σ, and c can, nevertheless, be obtained by considering a test region of radius X as isolated from the rest of the disk. In a uniformly rotating disk, any region viewed from its center rotates with angular velocity Ω. Particles at a distance X thus have an orbital velocity with respect to the region's center of $V \sim X\Omega$. The critical distance at which contraction is just possible is, to an order of magnitude, given by equating the centripetal acceleration to the gravitational acceleration due to the mass inside the region, i.e.,

$$V^2/X \sim GM/X^2. \tag{1-30}$$

However, the mass is related to the size of the region by $M = \pi X^2 \sigma$. Eliminating V and M from equation (1-30), one finds that contraction is possible for regions of radii

$$X \lesssim \pi G\sigma/\Omega^2. \tag{1-31}$$

An alternative way of looking at this is through consideration of the energy requirements. The kinetic energy associated with the rotation is of order $E_R \sim (MV^2)/2$ and the angular momentum is $L \sim MXV$. When a region contracts its angular momentum is conserved. Hence the rotational energy

$$E_R \sim \tfrac{1}{2}L^2/MX^2 \tag{1-32}$$

must increase with decreasing X. The only source for this new energy is the gravitational potential energy released during contraction. The gravitational binding energy of a region (the negative of the potential energy and, therefore, a positive quantity) is of order

$$U \sim GM^2/X. \tag{1-33}$$

Contraction is possible if more potential energy is released than that necessary to supply the rising needs of the rotational energy, the excess then appearing as a bulk contraction velocity of the material, i.e.,

$$-dU/dX > -dE_R/dX. \tag{1-34}$$

Combining equations (1-32)–(1-34) and eliminating L, M, and V again leads to equation (1-31).

Next consider a region containing particles with a characteristic dispersion velocity c. For such a region to contract, it must be able to "hold on" to its component particles. Hence, a particle must feel a deceleration (GM/X^2), during the time (X/c) it takes to cross the region, sufficient to reverse its velocity before it escapes, i.e.,

$$(GM/X^2)(X/c) \gtrsim c. \tag{1-35}$$

Eliminating M leads to the condition

$$X \gtrsim c^2/\pi G\sigma \tag{1-36}$$

for contraction.

We can again consider the energetics of the process. The kinetic energy associated with the random velocities is $E_T \sim Mc^2/2$ and the energy per unit volume is $\mathscr{E} = \rho c^2/2$. The particles behave as a monatomic gas with an adiabatic index of 5/3. The equivalent pressure is $P = 2\mathscr{E}/3$ and for an adiabatic contraction

$$P/P_0 = (\rho/\rho_0)^{5/3} = (X_0/X)^5, \tag{1-37}$$

where P_0, ρ_0, and X_0, are the values at the start of contraction and the thickness-to-diameter ratio of the disk is assumed to remain constant. Hence, the energy $E_T \sim \mathscr{E}X^3$ and

$$E_T = E_{T,0}(X_0/X)^2. \tag{1-38}$$

As before, contraction requires

$$-dU/dX > -dE_T/dX. \tag{1-39}$$

Combining equations (1-33), (1-38), and (1-39) and eliminating M recovers equation (1-36).

For the general case equations (1-34) and (1-39) can be combined into

$$dU/dX - d(E_T + E_R)/dX < 0, \tag{1-40}$$

which, after substitution of the appropriate expressions, becomes

$$c^2 - \pi G\sigma X + X^2\Omega^2 < 0. \tag{1-41}$$

This agrees, except for numerical factors, with the stability condition obtained from perturbation analysis (see appendix), which is

$$F(\lambda) = 4\pi^2 c^2 - 4\pi^2 G\sigma\lambda + \lambda^2\Omega^2 < 0 \qquad (1\text{-}42)$$

where λ, the perturbation wavelength, is roughly equivalent to the diameter of the collapsing region, $2X$. In equation (1-42) the differential rotation of the disk has also been taken into account.

Large wavelengths are stabilized by angular momentum, small wavelengths are stabilized by random particle motions. If these regions of stability overlap, the disk as a whole is stable against collapse due to self-gravity. Equation (1-42) is plotted in Figure 1-8 for various values of c. For a given

Figure 1-8.   Stability criterion $F(\lambda)$ for various values of the dispersion velocity c. Instability sets in for $F(\lambda) < 0$.

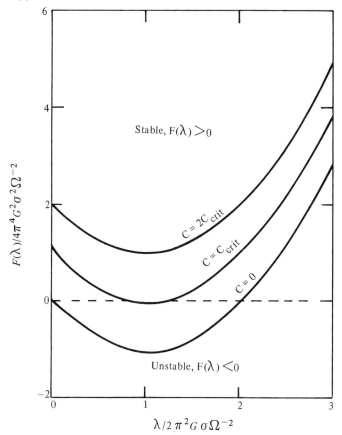

Figure 1-9. Planets begin to form when interstellar dust grains collide and stick to one another, forming ever larger clumps (a). The clumps fall toward the midplane of the nebula (b) and form a diffuse disk there. Gravitational instabilities collect this material into millions of bodies of asteroid size (c), which collect into gravitating clusters (d). When clusters collide and intermingle (e), their gravitational fields relax, and they coagulate into solid cores, perhaps with some bodies going into orbit around the cores (f). Continued accretion and consolidation may create a planet-sized body (g). If the core gets larger, it may concentrate gas from the nebula gravitationally (h). A large enough core may make the gas collapse into a dense shell that constitutes most of the planet's mass (i). From Cameron (1975b), The Origin and Evolution of the Solar System, copyright 1975 by Scientific American Inc., all rights reserved.

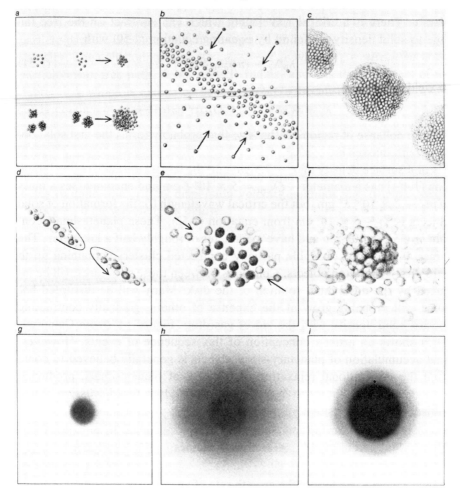

material:

$$dm/dt \sim (4\pi S^2)(\rho v). \tag{1-53}$$

(Many authors also include some dimensionless efficiency factor but we will assume such terms to be understood.) For a spherical body of radius R and density ρ_p,

$$dR/dt \sim v(\rho/\rho_p)(S/R)^2. \tag{1-54}$$

The physics of the problem resides in determining appropriate values or expressions for the effective collision radius S, and mass flux (ρv).

For an isolated two-body encounter, the effective radius can be easily related to the physical radius through the conservation of energy and angular momentum. Figure 1-10 shows a grazing impact of a projectile of mass m much less than the target mass M. By conservation of energy,

$$E = \tfrac{1}{2}mv_\infty^2 = \tfrac{1}{2}mv_I^2 - GMm/R, \tag{1-55}$$

where v_∞ is the approach velocity and v_I is the velocity at impact. By conservation of angular momentum,

$$mv_\infty S = mv_I R = L. \tag{1-56}$$

Combining these, one obtains

$$S^2 = R^2[1 + (v_e/v_\infty)^2], \tag{1-57}$$

where $v_e = (2GM/R)^{1/2}$ is the escape velocity from the target. Similar expressions can be derived for the case of comparable target and projectile masses. Equation (1-57) identifies the effective collision radius with the impact parameter that will result in a grazing blow. An alternative (but equivalent) inter-

Figure 1-10.   Geometry of a two-body encounter leading to a tangential impact.

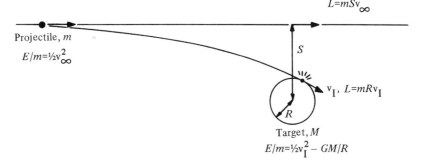

pretation is to take the target's radius R as the collision radius but consider the local particle flux as increased to $\rho v = (\rho v)_\infty[1 + (v_e/v_\infty)^2]$ by gravitational focusing. These are remote and local vantage points, respectively. Substitution of expression (1-57) into (1-54) yields

$$dR/dt \sim v_\infty(\rho_\infty/\rho_p)\left(1 + \frac{8\pi}{3}\frac{G\rho_p R^2}{v_\infty^2}\right). \tag{1-58}$$

For approach velocities $v_\infty \ll v_e$, the cross section varies as R^4. On the other hand, for $v_\infty \gg v_e$ accretion will not in general occur at all because the gravitational field of the target will not be able to hold on to the very energetic ejecta, i.e., the efficiency factor becomes negative.

Of course, impacts in the solar nebula are not strictly two-body collisions since both target and projectile orbit the primary. There *is* a region about the target object where its gravity field dominates the tidal forces of the sun and the target-projectile pair can be considered nearly isolated. The radius of this zone is found approximately by the L_2 point (Danby, 1962),

$$L_2 \sim r(m/3M_\odot)^{1/3}. \tag{1-59}$$

Far from this zone the two bodies orbit the primary independently. Since each conserves its orbital angular momentum, their relative angular momentum must vary as they drift nearer each other because of the differential rotation of the disk, i.e., equation (1-56) is not valid until near approach. In general, particles will not enter the zone with zero angular momentum; in other words, the equivalent of $v_\infty \to 0$ in equation (1-56) is not in general possible.

The mutual or relative angular momentum of the projectile with respect to the target resulting from orbital motion is easily calculated (e.g., Lyttleton, 1972). Let

$$\mathbf{r} = r(\cos\theta\hat{\mathbf{x}} + \sin\theta\hat{\mathbf{y}}), \tag{1-60}$$

$$\mathbf{r}' = r'(\cos\theta'\hat{\mathbf{x}} + \sin\theta'\hat{\mathbf{y}}), \tag{1-61}$$

be vectors locating the target and projectile with respect to the sun (Figure 1-11). We assume that both objects are in circular orbits of zero inclination so that

$$d\mathbf{r}/dt = r\dot\theta(-\sin\theta\hat{\mathbf{x}} + \cos\theta\hat{\mathbf{y}}), \tag{1-62}$$

$$d\mathbf{r}'/dt = r'\dot\theta'(-\sin\theta'\hat{\mathbf{x}} + \cos\theta'\hat{\mathbf{y}}) \tag{1-63}$$

are their velocities (neglecting mutual perturbations). The angular momentum per unit mass of the projectile with respect to the target is

$$\mathbf{l} = (\mathbf{r}' - \mathbf{r}) \times (\dot{\mathbf{r}}' - \dot{\mathbf{r}}), \tag{1-64}$$

which upon substitution of equations (1-60)–(1-63) becomes a vector of magnitude

$$l = \dot{\theta}r^2 + \dot{\theta}'r'^2 - rr'(\dot{\theta} + \dot{\theta}') \cos(\theta - \theta') \tag{1-65}$$

in the Z-direction. The orbital mean motions are $\dot{\theta} = (GM_\odot/r^3)^{1/2}$ and $\dot{\theta}' = (GM_\odot/r'^3)^{1/2}$. We have normalized l to the orbital angular momentum per unit mass of the target, $\tilde{l} = l(GM_\odot r)^{-1/2}$, and let $r = r'/r$ and $\theta = 0$ without loss of generality to obtain

$$\tilde{l} = 1 + \tilde{r}^{1/2} - [1 + \tilde{r}^{-3/2}]\tilde{r} \cos \theta'. \tag{1-66}$$

Curves of constant $\tilde{l}$ have been plotted in Figure 1-12. The lines of zero angular momentum separate the flow into two areas of negative and two areas of positive angular momentum. The critical angular momentum below which collision would occur if the target and projectile were isolated is

Figure 1-11. Target and projectile coordinates.

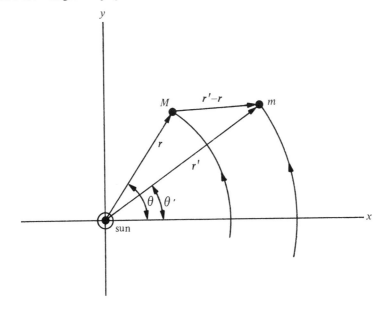

$l_{\text{crit}} \sim \pm (GMR)^{1/2}$. These values define thin bands about $\tilde{l} = 0$. The figure has been constructed with an exaggerated target to solar mass ratio of ~ 0.003 so that the distance to the L_2 point is about 10% of the target's semi-major axis. The circle surrounding the target indicates its approximate zone of gravitational dominance. Clearly much material penetrates this zone with too much angular momentum to be accreted. Although this picture is over-simplified, numerical integrations of collisions in a situation similar to that of Figure 1-12 do reveal that impacting particles arrive from four bands located in roughly the positions indicated here (Giuli, 1968a, 1968b). Hence, one must be careful in the application of equation (1-58), which is probably reliable only for v_∞ somewhat less than v_e.

Unless most of the accreting debris remains very small compared to the growing target object, it seems much more likely that orbits become quite eccentric rather than circular. Orbital eccentricities e tend to increase owing to gravitational relaxation and to decrease owing to inelastic particle collisions. The particles' dispersion velocity is of order $v \sim eV_0$, where V_0 is the

Figure 1-12. Angular momentum of a projectile with respect to the target as a function of relative position. Both objects are in circular Keplerian orbits about the sun.

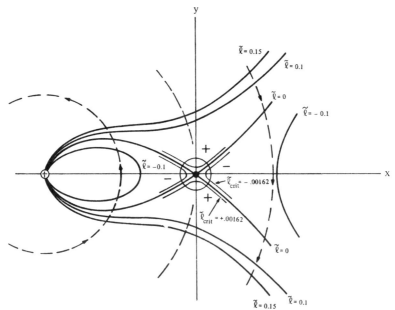

orbital velocity. The collision interval is given by the mean free path $\ell = m/\rho\pi R^2$ divided by the dispersion velocity. The spatial density of the debris $\rho = \sigma/h$, where $h \sim v/\Omega$ is the thickness of the disk of planetesimals (which is simply the height a particle with vertical velocity v can rise before the vertical component of the solar gravity forces it to return to the disk). If we assume that at each collision a particle loses β of its relative kinetic energy, the damping time scale due to collisions is

$$\tau_{\text{damping}} \sim \frac{4}{3} \frac{\rho_p R}{\sigma} \frac{1}{\beta\Omega}, \tag{1-67}$$

which is independent of v.

The relaxation time scale can be estimated from techniques employed in stellar dynamics (e.g., Woltjer, 1967; Safronov, 1969). A test particle traveling past a field particle (assumed to be stationary for simplicity) with velocity v will experience a velocity perturbation of order

$$\delta v \sim \frac{m_f v}{m_t + m_f} \sin 2\chi, \tag{1-68}$$

where

$$\sin \chi = \left(1 + \frac{D^2 v^4}{G^2(m_t + m_f)^2}\right)^{-1/2} \tag{1-69}$$

and D is the impact parameter as shown in Figure 1-13 and 2χ is the deflection angle experienced by the test particle in the center-of-mass frame. If we

Figure 1-13. Gravitational encounter of field and test particles producing a deflection of 2χ. From Woltjer (1967).

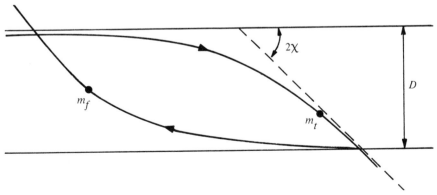

assume the dispersion velocity undergoes a random walk with a step size given by equation (1-68), then

$$dv^2/dt \sim \int v^2 \sin^2 \chi \cos^2 \chi n(D) \, dD, \qquad (1\text{-}70)$$

where $n(D) \, dD$ are the number of encounters per unit time with field particles at impact parameters between D and $D + dD$. [In equation (1-70) we have taken $m_t = m_f$.] The most important contributors are those within a distance $h = v/\Omega$ equal to the thickness of the planetesimal disk. Thus

$$n(D) \, dD = 2\pi D\sigma\Omega \, dD/m. \qquad (1\text{-}71)$$

Combining equations (1-68)–(1-71) and integrating from zero to h, we have

$$dv^2/dt \sim \frac{8\pi G^2 m\sigma\Omega}{v^2} \ln \Lambda, \qquad (1\text{-}72)$$

where $\Lambda = hv^2/2GM$.

The dispersion velocity will grow until its characteristic growth time

$$\tau_{\text{dispersion}} \sim \left(\frac{1}{v}\frac{dv}{dt}\right)^{-1} \sim \frac{v^4}{4\pi G^2 m\sigma\Omega \ln \Lambda} \qquad (1\text{-}73)$$

is no longer less than τ_{damping}, i.e., until

$$v \sim (GM/\Theta R)^{1/2}, \quad \Theta = (\beta/4 \ln \Lambda)^{1/2}. \qquad (1\text{-}74)$$

Numerically, $\Lambda \sim v^3/2GM\Omega \sim (\pi\rho_p G/3\Theta^3)^{1/2}/\Omega \sim 2 \times 10^3$ and $\Theta \sim 0.2 \, \beta^{1/2}$. By a more detailed treatment, Safronov (1969) obtains a value of order unity. He also argues that including gas drag effects further increases Θ to of order ~ 3–5.

If the characteristic dispersion velocity of equation (1-74) is substituted for v_∞ in equation (1-58),

$$dR/dt \sim \sigma\Omega(1 + 2\Theta)/\rho_p. \qquad (1\text{-}75)$$

Finally, the surface density must be related to the remaining unaccreted mass. The usual technique is to assume the unaccreted mass to be evenly distributed throughout a zone Δa, with boundaries lying halfway to each adjacent planet, and to set

$$\sigma \sim (\alpha m_{\text{final}} - m)/2\pi a \, \Delta a, \qquad (1\text{-}76)$$

where m_{final} is the final mass. In this case the accretion equation is easily integrated. The characteristic growth time required to achieve planetary dimensions by accretion is

$$\tau_{\text{growth}} \sim \frac{3}{2} \Omega^{-1} \left(\frac{a\,\Delta a}{R_{\text{final}}^2} \right) (1 + 2\Theta)^{-1} \alpha^{-1}. \tag{1-77}$$

Estimates of this quantity for the terrestrial planets are listed in Table 1-4. The total planetary accumulation time will be several times this value.

A difficulty of this model is the impossibly long accretion times predicted for the outer planets Uranus and Neptune, i.e., nearly two orders of magnitude greater than the age of the solar system. Various alternative theories have been suggested which can lead to more rapid accretion times.

1. Safronov (1969) suggests that Θ may have been of order $\sim 10^2$ during the final growth stage of the outer planets. The implied dispersion velocity would be only $\sim 10\%$ of the planet's escape velocity, but it is argued that because the escape velocities from Uranus and Neptune are greater than their orbital velocities, particles would leave the solar system before obtaining such a velocity by relaxation. It would seem to follow that Uranus and Neptune constitute only a fraction of the original mass in the zone. One must be careful here, however, since the ejection of a large mass will radically alter the orbits of the remaining material, primarily causing an inward orbital drift. Safronov discusses this problem and concludes that, as of yet, no satisfactory theory exists for the growth of the giant planets.

2. Cameron (1973) prefers a much larger surface density. If α is of order $\sim 10^2$, accretion times are correspondingly shorter. Again there is the problem of eliminating a large excess of material. Cameron also uses the differential orbital velocities between particles and gas (see appendix) to induce orbital decay of small particles via gas drag. This causes them to drift into

Table 1-4 Characteristic growth times for the terrestrial planets.

Planet	Orbit period (years)	Semimajor axis (AU)	Radius (cm)	$\tau \times (1 + 2\theta) \times \alpha$ (years)
Mercury	0.241	0.387	2.43×10^8	2×10^7
Venus	0.615	0.723	6.05×10^8	1×10^7
Earth	1.000	1.000	6.38×10^8	4×10^7
Mars	1.881	1.524	3.39×10^8	7×10^8

the accretion zone of a growing planet. Accretion times tend to be quite short ($\sim 10^3$ years for the earth).

3. Lyttleton (1972) argues that the correct capture radius for a growing planet is its L_2 point. Mutual collisions of debris while passing within this distance will result in capture. Such a large accretion cross section also leads to rapid growth ($\sim 10^4$ years).

4. Alfvén and Arrhenius (1970a, 1970b) contend that the relevant Δa was much smaller than present planetary separations because of the formation and contraction of jet streams. It is claimed that inelastic collisions cause radial clustering of particles, focusing them into a stream with low relative velocities and high particle density. A prerequisite appears to be that solid grains be formed in highly eccentric (~ 0.25) orbits. It is proposed that this situation is brought about by chemical condensation from a plasma partially corotating with the sun.

5. Still another possibility is that the sun's mass was somewhat larger prior to the removal of the solar nebula. Indeed, the dispersal of the nebula is most often attributed to the passage of the sun through a T Tauri stage during which several tenths of a solar mass may have been lost by the sun (e.g., Cameron, 1973). Since the planetary orbital angular momentum is conserved during solar mass loss, the semimajor axis $a \propto M_\odot^{-1}$. From this it follows that $\tau_{\text{growth}} \propto \Omega^{-1} a \, \Delta a \propto M_\odot^{-4}$ so that an increase of the solar mass by ~ 1.5 implies a decrease in τ_{growth} by a factor of ~ 5.

The correct answer may, of course, turn out to be some combination of the above (or even none of the above). The question is by no means a closed one.

Concluding Remarks

This chapter is no more than a cursory introduction to a complex, multifaceted subject. I have chosen three important problems to discuss, but there are many more that could have served the introductory purpose of this chapter just as well. Some unsolved or partially solved problems that are currently the subject of research efforts include the following:

1. Gas accretion by the giant planets. Perri and Cameron (1974) have found that when an accreting solid object exceeds a critical size, the hydrogen envelope surrounding it becomes hydrodynamically unstable to col-

lapse and the planet becomes a hydrogen sink. The critical size at the orbit of Jupiter is estimated to be a few tens of earth masses.

2. Planetary rotation. Only a partial explanation for isochronism (similar spin periods for bodies ranging from 10^{18} to 10^{30} gm) has been provided. Giuli (1968a, 1968b) has found that a planet capturing debris from circular orbits will acquire a retrograde rotation, but if accretion takes place from eccentric orbits as well, the rotation will be prograde and of the right order of magnitude. As we saw earlier, the sun may play an important role in determining the angular momentum added to the planet by the projectile. Giuli's investigations have thus involved numerical integrations of this three-body problem to determine which projectiles actually undergo impact.

3. Formation of satellite systems. We suspect that the formation of the satellite systems of the giant planets parallels in many respects the formation of the planets. However, many satellites exhibit orbital resonances —commensurabilities of orbital periods (Gold, 1974)—and it is not clear what role (if any) such phenomena play in the accretion process. Lunar origin has been the subject of considerable effort in past years. The moon is especially puzzling because of its large mass relative to the earth and its chemical peculiarities, such as low iron and volatile content (see Kaula and Harris, 1975).

4. Comets. The origin of the Oort cloud is not understood. Whether comets formed at the distance of the cloud ($\sim 50,000$ AU) or were formed in the solar system and then ejected is a topic of debate (Öpik, 1973; Cameron, 1973).

5. The nature and origin of the asteroid belt. Chapman and Davis (1975) have argued that the asteroids are a remnant of an initially much larger population containing a mass perhaps comparable to that of a small planet. This population has been depleted over geological time because of impact fragmentation and removal of small-sized debris by various nongravitational forces, i.e., Poynting-Robertson effect, Yarkovsky effect, solar radiation pressure, etc.

6. The dissipation of the solar nebula. It is commonly hypothesized that an intense solar wind during a T Tauri stage of the sun removed the nebula, but a detailed description of the physics involved and an estimate of the time scale has not yet been given.

The above list of subjects and references is by no means intended to be complete, but merely to furnish some illustrative examples. (Further references can be obtained from the bibliographies in these papers.) Research on the solar system has intensified enormously in the last two decades but many problems remain unsolved and many new ones will undoubtedly be generated by the continued acquisition of planetary probe data and future analyses of planetary and meteoritic samples. Thus, solar system research promises to be an exciting area of activity in the years ahead.

Appendix

We are interested in the behavior of small perturbations v', ρ', P', Φ'. Equations (1-26)–(1-29) can be linearized in these quantities to give (in cylindrical coordinates)

$$\frac{\partial v'_r}{\partial t} + \Omega \frac{\partial v'_r}{\partial \theta} - 2\Omega v'_\theta = \frac{\partial \Phi'}{\partial r} - \frac{1}{\rho} \frac{\partial P'}{\partial r} + \frac{1}{\rho} \left(\frac{\rho'}{\rho}\right) \frac{\partial P}{\partial r}, \tag{A-1}$$

$$\frac{\partial v'_\theta}{\partial t} + \Omega \frac{\partial v'_\theta}{\partial \theta} - 2B v'_r = \frac{1}{r} \frac{\partial \Phi'}{\partial \theta} - \frac{1}{r} \frac{1}{\rho} \frac{\partial P'}{\partial \theta}, \tag{A-2}$$

$$\frac{\partial v'_z}{\partial t} + \Omega \frac{\partial v'_z}{\partial \theta} = \frac{1}{\rho} \frac{\partial P'}{\partial z} + \frac{1}{\rho} \left(\frac{\rho'}{\rho}\right) \frac{\partial P}{\partial z} + \frac{\partial \Phi'}{\partial z}, \tag{A-3}$$

$$\nabla^2 \Phi' = -4\pi G \rho', \tag{A-4}$$

$$\frac{\partial \rho'}{\partial t} + \Omega \frac{\partial \rho'}{\partial \theta} = -\frac{1}{r} \frac{\partial}{\partial r} (r \rho v'_r) - \frac{\rho}{r} \frac{\partial v'_\theta}{\partial \theta} - \frac{\partial}{\partial z} (\rho v'_z), \tag{A-5}$$

$$\frac{\partial P'}{\partial t} + v'_r \frac{\partial P}{\partial r} + \Omega \frac{\partial P'}{\partial \theta} + v'_z \frac{\partial P}{\partial z}$$
$$= \gamma(P/\rho) \left(\frac{\partial \rho'}{\partial t} + v'_r \frac{\partial \rho}{\partial r} + \Omega \frac{\partial \rho'}{\partial \theta} + v'_z \frac{\partial \rho}{\partial z}\right), \tag{A-6}$$

where $B = -(\Omega + d(r\Omega)/dr)/2$.

Since we are primarily interested in the propensity of the disk to clump in the r and θ directions, the problem can be simplified if, in a manner similar to Toomre (1964), we consider a thin disk, neglecting variations and velocities in the Z-direction. In this case, equation (A-3) is deleted, terms containing v'_z are set equal to zero, equation (A-4) is replaced by

$$\nabla^2 \Phi' \sim -4\pi G \sigma' \delta(z), \tag{A-7}$$

where σ is the particle surface density, $\delta(z)$ is the Dirac delta function, and finally $\rho = \sigma/h$, $\rho' = \sigma'/h$, $h \to 0$ being the thickness of the disk. It is also customary to write

$$\partial P/\partial r = (\Gamma P/\sigma)\partial\sigma/\partial r, \quad c^2 = \gamma Ph/\sigma, \tag{A-8}$$

where Γ is a polytropic index describing the radial structure of the disk, and γ the adiabatic index, which changes equation (A-6) to

$$\frac{\partial P'}{\partial t} + \Omega \frac{\partial P'}{\partial \theta} = \frac{c^2}{h}\left(\frac{\partial\sigma'}{\partial t} + \Omega\frac{\partial\sigma'}{\partial\theta}\right) + v_r'\frac{\partial P}{\partial r}\left(\frac{\gamma}{\Gamma} - 1\right). \tag{A-9}$$

In order to take into account, in a simple manner, the effect of a nonzero dispersion velocity, we retain terms involving the pressure gradient (i.e., $h \neq 0$). Our derivation lacks rigor in this regard. It is inconsistent to use equation (A-7) for an infinitely thin disk in combination with a nonzero isotropic particle dispersion velocity. Toomre circumvents this difficulty by using a nonisotropic dispersion velocity with no component in the Z-direction. Goldreich and Lynden-Bell (1965), on the other hand, retain the equations relating to the Z-direction and treat various polytropic disk models in all three dimensions. Both analyses lead to dispersion relations that predict unstable wavelengths comparable to those obtained from (A-16), which is considerably simpler and sufficiently accurate for our purposes. Indeed, for a nonisotropic pressure, the results of Goldreich and Lynden-Bell reduce to ours.

First consider radial disturbances in the surface density of the form $\sigma' = \sigma_0' e^{ik_r r + i\omega t}$, i.e., of wavelength $\lambda = 2\pi/k_r$ and frequency ω. An approximate, local solution to equation (A-7) has been shown by Toomre (1964) to be

$$\Phi' \approx \left(\frac{2\pi G\sigma_0'}{k_r}\right) e^{ik_r r + i\omega t} e^{-k_r|z|}, \tag{A-10}$$

from which $\partial\Phi'/\partial r|_{z=0} \approx 2\pi i G\sigma_0' e^{ik_r r + i\omega t}$. Assigning the same exponential dependence to the other perturbed quantities and substituting them into the linearized equations leads to the following set of homogeneous linear equations:

$$i\omega v_r' - 2\Omega v_\theta' - \left(2\pi iG + \frac{h}{\sigma^2}\frac{\partial P}{\partial r}\right)\sigma_0' + ihP'k_r/\sigma = 0, \tag{A-11}$$

$$-2Bv_r' + i\omega v_\theta' = 0, \tag{A-12}$$

$$\left[i\sigma k_r + \frac{1}{r} \frac{\partial}{\partial r} (r\sigma) \right] v_r' + i\omega\sigma' = 0, \tag{A-13}$$

$$\frac{\partial P}{\partial r} \left(\frac{\gamma}{\Gamma} - 1 \right) v_r' + i\omega c^2 \sigma'/h - i\omega P' = 0. \tag{A-14}$$

Since these equations are not linearly independent, there exists a nontrivial solution only if the characteristic equation (i.e., the determinant of the coefficients) equals zero. This eventually leads to the dispersion relation,

$$\omega^2 = k_r^2 c^2 - 2\pi G\sigma k_r - 4\Omega B$$

$$+ \frac{h}{\sigma} \frac{\partial P}{\partial r} \frac{\partial}{\partial r} \ln (r\sigma) - ik_r c^2/r + 2\pi iG\sigma \frac{\partial}{\partial r} \ln (r\sigma). \tag{A-15}$$

Provided we restrict our attention to wavelengths $\lambda = 2\pi/k_r \ll r$, the fourth and fifth terms, of order $\mathcal{O}(c^2/r^2)$ and $\mathcal{O}(c^2 k_r/r)$ respectively, are much smaller than the leading term, while the last term, of order $\mathcal{O}(G\sigma/r)$, is negligible compared to the second. Hence, as a good approximation we shall use

$$\omega^2 = k_r^2 c^2 - 2\pi G\sigma k_r + K^2 \tag{A-16}$$

where $K^2 = -4\Omega B$. (Note that for a Keplerian disk, $K^2 = \Omega^2 = GM_\odot/r^3$.)

Instability sets in if $\omega^2 < 0$ so that $\omega = \pm i\sqrt{-\omega^2}$ and the assumed exponential form contains a growing part. For small wavelengths $k_r \to \infty$, the first term dominates the second and stability is maintained by the particles' dispersion velocity. For large wavelengths $k_r \to 0$, and the vorticity, K, ensures stability. However, for sufficiently small c there is a range of unstable wavelengths. The critical dispersion velocity at which instability sets in is given in equation (1-44) of the text.

For a uniformly rotating disk, a similar dispersion relation can be obtained for tangential disturbances of the form $\sigma' = \sigma_0' e^{irk_\theta (\theta - \Omega t) + i\omega t}$. In the case of differential rotation, tangential waves are sheared and tend to be converted to radial ones on a time scale $\sim (r\partial\Omega/\partial r)^{-1}$. However, even for a Keplerian disk this time is long compared to the collapse time for wavelengths somewhat smaller than λ_{crit}. Hence we conclude that the debris disk will disintegrate into fragments with masses of order $\xi^2 \sigma \lambda_{crit}^2$.

It is possible that wavelengths longer than λ_{crit} of equation (1-46) may play a role in the accumulation of solid material since such modes *are* destabilized on a longer time scale by the dissipative effects of viscous drag in the presence of the remnant gaseous solar nebula (Goldreich and Ward, unpublished).

Let us define a characteristic time scale for the decay of a particle's velocity by drag as

$$\nu^{-1} \sim m\nu/F_D, \qquad (A\text{-}17)$$

where F_D is the drag force. For particles smaller than the gas mean free path, F_D is determined by Epstein's law, equation (1-24), while for larger particles (with a Reynold's number $\mathcal{R}e = R\nu/\ell_g c_g \gg 1$), the force arises from the formation of a turbulent wake trailing the particle and

$$F_D \sim \tfrac{1}{2} C_D \pi R^2 \rho_g \, \Delta V \nu, \qquad (A\text{-}18)$$

where C_D is the drag coefficient (Whipple, 1971). In regard to equation (A-18) we should point out that for any reasonable model of a nebula for which the density and/or temperature fall with increasing distance from the sun, there is a radial pressure gradient which partially supports the gas in that direction,

$$r\Omega_g{}^2 = \frac{GM_\odot}{r} + \frac{1}{\rho_g}\frac{dp}{dr}, \qquad (A\text{-}19)$$

so that the gas orbital angular velocity is less than the local Keplerian velocity, i.e.,

$$\Omega_g \sim \Omega - c_g{}^2/2a^2\Omega \qquad (A\text{-}20)$$

where we have approximated $dp/dr \sim -\rho_g c_g{}^2/r$. The systematic velocity difference between particles and gas is thus

$$\Delta V \sim c_g{}^2/2a^2\Omega. \qquad (A\text{-}21)$$

To determine the effect of friction on the stability analysis we add a term of the form $-\nu v'$ to the right-hand side of the linearized version of equation (1-26). Making use of the same approximations as before, one arrives at a dispersion relation for the radial mode

$$i\omega(i\omega + \nu)^2 + (i\omega + \nu)k_r{}^2c^2 - 2\pi G\sigma k_r(i\omega + \nu) + i\omega K^2 = 0. \quad (A\text{-}22)$$

Note that in the limit of $\nu \to 0$ we recover equation (A-16). However, suppose $\nu \gg i\omega$. Then

$$i\omega \sim \frac{2\pi G\sigma k_r - k_r{}^2c^2}{\Omega}\left[\frac{\nu/\Omega}{1 + (\nu/\Omega)^2}\right] \qquad (A\text{-}23)$$

and wavelengths greater than $\lambda \sim c^2/G\sigma$ are unstable. The condition $\nu \gg i\omega$ is satisfied by all wavelengths greater than λ_{crit}. The growth time is of order

$$\tau_\lambda \sim (\nu^2 + \Omega^2)\lambda/4\pi^2 G \sigma \nu. \qquad \text{(A-24)}$$

For objects the size of the first-generation planetesimals, $R = R_4 \times 10^4$ cm and

$$\tau_\lambda \sim 10^3 R_4 (\lambda/\lambda_{crit})(\text{years}), \qquad \text{(A-25)}$$

while for objects less than a few centimeters,

$$\tau_\lambda \sim 30 R^{-1} (\lambda/\lambda_{crit})(\text{years}). \qquad \text{(A-26)}$$

Hence, if the particles do not participate in the local dynamic instabilities originally described (perhaps inhibited by gas turbulence), the growth rate of long-wavelength instabilities is enhanced by a factor of order 10^2. In particular, growth over an astronomical unit, 1.5×10^{13} cm, would require only $\sim 9 \times 10^5 R^{-1}$ years. Of course, as material accumulates, the local surface density increases and the material becomes more susceptible to the shorter-scale local dynamic instabilities. Progressively, larger objects substantially increase the time scale for collapse over still longer wavelengths so that the material decouples from the gas long before planetary-sized objects are approached.

References

Aflvén, H., and Arrhenius, G. 1970a. Origin and evolution of the solar system. I. *Astrophys. Space Sci. 8*, 338–421.

Alfvén, H., and Arrhenius, G. 1970b. Origin and evolution of the solar system. II. *Astrophys. Space Sci. 9*, 3–33.

Alfvén, H., and Arrhenius, G. 1973. Origin and evolution of the solar system. III. *Astrophys. Space Sci. 21*, 117–176.

Alfvén, H., and Arrhenius, G. 1974. Origin and evolution of the solar system. IV. *Astrophys. Space Sci. 29*, 63–159.

Anders, E. 1971. Meteorites and the early solar system, *Annu. Rev. Astron. Astrophys. 9*, 1–34.

Cameron, A. G. W. 1973. Accumulation processes in the primitive solar nebula, *Icarus 18*, 407–450.

Cameron, A. G. W. 1975a. Clumping of interstellar grains, *Icarus* 24, 128–133.

Cameron, A. G. W. 1975b. The origin and evolution of the solar system, *Sci. American 233*, 32–41.

Cameron, A. G. W., and Pine, M. R. 1973. Numerical models of the primitive solar nebula, *Icarus 18*, 377–406.

Chapman, C. R., and Davis, D. R. 1975. Asteroid collisional evolution: Evidence for a much larger early population, *Science 190*, 553–556.

Danby, J. M. A. 1962. *Fundamentals of Celestial Mechanics*, MacMillan, New York.

Giuli, R. T. 1968a. On the rotation of the earth produced by gravitational accretion of particles, *Icarus 8*, 301–323.

Giuli, R. T. 1968b. Gravitational accretion of small masses attracted from large distances as a mechanism for planetary rotation, *Icarus 9*, 186–191.

Gold, T. 1974. The movement of small particulate matter in the early solar system and the formation of satellites, in G. Contopoulous, ed., *Highlights of Astronomy*, pp. 483–485, Reidel, Dordrecht-Holland.

Goldreich, P., and Lynden-Bell, D. 1965. I. Gravitational stability of uniformly rotating disks, *M.N.R.A.S. 130*, 97–124.

Goldreich, P., and Ward, W. R. 1973. The formation of planetesimals, *Astrophys. J. 183*, 1051–1061.

Grossman, L. 1972. Condensation in the primitive solar nebula, *Geochim. Cosmochim. Acta 36*, 597–619.

Grossman, L., and Larimer, J. W. 1974. Early chemical history of the solar system, *Rev. Geophys. Space Phys. 12*, 71–101.

JANAF Thermochemical Tables, 1960 et seq. Compiled by the Thermal Research Laboratory, Dow Chemical Company, Midland, Mich.

Kaula, W. M.   1968. *An Introduction to Planetary Physics*, John Wiley and Sons, New York.

Kaula, W. M., and Harris, A. W.   1975. Dynamics of lunar origin and orbital evolution, *Rev. Geophys. Space Sci. 13*, 363–371.

Larimer, J. W.   1967. Chemical fractionations in meteorites. I. Condensation of the elements, *Geochim. Cosmochim. Acta 31*, 1215–1238.

Lewis, J. S.   1972a. Low temperature condensation from the solar nebula, *Icarus, 16*, 241–252.

Lewis, J. S.   1972b. Metal/silicate fractionation in the solar system, *Earth Planet. Sci. Letters 15*, 286–290.

Lord, H. C. III   1965. Molecular equilibria and condensation in a solar nebula and cool stellar atmospheres, *Icarus 4*, 279–288.

Lyttleton, R. A.   1972. On the formation of planets from a solar nebula, *Monthly Not. Roy. Astron. Soc. 158*, 463–483.

Öpik, E. J.   1973. Comets and the formation of planets, *Astrophys. Space Sci. 21*, 307–398.

Perri, F., and Cameron, A. G. W.   1974. Hydrodynamic instability of the solar nebula in the presence of a planetary core, *Icarus 22*, 416–425.

Reeves, H.   1972. Ed., *Symposium on the Origin of the Solar System*, Centre National de la Recherche Scientifique, Paris.

Safronov, V. S.   1969. *Evolution of the Protoplanetary Cloud and Formation of the Earth and the Planets*, Moscow; English translation 1972, Israel Programs for Scientific Translations, Jerusalem.

Toomre, A.   1964. On the gravitational stability of a disk of stars, *Astrophys. J. 139*, 1217–1238.

Ward, W. R.   1975. Cosmogony of the solar system, *Rev. Geophys. Space Sci. 13*, 422–424.

Whipple, F. L.   1971. On certain aerodynamic processes for asteroids and comets, in A. Elvius, ed., *From Plasma to Planet*, pp. 211–232, John Wiley and Sons, New York.

Williams, I. P., and Cremin, A. W.   1968. A survey of theories relating to the origin of the solar system, *Quart. J. Roy. Astron. Soc. 9*, 40–62.

Woltjer, L.   1967. Structure and dynamics of galaxies, in J. Ehlers, ed., *Relativity Theory and Astrophysics, 2, Galactic Structure*, pp. 1–65, American Mathematical Society, Providence, R.I.

Woolfson, M. M.   1969. Evolution of the solar system, *Rep. Progr. Phys. 32*, 135–185.

2

New Developments in Solar Research

Robert W. Noyes

The sun is a typical main-sequence star, of spectral type G2V. As such, it has no unique claims to set it apart from the 10^{11} other main-sequence stars in our Galaxy or the perhaps 10^{20} main-sequence stars in the observable universe. It is perfectly normal for its type in terms of the usual stellar parameters (see Table 2-1). The only apparently remarkable aspect is that its third planet has evolved a biology including intelligent life, and we have no evidence whether that aspect is unusual or not! However, the existence of that life form is critically dependent upon the sun and its variations, so as interested parties we naturally feel the need to understand solar phenomena.

Thus we study the sun both for what it can tell us about the fascinating phenomenon called a star and for what it can tell us about our own environment. But further, in studying the sun as a star, we also study many general astrophysical processes. We learn about nuclear energy generation and about the basic ways that energy flows, in the form of convection, radiation, or conduction. We observe close at hand the interaction of magnetic fields with matter, both in regions of weak field where motion of ionized plasma carries the field with it (the solar wind), and in regions of strong field where the matter is "frozen" to the field (sunspots). Fundamental phenomena of high-energy astrophysics, including particle acceleration to GeV energies, reveal themselves in an astrophysically modest scale in solar flares. And many of the basic data of astrophysics, such as wavelengths and identification of spectral features from the X-ray regions to the near infrared, come from solar observations.

It is because of the sun's closeness to us that solar research has such great usefulness. The sun is unique among stars in that it may be observed in two dimensions, and with rapid time resolution. It has been studied over an enor-

Table 2-1   The sun as a star.

Distance from earth: 1 astronomical unit (AU) = 1.496×10^{13} cm = 4.848×10^{-6} pc
Gravity at surface: $g_\odot = 2.74 \times 10^4$ cm/sec^2
Luminosity: $L_\odot = 3.86 \times 10^{33}$ erg/sec
Absolute visual magnitude: $M_v = 4.8$
Mass: $M_\odot = 1.991 \times 10^{33}$ gm
Radius: $R_\odot = 6.966 \times 10^{10}$ cm
Effective surface temperature: $T_{eff} = 5800$ K
Spectral type: G2V
Age: $\sim 4.7 \times 10^9$ years

mous range of the electromagnetic spectrum, spanning more than 14 decades of wavelength. In addition, we are able to measure the sun's particle emission and map its magnetic field. Apart from the earth and the moon, the sun is certainly the most thoroughly studied of all astronomical objects.

And yet many important problems remain unsolved. As we probe deeper into the physical interpretation of many solar phenomena that have become foundations for all of astrophysics, we discover unsettling chinks in those foundations. Thus the neutrino flux from the solar interior appears to be at least an order of magnitude below that predicted by models of stellar interiors, leading to a crisis of confidence in basic theories of stellar structure and evolution. The interaction of magnetic fields with matter, which is clearly basic to nearly all areas of astrophysics, is certainly not well understood when, after decades of study, we cannot explain the existence of a sunspot.

The above are two examples of areas of great current interest and research activity in solar physics. Other examples include understanding the 11 year cycle of solar activity, the mechanisms of solar flares, the detailed nature of coronal heating, the coronal source of the solar wind, and the impact of solar variability on the earth and its atmosphere. In subsequent sections we discuss some of these major current problems in turn, in an attempt to describe both what is known and what we may hope to learn from current research.

The Solar Interior

Research into the nature of the solar interior is a subject that a few years ago seemed well-tidied up, and theorists specializing in stellar interiors were turning to more challenging problems. The energy source of the solar interior was understood as being about 98% due to the proton-proton cycle and 2% due to the carbon-nitrogen-oxygen cycle, both of which have the effect of converting four hydrogen nuclei into one helium nucleus, and releasing the mass difference in the form of energy. The evolution of the sun since it arrived on the main sequence some 5 billion years ago was well described, and its ultimate fate when it will evolve through the red giant stage in another 5 billion years was completely predictable. Detailed models exist for the variation from surface to core of parameters such as temperature density, pressure, convective transport, opacity, and composition. (For a lucid description of the interior of the sun and other stars, see Schwarzschild, 1958).

Figure 2-1 shows schematically the variation of temperature, density, and energy flow with radius in the sun, as based on such models. The models are theoretically self-consistent, agree with known aspects of stellar evolution as shown by stars of other types and ages, and agree with all known parameters observed on the sun.

Except one. That is the now famous case of the missing neutrinos, and as of this writing it is threatening to topple the entire structure of stellar interior theory.

Solar neutrinos would seem to be an ideal probe of the solar interior. Their cross section for interaction with matter is so low that the chance of a neutrino being absorbed by the sun on its way out of the interior is completely negligible. They are certainly our best clue to the energy-generating processes in the interior, and one of a very small number of means we have of probing beneath the surface at all. There are a few others, which we discuss later; these are also now challenging any complacency we may still have about our understanding of the interior.

Briefly, standard theory predicts that two neutrinos are produced for every fusion of four protons into a helium nucleus. A unique detector, developed by R. Davis of Brookhaven National Laboratories, consisting of a tank car filled with cleaning fluid (C_2Cl_4) has been placed deep in a mine (to shield against reactions from cosmic ray protons that would overwhelm the experiment at the earth's surface) to detect the more energetic of these neutrinos by the reaction $\nu + {}^{36}Cl \rightarrow {}^{37}A + e^-$. The amount of radioactive ${}^{37}A$ produced is measured by sensitive radiochemical techniques. According to standard models, the sun in producing its measured luminosity of 4×10^{33} erg/sec by nuclear reactions in the interior will generate enough neutrinos to produce 5.6×10^{-36} neutrino captures per Cl atom per sec in the Davis de-

Figure 2-1. Idealized general solar properties, structure, and modes of outward energy flow. The features shown are not to scale and provide a qualitative picture only. From Gibson (1973).

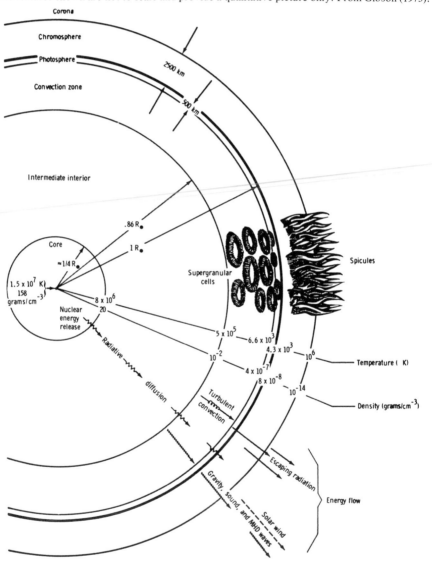

tector. This is 5.6 solar neutrino units, where the SNU is defined as 10^{-36} neutrino captures per sec per Cl atom. To date no neutrinos have been detected by the Davis experiment, and the current upper limit is 1.5 SNU (Bahcall and Davis, 1976), which is well below the lowest limit of any standard interior model. See Bahcall and Davis (1976) for further discussion of the experiment, and its implications for the structure of the interior.

What can be wrong? The possibility of errors in nuclear reaction cross sections, upon which the predicted number of SNU depends, has been carefully studied, and does not seem to be a likely way out. If the opacity of the interior were lower, the central temperature of the sun could be lower and still allow the observed radiative luminosity to escape, but the production rate of neutrinos energetic enough to be detected by the Davis experiment would decrease because of the decreased core temperature. However, it does not seem possible to decrease the opacity sufficiently below presently accepted values to bring the predicted count rate down to the observed upper limit (Bahcall et al., 1973). Fowler (1972) made the ingenious suggestion that the core is in a transient state which does not produce neutrinos while the surface reflects an earlier high-temperature, high-luminosity state; this is possible because the time for diffusion of radiative energy from the core to the surface is about 10^7 years. A specific mechanism for this was suggested by Dilke and Gough (1972, 1974) involving the periodic mixing into the core of ^{3}He, which burns rapidly, causing the core to expand and cool to a lower temperature. It would also cause the solar luminosity to decrease somewhat later, and has been suggested as a method of causing ice ages.

However, as recently discussed by Ulrich (1975), paleoclimate evidence concerning the long-term temperature variations on the earth suggests that the sun's luminosity has not changed from its present value by more than about 3% over the past million years, although the Dilke and Gough mechanism would induce at least a 10% change in the solar luminosity over that time scale.

Looking at the earth's temperature over even larger time scales, we find another source of uneasiness about the theory of the solar interior. The theory of stellar evolution demands that the sun's luminosity must have increased by perhaps 30% since it began its existence as a main-sequence star some 5 billion years ago. Yet it has been argued (Sagan and Mullen, 1973) that if the solar luminosity had been as much as 25% less than the present value, the oceans would have frozen and they probably would not have melted thereafter even at the present luminosity (because of the increased albedo of ice). It is premature to take this as a serious objection to a varying solar luminosity, however, until more sophisticated climate models are developed.

In what some might classify as desperation, other even more bizarre suggestions have been made to resolve the neutrino problem. Clayton, Newman, and Talbot (1975) point out that if a black hole, with mass $10^{-5}\,M_\odot$ and radius of a few centimeters, resides at the center of the sun, it would provide about half of the solar luminosity. The remainder could be provided by standard proton-proton reactions without producing more than one SNU of neutrino emissions. Hoyle (1975) proposed that the sun was formed in two stages: an initial inner core containing half the present sun's mass was first formed with a low helium abundance and a high abundance of metals. The exterior, with the composition that we observe today, was added by accretion about 4.7×10^9 years ago. Such an object would have a convective core, with reduced neutrino emission.

Aside from rather extreme suggestions of the type just described, then, no solution appears in sight for the neutrino problem. Table 2-2, from Bahcall and Davis (1976) presents the implication of the present observational limits. From the table we see that if the CNO cycle dominated the proton-proton chain, 35 SNU should be detected; the upper limit of 1.5 SNU clearly verifies that the CNO cycle is only a minor contribution to the sun's energy generation. In addition, however, the upper limit seriously threatens standard interior models, and basic ideas of steady evolution (without perturbations of the sort described above) are also under fire. Fortunately, the concept of nuclear fusion as the primary energy source (which is the only

Table 2-2 Levels of meaning in the ^{37}Cl experiment.

Counting rate (SNU)	Level of meaning
35	Expected if the CNO cycle produces the solar luminosity
~7 ± 3	Predictions of current models
1.5	Expected as a lower limit consistent with standard ideas of stellar evolution
0.3	Expected from proton-proton reaction. Hence a test of the basic idea of nuclear fusion as the energy source for main sequence stars.

From Bahcall and Davis (1976).

conceivable energy source given the near-constancy of the solar luminosity for at least 3×10^9 years) is not yet under challenge by the data.

Very recently two new observational programs have emerged to create a new and potentially very valuable method for probing the interior, and we may hope that this will combine with the neutrino experiment to lead to a resolution of the current dilemma. The first of these two programs arose from an entirely independent chain of observations and theory, sparked by speculations about the correct formulation of general relativity. This is the search for a small oblateness of the sun, undertaken by Dicke and Goldenberg (1967, 1974; see also Dicke 1974 for a review) in order to provide an alternative source for part of the relativistic precession of the perihelion of the planet Mercury. According to Dicke's scalar-tensor theory of general relativity, only about 93% of the observed excess rate of perihelion advance can be explained by relativistic effects (as opposed to Einstein's general theory of relativity, which neatly explains the entire excess perihelion advance.) The remainder could be supplied by the quadrupole moment that the sun would have if it had an oblateness of $\Delta R/R \sim 4.5 \times 10^{-5}$, where ΔR, the difference between equatorial and polar radii, is then 30 km. Of this only about 4 km is expected from rotation at the observed surface velocity. The needed excess oblateness *could* result if the core were rotating rapidly (some twenty times as fast as the surface) and as a by-product such a rapid rotation would decrease the central temperatures and pressure enough to lower the neutrino flux to near present upper limits (Demarque, Mengel, and Sweigart, 1973; Roxburgh, 1967). Determining the oblateness is extremely difficult because of the smallness of the expected effect and the confusing influence of surface phenomena such as active regions and small variations in atmospheric temperature structures. However, a recent set of measurements with a telescope specially built for the purpose (Hill and Stebbins, 1975) has indicated that the oblateness predicted by Dicke does *not* exist, that there is thus support for Einstein's general theory of relativity, that the sun does not have a rapidly rotating core, and finally that the neutrino problem is still with us.

It is almost a truism that a new sensitive instrument will uncover additional unexpected effects, and the telescope just mentioned is no exception. Hill and his colleagues have recently detected periodic oscillations of the brightness of the solar limb, with a fundamental period of about 52 minutes, as well as several higher-frequency harmonics (Hill, Stebbins, and Brown, 1975). These oscillations are thought to be due to radial pulsations at the fundamental and harmonic normal modes of pressure wave (so-called p-mode) fluctuations in the sun. Their frequencies are sensitively dependent on the distribution of temperature and density throughout the interior, and thus careful measurement of their frequencies should permit determination of the interior structure, just as the free oscillation of the earth after earthquakes gives us information on its interior. For a discussion of the inversion of such data to give the structure of the solar interior, see Christensen-Dalsgard and Gough (1976).

At the same time, two other groups, one in the Soviet Union (Severny et al., 1976) and one in England (Brookes et al., 1976) have independently dis-covered global velocity oscillations of an even longer period—2 hours and 40 minutes. The velocity amplitudes of these oscillations are exceedingly small—only about 2 m/sec—leading to the total displacement of the surface during the oscillation of about 10 km. The long period of the oscillation is very difficult to understand in terms of radial p-mode oscillations, which should have a fundamental period close to the 52 minutes observed by Hill et al. (1975). In particular, if the period of 2 hours and 40 minutes is a true radial oscillation, it is extremely difficult to see how the density of the core could be great enough to permit energy generation from the proton-proton reaction (Severny et al. 1976). While this would conveniently resolve the neutrino problem, it is extremely unlikely that nuclear reactions are not the main energy source for solar and stellar luminosity. A more likely explana-tion is that the oscillations are nonradial gravity modes (g-modes), rather than p-modes; g-modes should exist, but at many close-lying periods, and it is not clear why only one is actually selected out.

The observations are at this writing only in their earliest stages, and like all first observations of subtle phenomena require further verification before they can be accepted as definitive. Should they be verified, however, the new information they convey, together with further observational and theo-retical work on the neutrino problem, will place our knowledge of solar and stellar interiors on a far firmer footing.

Convection, Waves, and a Hot Corona

The rate of outward thermal diffusion of radiation from the interior varies directly with the temperature gradient dT/dr and inversely with the mean opacity κ according to the relation

$$L_r = 16\pi r^2 \frac{1}{3\kappa\rho} \frac{d}{dr} (acT^4), \qquad (2\text{-}1)$$

where L_r is the radiative flux through a shell of radius r (see Schwarzschild, 1958, Chapter II). At high temperatures such as exist in the deeper layers of the sun, the opacity is low enough that all of the energy produced in the core can easily be transported by thermal diffusion, even if the temperature gradient is rather small. However, toward the surface of the sun the opacity begins to increase, owing to the recombination of electrons and nucleii into ions that can absorb photons by photoionization or free-free absorption. In compensation, the temperature gradient increases to allow the radiation to flow unimpeded, according to equation (2-1). However, at a certain point ($r \sim 0.85\, R_\odot$ in the sun) the increased opacity drives the temperature gradient to a value so large that a new effect occurs, called convective instability, that completely alters the energy transport properties of the interior.

To understand convective instability, consider a blob of gas which is in horizontal pressure equilibrium with its surroundings. Then $P = (R/\mu)\rho T$ (where R is the gas constant) is constant across any horizontal surface. Neglecting changes in molecular weight μ, if the blob is hotter than its surroundings it will be less dense, hence more buoyant, and will rise. As the blob rises into regions of lower pressure, it will expand and, in the process, cool adiabatically. Whether the buoyant rise of the blob eventually stops depends on whether, after expanding adiabatically, the blob is now cooler (and hence denser and less buoyant) than its surroundings, or whether it is still hotter and more buoyant than its surroundings.

The onset of convective instability occurs, then, when the ambient temperature gradient dT/dr becomes greater than the *adiabatic* temperature gradient dT/dr_{ad}, i.e., that temperature gradient experienced by a blob rising and cooling adiabatically. The adiabatic gradient may easily be seen to be

$$\frac{dT}{dr_{ad}} = \frac{\gamma - 1}{\gamma} \frac{T}{P} \frac{dP}{dr},$$

where γ is the ratio of specific heats ($\gamma = \frac{5}{3}$ for perfect gases) and dP/dr is the ambient pressure gradient.

Above the level of onset of convective instability, the vast bulk of the energy flow is by convection, in which hot blobs of material rise and then give up their excess heat to their surroundings. Similarly, the falling material in the convective flow is cooler than its surroundings and also contributes to a net upward flow. Further details on the energy carried by convection may be found in Chapter 4, as well as in treatments by Schwarzschild (1958) or Spiegel (1971). We content ourselves here with noting that above $r \sim 0.85\,R$ in the sun a *convective zone* exists, in which convective transport is so efficient that nearly all of the sun's energy flow is by convection. (While the convective zone occupies some 30% of the volume of the sun, because of the low density near the surface it contains only 1–2% of the mass of the sun.)

Convection manifests itself at the surface through the well-known granulation, which we now discuss, and the somewhat less well known supergranulation, to be discussed later.

Granulation and Oscillations

Near the outer boundary of the sun, where radiation can escape directly, the opacity and temperature gradient decrease and the atmosphere becomes stable against convection once again. The granulation is commonly thought to represent the overshoot of the topmost convective elements as they penetrate the stable atmosphere. However, the situation is by no means so simple, as we see below.

Briefly, the granulation (Figure 2-2a) may be described as a pattern of quasi-polygonal cells, with characteristic size $\sim 10^3$ km, temperature fluctuations ~ 300 K, and lifetimes of a few minutes (Leighton, 1963). The bright granules appear to be rising, and the dark intergranular lanes falling; characteristic velocities are about 0.5 km/sec. The correlation of rising material with excess temperature implies upward transport of nonradiative energy, which may be directly calculated from observations (Leighton et al., 1962) as $\gtrsim 10^7$ erg sec^{-1} cm^{-2}. This is exceedingly small compared to the radiative flux ($\sim 6 \times 10^{10}$ erg sec^{-1} cm^{-2}), in distinction to the situation well inside the convective zone, where convection dominates the energy transport. In fact, the nonradiative energy flux in the granulation would be of little importance, were it not large enough to supply the very small amount of nonradiative heating ($\sim 10^7$ erg sec^{-1} cm^{-2}) necessary to maintain the high temperature of the chromosphere and corona. For this reason, the details of the nonradiative flux in the granulation become exceedingly interesting.

In 1960, it was discovered (Leighton et al., 1962) that a large part of the velocity field in and above the granulation was distinctly nonrandom, but instead was vertical and oscillatory in nature, with a well-defined average period of almost exactly 5 minutes (300 sec), and velocities about 0.5

Figure 2-2. Solar granulation. a, Observed with the Sacramento Peak Vacuum Tower Telescope through ¼ A filter in the continuum near Hα(6563 Å). b, The same region observed in the far wing (+2 Å from line center of Hα, showing filigree structures (see discussion of Photospheric Magnetic Knots later in this chapter). Photographs were taken about 1 minute apart, 10 arcsec corresponds to 7000 km on the sun. See Dunn and Zirker (1973) for details.

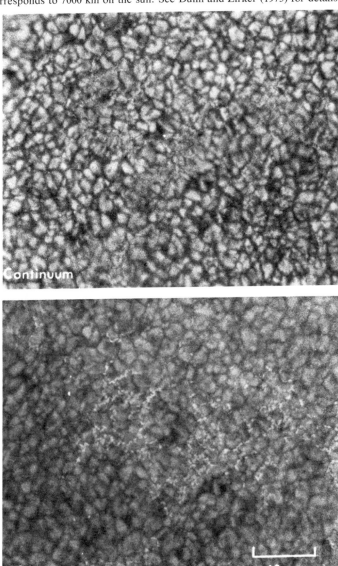

km/sec. In other words, virtually all points in the convectively stable layer above the top of the convection zone were moving up and down quasi-periodically with respect to the mean atmosphere, with an excursion of about ± 25 km (Figure 2-3). The horizontal coherence of this vertical oscillation was variously measured from about 10^3 km (granule size) to sizes as large as 5×10^4 km. It was immediately deduced that the oscillation was a standing wave (from the phase relations at various heights), and existing theoretical descriptions of waves in stratified, gravitating, compressible atmospheres (e.g., Tolstoy, 1963) were used to describe the phenomenon.

Two cutoff frequencies determine the propagation characteristics of waves in a gravitating atmosphere. An acoustic cutoff frequency is defined according to $\omega_{ac} = (c/2H)$, where c is the sound speed and H the density scale height, $RT/\mu g$ (this is the height over which the density decreases by the factor $1/e$). Above ω_{ac} ordinary sound waves (modified by gravity) can propagate, but sound wave propagation is impossible below ω_{ac}. There is another critical frequency N, known as the Brunt-Vaisala frequency, where $N = (\gamma - 1)g/c$ for an isothermal atmosphere. Below this frequency propagation is possible in the form of gravity waves, which are ones in which the restoring force is provided by gravity rather than by compressibility, as for acoustic waves; a familiar example of one type of gravity wave is ocean waves.

Figure 2-3. Time variation of the velocity of a typical point on the solar disk, showing the growth, decay, and phase change of the "5-minute" oscillation. From White and Cha (1973).

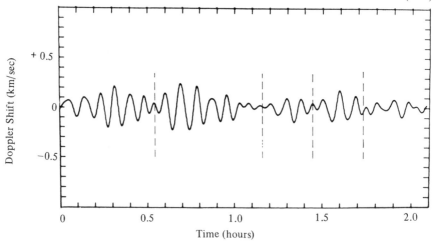

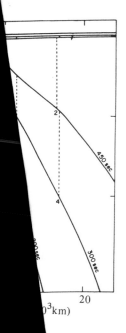

ecting boundary as a function of horizontal wave-
ds. For each period, vertical dashed lines are in-
mber of wavelengths just fits between the bounda-
lrich (1970). b, Diagnostic diagram, in which the
ch resonance occurs at the labeled modes, from
ved power density in the oscillation, where each
revious one, and exhibit the predicted normal

The *dispersion relation* defining the propagation characteristics is given by

$$k_z^2 = \frac{\omega^2 - \omega_{ac}^2}{c^2} - k_H^2\left(1 - \frac{N^2}{\omega^2}\right), \qquad (2\text{-}2)$$

where k_z and k_H are the vertical and horizontal wavenumbers (i.e., $2\pi\lambda_z$ and $2\pi/\lambda_H$) and ω is the temporal frequency $2\pi/T$ where T is the wave period. For a derivation and discussion of this equation, see Tolstoy (1963.)

From equation (2-2), one may deduce the regions of ω and k_H where propagation is possible, i.e., where $k_z^2 > 0$; for $k_z^2 < 0$ one has vertical attenuation, not propagation. These regimes are shown in Figure 2-4. In this figure the horizontal axis is k_H and the vertical axis is ω. In the shaded portion at the upper left, acoustic waves, modified by gravity, can propagate. For $\omega \gg \omega_{ac}$ the waves are ordinary sound waves (the wavelength is so small compared to the scale height that stratification exerts no influence) but for frequencies only slightly above the cutoff frequency ω_{ac}, only near-vertical propagation is possible ($k_H \sim 0$). In the shaded portion at the lower right of the graph, gravity waves (modified by compressibility) can propagate. These waves propagate primarily horizontally. Waves in the nonshaded region cannot propagate, i.e., they can exist only as standing waves without transporting energy.

Until recently, it was commonly thought that the observed oscillations (Figure 2-3) were simply this nonpropagating component, which was perhaps excited when the base of the stable zone was perturbed by the impact of a rising hot granule. Those frequencies which were able to propagate away by gravity or acoustic waves did so, leaving the atmosphere "ringing" in its trapped frequencies. However, there were nagging discrepancies associated with this picture: The oscillating regions were often coherent over sizes much larger than a granule; there did not seem to be clear evidence that rising granules were followed by oscillations; and the observed frequencies were not quite correct.

Over the past few years a far more convincing and more interesting, picture has begun to appear. That is that the waves are produced below the surface in the convective zone at preferred frequencies that match the characteristics of a "resonant cavity" in the upper convective zone.

Ulrich (1970) pointed out that while acoustic waves modified by gravity (upper left portion of Figure 2-4) can propagate near the top of the convection zone, for each combination of k_H and ω there is a depth below which

propagation ceases, i.e., below which the boundary between propagation
and nonpropagation in Figure 2-4 has shifted to a higher ω. This is because
of the rapid increase of temperature and hence sound speed c with depth.
(From equation 2-2 we see that the minimum frequency for propagation is
always greater than $\omega = k_H c$.) The transition from propagating to nonprop
agating conditions is equivalent to a reflecting boundary. At the sam

Figure 2-4.   Diagnostic diagram for oscillations in an isothermal atmosphere with $T = $
the shaded region $k_z^2 > 0$, and propagation occurs. In the unshaded region $k_z^2 < 0$, and a
ation occurs. The dashed lines show the boundary between propagating acoustic wav
nonpropagating waves for temperature T_1 and T_2, where $T_1 < T_0 < T_2$. The intersectio
dashed curves occurs at that value of (ω, k_H) for which there is reflection at both up
lower boundaries at temperatures T_1 and T_2 respectively (see text).

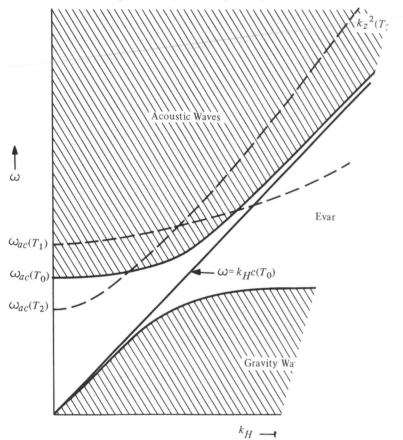

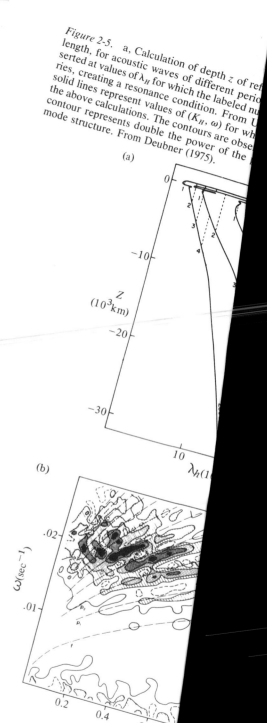

Figure 2-5.   a, Calculation of depth z of ref
length, for acoustic waves of different perio
serted at values of λ_H for which the labeled nu
ries, creating a resonance condition. From U
solid lines represent values of (K_H, ω) for wh
the above calculations. The contours are obse
contour represents double the power of the
mode structure. From Deubner (1975).

Coronal Heating

It should be noted that the 5-minute oscillations are actually observed in the upper photosphere, above the reflecting upper boundary of the propagation region discussed earlier. However, the observed region is less than one vertical wavelength from that boundary. Hence the waves are easily seen in the "evanescent" region overlying the upper reflecting boundary. Some of the energy in these waves can indeed be deposited in the upper atmosphere by means of shock dissipation if the velocity amplitude grows until it approaches the sound speed. (Such dissipation changes the reflecting boundary somewhat, allowing the necessary energy to propagate into the evanescent region.) In addition, waves of frequency larger than the maximum critical frequency in the photosphere (which occurs at the temperature minimum in the low chromosphere and corresponds to a period of about 190 sec) will suffer no reflection at all and will proceed unimpeded into the outer atmosphere.

Acoustic waves are thus a primary candidate for supplying the energy necessary to heat the solar chromosphere and corona. The mean temperature structure of the chromosphere and corona, and the energy flow associated with it, are shown in Figure 2-6. We see that about 10^7 erg sec^{-1} cm^{-2} must be supplied to the chromosphere and corona, of which about 10^6 erg/cm^2 must be deposited at the region of maximum temperature, probably located more than 50,000 km above the surface.

Much of the energy deposited in the corona flows *back* to the chromosphere by thermal conduction, whose flux $F_c = \kappa(dT/dr)$, where κ is the thermal conductivity and dT/dr the temperature gradient. In fact, until very recently (see below) observations indicated that thermal conduction greatly exceeded all other energy losses. If thermal conduction dominates the energy balance, energy conservation then requires that the conductive flux be nearly constant with height, and thus $dT/dr \sim 1/\kappa$ in those regions. Since for a fully ionized gas $\kappa \sim T^{5/2}$ (see, e.g., Spitzer, 1962) this implies that the temperature gradient at $T \sim 10^5$ K is 300 times steeper than at 10^6 K. Such a result is in good agreement with observational data from extreme ultraviolet emission lines formed at these temperatures: The *emission measure* $n_e^2 \, \Delta V$ of lines formed at a few hundred thousand degrees is very low compared to lines formed at 10^6 K, and implies that the volume ΔV of material emitting at several hundred thousand degrees is low. This is consistent with a very sharp *transition zone* between the 10^4 K chromosphere and the 10^6 K corona, as is sketched in Figure 2-6. It is also consistent with the very large temperature gradient at 10^5 K implied by constancy of conductive flux. Indeed the model sketched in Figure 2-6 satisfies, for 10^5 K $< T <$ 10^6 K, the condition that $F_c = \kappa(dT/dr) = 6 \times 10^{-7} T^{5/2}(dT/dr) =$ constant $= 6 \times 10^5$ erg sec^{-1} cm^{-2}.

Although we shall not deal with it further here, models may be con-

Figure 2-6. Model of the mean temperature and density structure of the quiet solar atmosphere. Also labeled are the approximate heights of EUV emission lines, as seen in Figure 2-8.

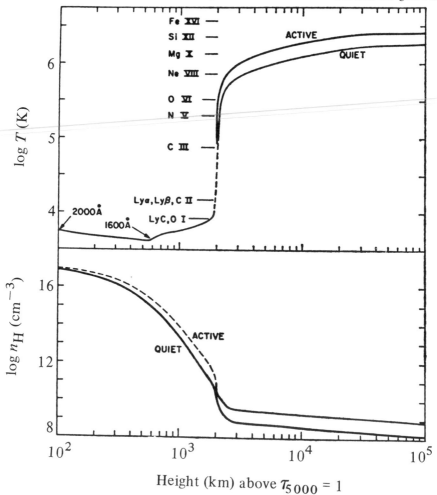

Height (km) above $\tau_{5\,000} = 1$

structed (Withbroe, 1970a, 1970b, and others; see Noyes, 1971, for a review) that satisfy the relation of hydrostatic equilibrium, match observed intensities of extreme ultraviolet lines, and preserve the condition that F_c is approximately constant in the transition zone. Such a model in fact is the basis of Figure 2-6. Given the model one can calculate other loss rates such as radiative losses from each height.

In principle, one might hope to predict a priori the structure and energy balance of the corona, knowing only the details of the convection zone and the waves that pass through it, if such energy inputs were the sole factor determining the structure. In practice, however, this must be dismissed as impossible. Not only are the processes described above already so complicated in detail as to defy accurate prediction with our present understanding, but there is a major additional complication, which is the influence of magnetic fields. The overriding influence of magnetic fields in the corona is just now becoming understood as we obtain high-resolution data on the upper chromosphere and corona from space vehicles. In the light of these data, to be discussed below, it seems that our best hope of understanding coronal heating is to establish empirically the detailed energy losses from the corona, and hence the requirements for sustaining it, and ultimately to use those requirements to determine the physics of the energy transport into the corona. For this purpose we must now look at the detailed structure of the chromosphere and corona. To do so, however, we first go back to another regime of motions in the photosphere—the supergranulation.

Supergranulation Flow and Its Effects

Coexisting with the motions associated with the granulation and five-minute oscillations of the solar photosphere is a completely different type of motion. This is the supergranulation (Figure 2-7a), so named because it looks like a convection pattern whose elements have sizes more than an order of magnitude larger (30,000 km) than those of the granulation (1000 km). However, it may be that a similarity of names is *all* that the two types of flow have in common. The supergranulation has a very regular structure of upwelling and outflowing convective elements, in contrast to the much less regular motion of the granulation. As opposed to the granulation, there are no bright and dark lanes because of temperature differences associated with the supergranulation. (For this reason the supergranulation is seen only through its pattern of motion, and was unrecognized until 15 years

Figure 2-7. a, "Velocity spectroheliogram," from Mount Wilson Observatory, in which velocities of approach are indicated by lighter-than-average areas and velocities of recession appear
darker. Outflow of gas occurs from the centers of supergranulation cells, with predominantly
horizontal velocities of 0.5 km/sec and scale size of 30,000 km. This produces the mottled
appearance of the image. See Simon and Leighton (1964) for details. b, Ca^+K spectroheliogram
(not made at the same time as 2-7a), showing the chromospheric emission network that overlies
the boundary of supergranulation cells. Also shown is the strong Ca^+K emission overlying the
strong magnetic fields in active regions. Courtesy Sacramento Peak Observatory, Air Force
Cambridge Research Laboratories.

(a)

(b)

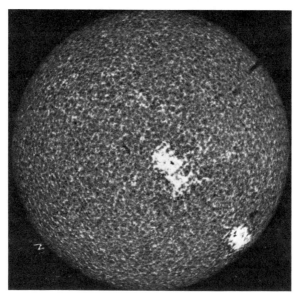

ago.) The lifetime of supergranulation elements is about a day, rather than a few minutes as for the granulation. Most important from the point of view of the structure of the sun's outer layers, the supergranulation is associated with, and presumably causes, a network of concentrated magnetic fields, concentrated at the boundaries of the supergranulation cells. This network appears to exert fundamental control over the structure of the upper atmosphere. The granulation and its associated waves may perhaps produce the heating of the corona, but certainly the supergranulation and its associated magnetic fields produce the shaping of the corona.

Figure 2-7b shows the related (and much longer known) pattern of emission in the chromospheric Ca^+K line known as the *chromospheric emission network*. Shortly after the discovery of the supergranulation it was found (Simon and Leighton, 1964) that the chromospheric network overlies the outer perimeter of the supergranulation cells. Detailed measurements of the velocity field in the cells also showed that the gas, having risen at the center of cells and flowed outward from the center, sank again at the edges. Because it was already known that the Ca^+K line mapped out the magnetic field in a one-to-one correspondence (Leighton, 1959), it was suggested (Leighton, 1963) that magnetic fields were being concentrated at the downflowing boundaries of the supergranules, presumably being swept there by the motions themselves.

How far upward in the atmosphere does the chromospheric network persist? This important question has been recently answered by the Harvard experiment on the Skylab Apollo Telescope Mount. Figure 2-8 shows simultaneous observations of the structure of the network in six extreme-ultraviolet emission lines, characteristic of the chromosphere, chromosphere-corona transition zone, and low corona. The approximate height of these emissions is marked on Figure 2-6. We note that the network is clearly visible up through the transition zone, but becomes "washed out" at coronal levels. A reasonable inference (not yet proven, however) is that the intensity and contrast of the network reflect the intensity and fluctuation of the magnetic field in the transition zone and corona, just as it does in the chromospheric Ca^+K line. This supports the schematic pictures of the network and its associated magnetic fields shown in Figure 2-9 (from Gabriel, 1975). In this picture, as we have already suggested, photospheric convection currents associated with the supergranulation outflow sweep the magnetic field lines to the supergranular boundaries. This is possible in the photosphere, where the gas density is so high that the gas pressure exceeds the magnetic pressure.

(In the photosphere the gas pressure $P_g \sim 10^5$ dyne/cm², which is balanced by the magnetic pressure $B^2/8\pi$ for a magnetic field $B \sim 1500$ gauss. Present indications are that the actual field strength in the photosphere may be bunched into tight knots of this strength or even greater; we return to this point in a later section.)

However, because of the very small pressure scale height (~ 100 km), the magnetic pressure exceeds the gas pressure above a level in the low chromosphere, so that the gas is unable to push the fields around, and the field remains nearly uniform.

Even this picture is seen to be too simple when compared with very high resolution data, as we mention below. Nevertheless, it represents a major

Figure 2-8. The chromospheric emission network in the extreme ultraviolet, showing its extension to higher temperatures in the transition zone and low corona. Approximate temperatures and heights of emission are given in Figure 2-6. Harvard College Observatory photograph.

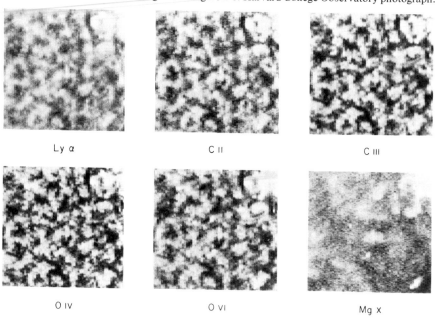

Ly α   C II   C III

O IV   O VI   Mg X

QUIET CHROMOSPHERIC NETWORK

AUG 13,1973 16:00 UT

advance over plane-parallel models of the corona, and for two reasons drastically reduces the energy requirements to support the corona. These are:

1. The energy lost from the corona by thermal conduction is forced to flow along field lines (the small gyro radius of electrons in magnetic fields effectively chokes off their ability to transport thermal energy normal to field lines). Thus the downflow of heat is confined to the network boundaries, covering only a small fraction of the total solar surface.

2. The thickness of the chromosphere-corona transition zone in the network must be several times greater than the thickness of the transition zone that one infers for a homogeneous atmosphere, such as modeled in Figure 2-6. The reason is that, as we mentioned above, the thickness is derived from the observed emission measure $n_e^2\,\Delta V$ and hence from the volume ΔV that emits extreme-ultraviolet line radiation at transition zone temperatures. Constricting the volume elements horizontally as in Figure 2-9 requires increasing their thickness vertically, if ΔV is to continue to match the value inferred from observations (see Gabriel, 1975). The importance of this geo-

Figure 2-9. Structure of magnetic fields in the chromosphere and corona, according to calculations of Gabriel (1975). The convective flow of the supergranulation is assumed to carry photospheric magnetic fields to the periphery of the supergranules, as originally suggested by Leighton (1963). The fields assume a force-free configuration, subject to the boundary condition that they become uniform in the corona. Thermal conduction downward along the field lines establishes the temperature structure indicated by the isotherms shown.

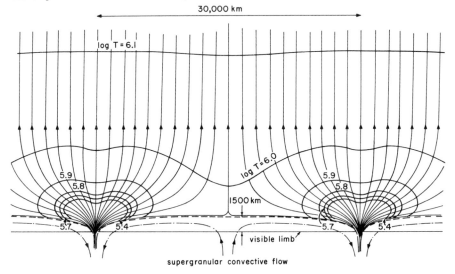

metrical change for the energy flow is that the thermal conduction rate per unit area is proportional to the temperature gradient dT/dz and will vary inversely as the thickness of the transition zone.

The two effects just described will decrease the thermal conduction back from the corona by at least an order of magnitude, as first pointed out by Kopp (1972). Since thermal conduction dominates the loss of energy from a plane-parallel corona, the requirements for heating the corona are correspondingly reduced. At present, the new ultraviolet data from Skylab are being intensively studied to pin down as precisely as possible the actual energy balance of the corona, given the complicated structure that exists there.

In closing this section, we must point out that even the high-resolution ultraviolet data from Skylab are not sufficient to resolve the finest details of coronal structure, and as a result major physical processes are still poorly understood. Chief among these are the dynamics of *spicules*—small, very dynamic jets of chromospheric material shot up into the corona with velocities of about 20 km/sec from the region of the chromospheric network. Figure 2-10 shows recent superb observations from Sacramento Peak Observatory that show how the network in Hα leads up into a pattern of spicules when observed in very high resolution. The kinetic energy carried by the spicules is comparable to other flows of energy in the chromosphere and corona (Kuperus and Athay, 1967), and surely must be included as a significant part of the entire picture of energetics of the chromosphere and corona.

Large-Scale Structure of Corona, Coronal Holes, and Solar Wind

In the previous sections we saw that magnetic fields associated with the supergranulation shaped the small-scale structure of the low corona. It has long been known that the outer corona consists of large-scale structures that also appear to be shaped by magnetic fields, as seen by eclipse data such as the beautiful example in Figure 2-11 (Newkirk, 1968). Here the density structure of the outer corona, revealed by the Thomson scattering of free electrons, shows structures clearly suggesting entrapment of matter by magnetic fields.

The true richness of coronal structures was revealed, however, only through X-ray and extreme ultraviolet observations obtained on the Skylab ATM mission, which showed the corona face on, rather than only from the side, as is required by eclipse observations. Further, X-ray emission is sensitively dependent on temperature and also varies as n_e^2, while white light

Figure 2-10.   Spicules at the edge of supergranulation cells, as observed 0.8 Å to the red of center of Hα 6563 Å, with a ¼ A Zeiss filter at Sacramento Peak Observatory's vacuum tower telescope. Small bright mottles mark the base of spicule clumps, which occur in the chromospheric emission network. Photograph courtesy Sacramento Peak Observatory, Air Force Cambridge Research Laboratories.

emission depends only on n_e. Thus hot or dense features show up with greater contrast in X-rays than in white light. Figure 2-12 shows an image of the sun in broadband X-rays (Vaiana et al., 1976), along with a photospheric magnetogram; in the latter, light and dark areas represent fields of opposite polarity.

In Figure 2-12 we see a wealth of X-ray structures in the corona. Prominent among them are:

1. Active regions. These large bright areas of high density and temperature in the corona are seen to overlie photospheric regions characterized by strong bipolar or more complex magnetic fields. Many loop-like structures

Figure 2-11.   The solar corona, as seen during the 1966 solar eclipse, is seen in white light. This photograph, made with a radially graded, neutral density filter which compensates for the rapid decline of coronal brightness with increasing height above the sun, displays the fine structure of the solar corona out to several solar radii. Note the ''helmet'' streamers, prominences, plumes, and the image of Venus overexposed at the left-hand side of the photograph. Photograph courtesy of G. Newkirk, High Altitude Observatory.

Figure 2-12. Soft X-ray image of the sun, with a corresponding Kitt Peak magnetogram. The soft X-ray image, obtained with a grazing-incidence, spectroscopic telescope aboard Skylab, delineates the low corona, with active region loops at million-degree temperatures, bright points, and coronal holes. A comparison with the magnetogram, a map of the longitudinal magnetic field, indicates the correspondence of magnetic field with the coronal structures. From Vaiana (1976).

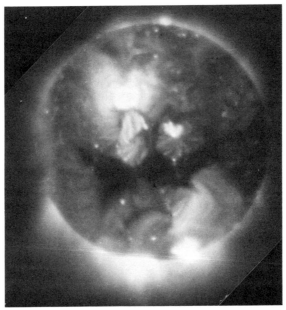

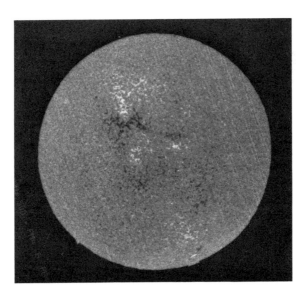

are seen, invariably connecting regions of opposite magnetic polarity in the photosphere. Apparently the greater X-ray emission (i.e., the high density and temperature) is related to the fact that the associated magnetic fields are both rather strong and represent a *closed* configuration; i.e., they may be traced without interruption from one end to the other. We shall discuss active regions in more detail in the next section.

2. Large-scale closed structures of medium brightness often interconnecting separate active regions, and often joining much smaller-scale features remote from active regions.

3. Small bright points of X-ray emission, which are associated with tight bipolar magnetic regions (adjacent small black and white regions on the magnetograph). We shall discuss them in detail later in this chapter.

4. "Open" structures such as the linear structures lying more or less on the 60° S parallel of the figure, in which the X-ray features diverge. Such features often underlie streamers such as that seen in the white light coronal photograph of Figure 2-11.

5. Coronal "holes," large areas that are essentially devoid of X-ray emission, which are seen in Figure 2-12 at the center and at the north pole. The diverging nature of the emission features, and the predominantly unipolar nature of the underlying photospheric field, suggest that these regions are open.

Coronal holes were first thoroughly studied by use of lower-resolution ultraviolet data (Munro and Withbroe, 1972), from which it was inferred that (a) the density of the corona immediately above coronal holes is about a factor of 3 lower than in the average "quiet" sun. (b) The coronal temperature is reduced to about 1.0×10^6 K over holes, as opposed to about 1.8×10^6 K over normal quiet areas. (c) The thickness of the chromosphere-corona transition zone is increased by about a factor of 3 in holes over that in the quiet sun; this implies that the temperature gradient, and hence the conductive flux, are decreased by a factor of 3 in holes. (d) Coronal holes are virtually invisible in radiation emitted from the chromosphere or photosphere, i.e., regions with temperatures less than 20,000 K.

It is interesting that coronal holes seem to have few if any manifestations in the photosphere or low chromosphere. The granulation, supergranulation, and oscillations referred to in the previous section seem to be the same inside or outside of coronal hole regions. The only apparent difference is in the topology (openness or closedness) of the magnetic field. Although not a proof, this suggests that the mechanical energy input passing up through the photosphere, as discussed earlier, may be substantially the same inside and outside of holes.

However, low coronal density and temperature in holes means less radiative loss (that is of course why holes appear dark in coronal emissions). And a lowered value of the temperature gradient in the transition zone means less thermal conduction from the corona back to the chromosphere in regions of holes. If these two major energy leaks are partially plugged and the input remains the same, where does the excess energy go?

Recent evidence strongly suggests that the excess energy goes into creating and accelerating a high-speed solar wind that emanates primarily from the regions of the holes. Calculations based on energy arguments such as just given (Noci, 1973; Pneuman, 1973) predict that in magnetically open regions such as holes there may occur outward expansion of the solar wind, accompanied by temperature decreases of a factor of 2 and density decreases of a factor of 10, which are similar to the conditions observed in coronal holes. In addition, Krieger, Timothy, and Roelof (1973) found a strong temporal correlation between the passage of coronal holes across the center of the solar disk and the arrival of high-speed solar wind streams at the earth. The correlation is best when the time of arrival at the earth of the high-speed stream is shifted backward by the several-day transit time required for the stream to reach the earth. The weight of the evidence, plus other data relating coronal holes to times of high geomagnetic activity (known to be related to solar wind fluctuations), strongly suggests that coronal holes do have a profound influence on the solar wind and therefore on geophysical phenomena. The suggestion is particularly interesting because independent studies indicate circumstantial evidence for a relation between solar-wind-related interplanetary magnetic field structures and terrestrial weather, in the sense that changes in the interplanetary field are correlated with changes in the "vorticity" of the earth's atmosphere (Wilcox et al., 1974; Hines, 1974). At this writing the evidence is not completely clear and no clearcut mechanism for the effect has been found, but the subject is of such great general interest that further study is being vigorously pursued.

One other interesting aspect of the relation of coronal holes to the solar wind follows from the appearance of coronal holes at the poles, as seen in Figure 2-12. These features appear identical in their measurable characteristics to low-latitude holes, but differ in that they are larger and longer lived. This is illustrated by Figure 2-13, which shows the projected appearance of a long-lived polar coronal hole as it would have been observed from above the ecliptic for eight months during the flight of Skylab in 1973 (Sheeley et al., 1975). The size and persistence of polar holes suggests that a continuous stream of high-speed solar wind emanates from polar regions, conceivably dwarfing the rather modest solar wind seen from the earth's position near the sun's equatorial plane.

Indirect evidence for polar wind is provided by radio scintillation measurements of sources that pass behind the north or south poles of the sun. However, it will be extraordinarily interesting when, sometime in the next

Figure 2-13. Diagram of the evolution of a coronal hole as it would have been seen from above the south pole. Data are rectified from ATM Skylab observations, May 1973 to February 1974, made with the Naval Research Laboratory XUV monitor instrument. Data courtesy N. R. Sheeley, Naval Research Laboratory.

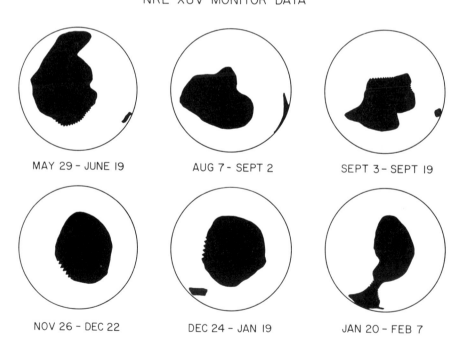

SOUTH POLAR HOLE, 1973-1974,
RECTIFIED FROM
NRL XUV MONITOR DATA

MAY 29 - JUNE 19   AUG 7 - SEPT 2   SEPT 3 - SEPT 19

NOV 26 - DEC 22   DEC 24 - JAN 19   JAN 20 - FEB 7

few years, a small spacecraft will be launched to fly close by Jupiter and pick up enough angular momentum to swing out of the ecliptic plane and over the solar pole. Direct in situ measurements of the density, velocity, and magnetic fields carried out in the polar wind will then be possible for the first time.

Solar Activity and Magnetic Fields

In previous sections we have noted how magnetic fields control both the structure and the flow of thermal conduction in the upper solar atmosphere, and in this and the next section we shall see the even more dramatic role of magnetic fields in solar activity. But first we should discuss the general nature of solar magnetism. It has long been known that sunspots vary in frequency and latitude of occurrence with a period of about 11 years. Sunspots, which generally occur in pairs, are regions of very strong magnetic field, whose sign depends on whether a spot is a "leading" or "following" member (in the sense of the direction of solar rotation) of a spot pair. The sense of polarity between leading and following spots is opposite for spot pairs in the northern and southern hemisphere. And finally, the sense within each hemisphere alternates in sign with alternating 11-year sunspot cycles. Thus the sun is a quasi-periodic magnetic variable star with a period of about 22 years.

This seems strange at first since it may be easily shown that if left to itself, the magnetic field in a conducting body should decay in a characteristic time $\tau \sim 4\pi\sigma L^2$ where σ is the electrical conductivity of the material and L the characteristic scale of the field. (For an old, but very lucid, description of basic concepts of solar magnetism, see Cowling, 1953). For the body of the sun, where $\sigma \sim 6 \times 10^{-4}$ emu (about that of copper) and $L \sim \frac{1}{3}R_\odot \sim 2 \times 10^{10}$ cm, this becomes 10^{10} years, which is greater than the age of the sun and of course much greater than the period of the 22-year magnetic cycle.

The answer, of course, is that the field is *not* left to itself. Motions of conducting plasma through magnetic fields, such as are caused by convective instability and differential rotation of the sun's surface layers, induce currents; these in turn cause a change in the field.

If the electrical conductivity is high enough, as it is almost everywhere in the sun, the field becomes "frozen" into the plasma, in the sense that the magnetic flux passing through any surface carried with the plasma remains constant during the motion. At low densities such as exist in the corona the thermal energy density of the gas nkT is generally much less than that of the field $B^2/8\pi$, and the gas is constrained to move with the field, as we have seen earlier. However, if the reverse is true and in particular the kinetic energy density of the plasma, $\rho v^2/2$, exceeds the energy density $B^2/8\pi$ of the field, the motions of the plasma will control and modify the magnetic field. This occurs in the high-density interior and permits motions to dictate how the field changes.

It is a basic problem of solar magnetism to explain how velocities in the solar interior, by interacting with the existing magnetic fields, can produce the 22-year oscillation of magnetic polarity and the accompanying detailed variations of surface fields. Although the general outlines are appreciated, the detailed nature of the process is far from understood, and is the subject of much current research. In broad outline, the interior is a reversing magnetic dynamo; i.e., motion of material across magnetic fields generates currents that in turn create an oppositely directed field. It is generally thought that the process involves distorting an initial poloidal (N-S oriented) general magnetic field into a predominantly toroidal (azimuthal) one by E-W stretching of magnetic lines of force through differential rotation of the interior. In turn, as convective motions carry the toroidal lines of force upward through the convective zone, coriolis forces on the rotating sun might cause the lines of force to bend poleward again, but in the opposite sense from that of the original poloidal field. Several detailed studies have shown that not only the overall character of the sunspot cycle, but also many of its detailed characteristics, can be explained in this way. The interested reader is referred to a review by Parker (1970) or a pioneering study by Babcock (1961) for more details.

This topic was high on the list of important solar physics problems a few years ago, because of the very successful models just mentioned, coupled with the discovery of an apparent 22-year periodic reversal of the general (dipole-like) field of the sun. At the time of this writing, interest is reaching an even higher peak, owing to an increase of the power of modern computers that enable researchers to model the motion of material and magnetic fields in the interior with considerable sophistication, coupled with several important very recent observational findings. These include:

1. The development of new tools to explore the structure of the interior of the sun, especially through neutrino observations (or nonobservations!) and through studies of the global oscillations of the sun, as we described earlier.

2. The observation that the absolute rotation and the differential rotation of the solar interior, as shown by the measured rotation of the surface layers, is much more complicated than once thought. For one thing, the sunspots and small-scale magnetic fields appear to rotate faster than the gas in which they are embedded; for example, at the equator, they complete their rotational circuit in 25 days, while the gas, as determined from its Doppler shift, completes one circuit in about 26 days. Larger-scale nonsunspot fields rotate at even a slightly different rate. In addition, the over-all rotation rate of the surface layers appears to change in an irregular fashion by significant amounts over a time scale of months. These results constitute an important constraint on theoretical calculations of the dynamics of the interior. For a recent review of solar rotation observations and their implications, see Gilman (1974).

3. The very recent historical research of Eddy (1976) has reaffirmed the existence of a prolonged period from 1645 to 1715, called the Maunder minimum, during which the solar activity cycle essentially disappeared. For nearly half this time (1672–1704) there were no spots at all seen on the northern hemisphere of the sun. Figure 2-14 shows a plot of reconstructed sunspot numbers, based on historical records. The striking disappearance of the 11-year cycle of spot numbers shows that the amplitude of the free-running oscillation of the solar dynamo has in the recent past varied enormously, and suggests that the presence of spots from a previous cycle is not necessary to trigger a new cycle. Eddy, Gilman, and Trotter (1976) have noted the interesting and perhaps highly significant fact that the differential rotation of the surface layers, generally accepted as related to the solar dynamo action, was a factor of 3 greater during the years just before the onset of the Maunder minimum. The interpretation of these new results is still unclear. As a side issue, it is interesting to note that the time of the ex-

Figure 2-14. Annual mean sunspot numbers, AD 1610–1974, from Eddy, Gilman, and Trotter (1976). The "Maunder minimum" occurred from 1645 to 1715.

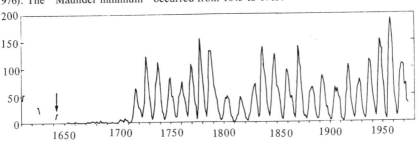

tended minimum of solar activity coincided very nearly with the "little ice age," a period of extreme cold weather on the northern hemisphere that is very well documented in historical records. Whether the decreased solar activity was somehow responsible for the little ice age is still unknown, but, not surprisingly, there is much current speculation about a possible connection.

In the remainder of this section and the next, we shall discuss some of the detailed manifestations of solar magnetic fields, as seen from observed surface fields and their associated structures.

Active Regions

Active regions on the sun are areas of strong magnetic fields that grow and decay over times of days to months; they occur in well-defined latitude zones called activity belts, symmetrically north and south of the equator. Their size ranges from about 10^4 km to up to a few times 10^5 km.

Active regions are in a sense the fundamental unit of solar activity and the most persistent and visible manifestation of solar magnetism. A detailed physical understanding of active regions and their evolution is thus basic to an understanding of solar activity. Probably all sunspots and all flares occur within active regions, which are the source of the slowly varying components of the soft X-rays, XUV, and UV which greatly influence the structure of the earth's upper atmosphere.

Most new solar magnetic flux is believed to emerge within active regions, from whence it is spread by supergranular velocity fields over the surface of the sun. Thus the synoptic study of active regions is important for understanding large-scale solar magnetism. In addition, however, active regions are of great physical complexity and they offer an unparalleled opportunity to study astrophysical processes.

Because their coronal manifestations (i.e., coronal condensations) are often very bright, they were the first coronal structures that could be observed outside of eclipse with the coronagraph. But they are also the most conspicuous sources of XUV emission and soft X-rays observed from space.

Observations carried out on satellites, especially the high-resolution ATM observations, have now revealed the detailed structure of active regions in the corona. The large-scale structure of active regions in visual and X-ray wavelengths was seen in Figure 2-12, above. Figure 2-15 shows some of the fine details of active regions in the corona. We see that the emission is concentrated in very fine loop structures, clearly showing the restraining influence of magnetic field lines. This implies that the gas pressure nkT must be much less than the magnetic pressure $B^2/8\pi$.

While loops at the limb have been well studied from ground-based observations of a few visual forbidden lines such as [Fe XIV] $\lambda5303$, the extreme ultraviolet and X-ray data permit study of loops at different projections on the disk, and at many different temperatures, as shown in Figure 2-15. These new data are under active study, to determine temperature gradients and pressure balance in magnetic loop structures.

Another observation of some interest is that the loops are discrete; i.e., there are neighboring regions that are not filled with X-ray or ultraviolet-

Figure 2-15. Active region near the solar limb, viewed by the Harvard spectrometer on ATM. Each image covers a 5-arcmin square field of view. Approximate temperatures of emission are: 2×10^4 K for Ly α, 8×10^4 K for C III, 3×10^5 K for O VI, 6×10^5 K for Ne VII, 1.5×10^6 K for Mg X, and 2.5×10^6 K for Si XII. From Noyes et al. (1975).

ACTIVE REGION

EAST LIMB

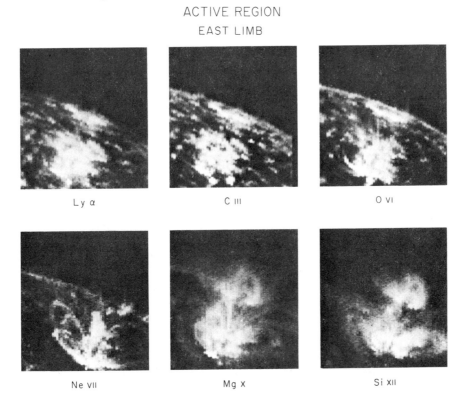

Ly α   C III   O VI

Ne VII   Mg X   Si XII

emitting material. This suggests that either there are no magnetic fields spanning regions in which no loops appear, or there are magnetic loops but for
some reason they are not filled with emitting material. The latter possibility
seems more likely, in view of magnetic field extrapolations from observed
photospheric fields (Rust and Roy, 1971). The interesting question of why
some magnetic flux tubes should be free of material and others full has, however, not been answered.

Given the extreme complexity introduced by magnetic fields, the problem
of mass and energy balance in the chromosphere, transition layer, and
corona of active regions is even more difficult than in the case of the quiet
sun. The fields not only determine the structure of active regions, but also
probably are instrumental in supplying the increased heating in active
regions as compared with quiet regions. How this is done is unclear: Is there
an increased mechanical energy flux carried upward from the convection
zone? Or does the presence of strong magnetic fields lead to more efficient
transmission of that energy into the chromosphere and corona? What is the
role of dissipation of energy stored in the fields themselves?

As in the case of the quiet sun, the first approach to this problem is to
study the structure and energy requirements of active region features. Earlier studies with low-resolution data (Noyes, Withbroe, and Kirshner, 1970)
had indicated that, on the average, active regions require several times as
much mechanical energy per unit area as the quiet sun, and have a coronal
temperature some 0.5×10^6 K greater than the 1.8×10^6 K temperature of
the quiet corona. However, the fine structure described above makes clear
that this is only the crudest of approximations.

Another approach is through the study of evolution of active-region
structures, including the emergence through the photosphere and subsequent growth of active-region loops. Some evidence has recently been presented (Sheeley et al., 1975) indicating that the magnetic flux of emerging active regions may *reconnect* to join up with pre-existing magnetic lines of
force. In the process the field structure relaxes to a lower-energy configuration, the energy differences being liberated as heat. The problem is that this
reconnection theoretically must occur on the time scale for diffusion of magnetic fields in an ionized plasma. As we stated earlier, this is $\tau_D = 4\pi\sigma L^2$
(e.g., Cowling, 1953), where σ is the electrical conductivity and L the characteristic length of the magnetic structure undergoing reconnection. For
$L \sim 10^4$ km and $\sigma = 10^{-7}$ emu, characteristic of the photosphere and low
chromosphere (Cowling, 1953), we find $\tau_D \sim 3 \times 10^4$ years! As opposed to

the case in the interior, here we can watch the detailed time evolution of the fields as outlined by the hot plasma, and perhaps find a clue to how fields circumvent the diffusion equation. It is worth noting that significant heating may be expected if magnetic field reconnection occurs; it is an open question whether this is an important part of the energy balance of active regions.

It may be that an understanding of active regions will ultimately be aided by detailed study of the recently discovered X-ray bright points (Golub et al., 1974), which have recently been extensively observed on ATM. Some of them are easily seen in the X-ray photograph of Figure 2-12. These small emitting features are very like miniature active regions in two ways:

1. Like active regions, they overlie bipolar magnetic field regions, but the regions are so compact that the tight loops that must exist are essentially unresolvable with ATM instrumentation (resolution limit about 3000 km). Related to their smallness is their relatively short life of a few hours, comparable to that of the supergranulation flow fields in which they are embedded. Temperatures and densities, hence mechanical heat input, of X-ray bright points are comparable to active regions. And, because they have similar, but far less complicated, magnetic field structures, bright points may hold important clues to active region heating.

2. Like active regions, bright points often show a rapid brightening and heating, which may well be essentially the same phenomenon as active-region flares. As we shall see in the next section, flares are extraordinarily complex; and once again the simpler geometry of bright points may ultimately lead to a better understanding of flares.

X-ray bright points differ from active regions in one important property, however (Golub et al., 1974): namely, their distribution in latitudes over the sun is uniform to first order, as opposed to active regions which we have seen to be confined to well-defined activity belts some 20° to 40° north and south of the equator. The total magnetic flux brought to the surface by bright points may equal that of active regions, so they are a significant component of solar activity. This large magnetic flux, and its quite different latitude distribution, must be considered in any dynamo theory of the solar magnetism.

Sunspots
Sunspots are among the easiest to observe and most difficult to understand of all solar phenomena. It is strange, when so much is known of

sunspots, that we cannot understand the first two things about them: why they exist at all and why they are black.

Sunspots (Figure 2-16) are black, of course, because they are cold. [Their effective surface temperature is only about 4100 K, some 1700 K cooler than the surrounding photosphere. By Planck's law their emission per unit area is $(4100)^4/(5800)^4 = 0.25$ as large as the photosphere, so they appear black by contrast; if seen against a black sky, a typical sunspot would have an apparent magnitude $m_v \sim -12$, about half as bright as the full moon.] But why are they cold? Clearly the magnetic field must be the cause; magnetic fields $B \sim 3000$ gauss are typically measured in sunspot umbrae. The field must be the cause, rather than the effect, of the cool umbra because at 3000 gauss the energy $B^2/8\pi$ in the field dominates the thermal energy nkT of the gas by a factor of 10.

It is often stated that sunspots are cold because the strong field inhibits convection by freezing the material to the field lines and thus preventing turnover of convective cells. In other words, the magnetic fields create a layer of insulation against the convective energy flow. This should have an important effect, since as we saw earlier most of the solar flux is carried by convection near the surface. However, as Parker (1974) has pointed out, thermal diffusion around the edges of such a plug would cause a gradual temperature increase at the edge of the umbra, of width comparable to the vertical thickness of the plug, rather than the extraordinarily sharp boundary seen in Figure 2-16. Parker suggests, rather, that energy is carried away from the spot umbra by hydromagnetic Alfvén waves, thus refrigerating the spot. In either case, there does not appear to be a visible sign of the missing energy, which might be expected in the form of a bright ring around the spot, in the case of a convection-inhibiting plug, or as a greatly enhanced heating of the corona over the spot in the case of Alfvén-wave energy transport. The only satisfactory picture is for all of the energy to be transported back *downward* into the photosphere, conveniently out of sight. Clearly the problem is by no means solved.

There are many other intriguing unresolved questions about sunspots, including the nature of the penumbra that surrounds the dark umbra as seen in Figure 2-15, the flow of material along the penumbral filaments, the stability of sunspots, and the mechanism of their growth and decay. The interested reader is referred to the monograph by Bray and Loughhead (1965) on the subject. We mention here, however, one new observational result of current

Figure 2-16. Sunspot obtained with a 12-inch, balloon-borne telescope. The dark umbra is surrounded by the gray penumbra, consisting basically of narrow, bright filaments on a dark background. The photospheric granulation remains unaltered right up to the boundary of the sunspot. Photograph courtesy of Project Stratoscope, Princeton University.

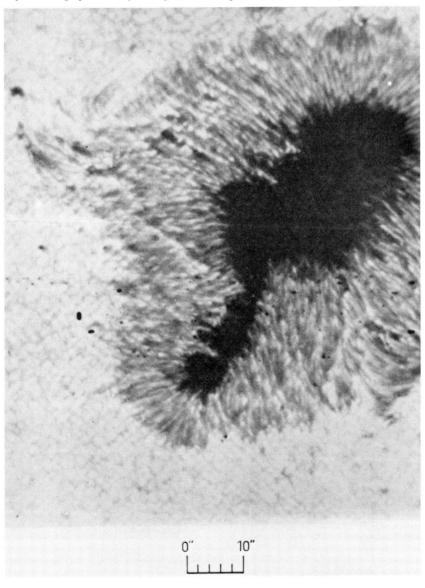

interest that will not be found in earlier reviews on the subject; namely, the appearance of sunspots in the overlying corona. Figure 2-17, from the recent Skylab ATM mission, shows that a bright "plume" of intense emission exists above sunspot umbrae. This is due to an extended region of relatively cool (2×10^5 K $< T < 8 \times 10^5$ K) gas immersed in the hotter surrounding corona (Foukal et al., 1974). Such plumes, which often bend over to become

Figure 2-17. Emission in Ne VII 465 Å (temperature about 600,000 K) above a sunspot. Note growth of a closed loop structure during the three days of observation.

SUNSPOT, JUNE 27-29, 1973

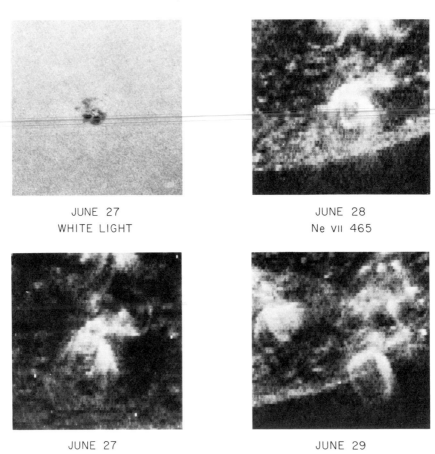

JUNE 27
WHITE LIGHT

JUNE 28
Ne VII 465

JUNE 27
Ne VII 465

JUNE 29
Ne VII 465

closed loops running between the sunspot umbra and a point of opposite polarity in the surrounding photosphere, seem to be associated with nearly all spots. Because they are relatively large and easy to resolve with existing spacecraft data, they are under close study to determine their energy balance and dynamics. The results may well be applicable to the smaller-scale loops mentioned above in connection with active regions.

Photospheric Magnetic Knots

Recent high-resolution magnetic observations (Frazier and Stenflo, 1972) strongly suggest that the general magnetic field, whose spatially averaged value may be only about 1 gauss, actually consists of numerous knots of much higher field strength—probably exceeding 10^3 gauss. This very recent finding suggests that the problem of the existence and stability of very large field strengths at the surface of the sun is common to magnetic regions elsewhere than in sunspots, and perhaps everywhere. The problems may be even more severe for the small photospheric flux knots since, unlike sunspots, they do not appear to be dark, or cool, at the surface. For a sunspot, the cool interior of the umbra at least allows horizontal pressure balance to exist, i.e., the magnetic pressure $B^2/8\pi$ plus the gas pressure $(nkT)_u$ in the umbra can balance the gas pressure $(nkT)_p$ of the photosphere. In a photospheric flux knot, unless the temperature is cooler below the visible surface, such an equilibrium seems difficult to achieve.

It may well be that the flux knots are reflected in the recently discovered solar "filigree" (Dunn and Zirker, 1973), seen in very high-resolution photographs taken through narrow-band filters positioned on the wings of chromospheric lines like Hα (Figure 2-2b). The filigree requires a combination of superb instrumentation and fine "seeing" to observe, and seems to be associated with strong fields. It is also clear that the filigree, presumably by virtue of these fields, completely alters the structure of the granulation, leading to the "washed-out" appearance seen in Figure 2-2a at the location of the filigree.

Flares

Solar flares are perhaps the most complex phenomena observed on the sun. They involve an astounding variety of physical processes, including (a) preflare storage of large amounts of energy in a magnetic field, (b) extremely rapid release of that energy (triggering), (c) acceleration of electrons to near-relativistic energies, (d) gyrosynchrotron radiation from the interaction of relativistic electrons with the magnetic field, (e) acceleration of nuclear particles to highly relativistic energies, up to GeV, (f) nuclear reactions produced by the impact of high-energy protons on the ambient solar material, (g) inexplicable anomalies in isotope abundances, (h) sudden deposition of heat in the chromosphere, resulting in explosive ejection of material, and (i) the generation of a blast wave carrying particles and energy into the interplanetary medium and eventually interacting with the earth and its magnetosphere. The conversion of the very small random energies of many individual particles into the very large energies of a few, by means of amplification and sudden destruction of magnetic fields, ranks as one of the most interesting phenomena in all of astrophysics. Similar phenomena may well be occurring in radio galaxies, in quasars, in supernovae remnants, in X-ray stars, in pulsars, and in flare stars; we are fortunate to have an example close at hand which we may study in detail.

From observations over the entire electromagnetic spectrum, and direct measurement of particle fluxes at the earth, a phenomenological picture of flares has evolved, and will be described in the following sections.

Flare Energy Buildup

Hours to days before the flare release, excess energy becomes stored in the magnetic field. By ''excess'' is meant the amount above the minimum energy associated with a potential field, which represents the lowest possible energy for a magnetic field. A potential field is one undistorted by currents; a familiar example is the field in the vicinity of a permanent magnet. Using the familiar analogy of magnetic fields and stretched rubber bands (both of which have tension), we may describe the energy in a potential field as like that in the stretched rubber band of a toy airplane before it is twisted up. The preflare energy storage is analogous to winding up the rubber band. Indeed the two situations may be quite similar geometrically, as a twisted magnetic field can store energy; twisted structures are often seen in the atmosphere near regions of strong magnetic field. Associated with a twisted field must be a current that supplies energy, for from one of Maxwell's equations, the curl, or shear, of the field at any point numerically equals the current density ($\nabla \times \mathbf{B} = 4\pi \mathbf{J}$).

The question of the precise geometry of the magnetic fields involved in the energy release is important, in order to pin down the nature of the instability. One very simple model is that of a sheet pinch (Sturrock, 1973), as shown in Figure 2-18a. In this geometry a neutral sheet exists, separating regions of oppositely directed open field lines overlying the closed magnetic loop structure of a bipolar magnetic region. A sheet current flows in the neutral sheet normal to the figure, thus satisfying the relation $\nabla \times \mathbf{B} = 4\pi\mathbf{J}$. In this model, the flare instability causes rapid reconnection of field lines in the current sheet, releasing about one-half the total magnetic energy $B^2/8\pi$ in the reconnection region in the form of rapid heating. Detailed modeling shows that this energy is sufficient to supply the energy of flares, and further that the reconnection of fields as seen in Figures 2-18b and 2-18c naturally provides for ejection into the corona of plasma that had been located above the instability, as demanded by the observations (Sturrock, 1973). However, it is not easy to understand how the reconnection can occur so rapidly as to deposit

Figure 2-18. a, The open, current-sheet model above a bipolar magnetic field. b. Reconnection at the neutral line results in particle acceleration away from the sun, which causes evaporation of material in the loops. c, A force-free configuration exists once the reconnection has occurred, and a large mass of material, in which field lines are frozen, is ejected into the corona. From Sturrock (1973).

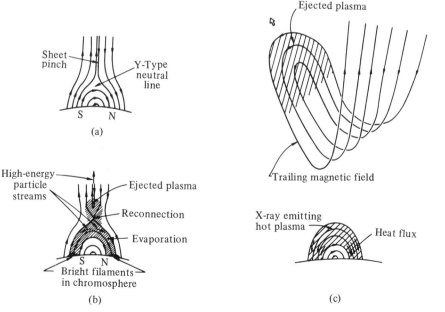

the entire energy requirement (see below) in the few hundred seconds required of large flares. Further, the recent high-resolution ATM data seem to indicate (Brueckner, 1976) that flares actually occur in the top of loop structures rather than in current sheets as in Figure 2-18.

Energy Release

In order for the preflare energy to be stored without premature release, the magnetic field configuration must be *metastable;* i.e., even though there exists a lower-energy state, the field will not relax to it, even in the presence of small fluctuations. However, at some instant of time, the nonpotential energy in the field appears to reach a level where a small disturbance *can* trigger the onset of instability. At this point a very rapid conversion of the excess energy occurs, converting as much as 10^{30} to 10^{32} erg to kinetic energy of high-speed particles in a matter of seconds to minutes. A large fraction and perhaps essentially all of this energy goes into acceleration of electrons to energies of up to about 100 keV, or velocities of up to about 0.5 c. Figure 2-19 (Kane, 1973) shows some of the immediate effects produced by these electrons.

1. A hard X-ray burst appears instantly, with X-ray energies rising in times of less than one minute; this is bremsstrahlung ("braking radiation") produced by the deceleration of the electrons by the ambient gas. Very likely most of the X-radiation comes from the dense chromosphere, and it appears that the electrons may be channeled down to the chromosphere by the magnetic fields themselves (Figure 2-20).

2. As the relativistic electrons move through the magnetic field in the flare region, they are also accelerated (although more gently than in the case of bremsstrahlung described above) to spiral around the magnetic field lines. The centripetal acceleration causes *gyrosynchrotron radiation* to occur in the radio (microwave) region (see Figure 2-19). Theoretical calculations of the exact microwave spectrum expected have been carried out by Takakura (1973) and give results dependent on the size of the emitting region, the field strength, and the optical thickness for self-absorption at microwave frequencies. Further, the ratio of the microwave radiation to the X-ray bremsstrahlung radiation is dependent on the density in the emitting region, which is thus calculated to be about 10^{10} cm^{-3}. This places the source of the microwave emission in the high chromosphere or chromosphere-corona transition zone.

Figure 2-19.   An example of time variations in a flare as seen in various regions of the electromagnetic spectrum. Note the two component portions of the light curves (impulsive and slow)
and their differences among the X-ray, EUV, and microwave regions of the spectrum. From
Kane (1973).

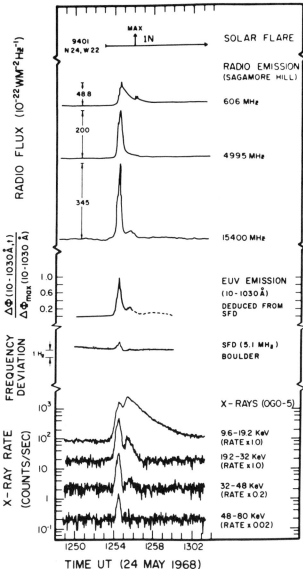

3. Some of the fast electrons move outward through the solar atmosphere, rather than inward as in (1) above. As they pass through successively higher layers in the atmosphere, they excite radiation from the various layers in turn. This radiation, known as a type III radio burst, is due to plasma oscillations, i.e., a "ringing" of the atmosphere after it is struck by the beam of fast electrons; the frequency of the oscillation at each height is proportional to the square root of the electron density. From a knowledge of the electron density distribution with height, then, one can use the observed time of arrival of radiation at different frequencies to determine the speed of the exciting electron beam. In this way it is found that the electrons are moving at about one-third to one-half the velocity of light, in good accord with the energy of about 100 keV deduced from the X-ray bremsstrahlung spectrum in (1) above.

Thermal Phase

It can be shown (Kane and Donnelly, 1971) that the amount of radiation produced by bremsstrahlung and by gyrosynchrotron radiation from the high-energy electrons is quite small—only about 10^{-4} of their total energy. What becomes of the vast bulk of the remainder? It appears to be shared among the ambient nonaccelerated material ("thermalization"), thus heating the gas. The temperature to which the gas is heated depends on the number of particles with which the electrons must share their energy, and thus on the density. Thus thermalization in the chromosphere may raise the temperature a few thousand degrees and produce enhanced $H\alpha$ and ultravio-

Figure 2-20. A model of the relative positions of the emission in the X-ray, microwave, and EUV regions of the impulsive phase of a solar flare. From Kane (1973).

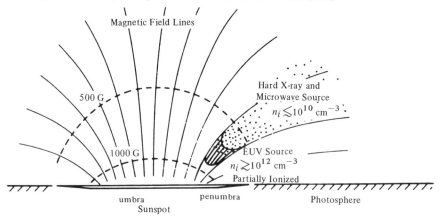

let emission (Figures 2-19 and 2-21), whereas thermalization in the low corona may produce a 10^7 K soft X-ray emitting thermal plasma (Figures 2-19 and 2-22).

This thermal phase of the flare may last for minutes to hours, depending on the size of the flare. In fact, in some cases the very hot (10^7 K) plasma mentioned above lasts so long that one must postulate a continuing source of energy input to keep it from cooling off; such an energy input has however not been discovered, and the present situation is unclear.

Secondary Acceleration

In the chromosphere, the deposition of energy often causes an explosive heating that gives rise to material ejection, at velocities of several hundred km/sec. In fact, the expanding blast wave may exceed escape velocity, and eject as much as 10^{16} gm of material into the interplanetary medium.

It appears that as the blast wave moves outward it carries magnetic fields along with it, and that these moving magnetic fields may cause a secondary acceleration of particles, perhaps through the Fermi mechanism. While the details are obscure at this time, it is clear that acceleration of nucleons to very high energies does occur, and apparently at a time associated with the occurrence of the above-mentioned blast wave. The evidence for this second acceleration takes the form of direct observation of electrons, protons, and heavier nuclei at the orbit of the earth with energies ranging up to a hundred MeV (see Fichtel and McDonald, 1967, for a review), as well as gamma rays (Chupp, Forrest, and Suri, 1973) from nuclear reactions in large flares. These gamma rays were identified as the 511-keV emission produced by electron-positron annihilation, the 2.2-MeV emission produced by de-excitation of deuterium following its creation by proton-neutron reactions, and other line emission resulting from reactions of neutrons with abundant elements in the atmosphere.

We conclude this brief description of the flare process with a list of outstanding current questions on flares, which will be under intensive investigation over the coming few years:

1. How is the energy stored in the preflare magnetic field?

2. How is the energy released? Where is the precise release point relative to the magnetic field? What is the energy spectrum of the accelerated particles? How does the magnetic field change during the energy release? What event triggers the flare?

3. What detailed mechanisms occur during the period following the initial

Figure 2-21. A solar flare, photographed in Hα, seen at its peak intensity. The structure of the surrounding chromosphere indicates the complexity of the magnetic field in this active region. Photograph courtesy of Big Bear Solar Observatory.

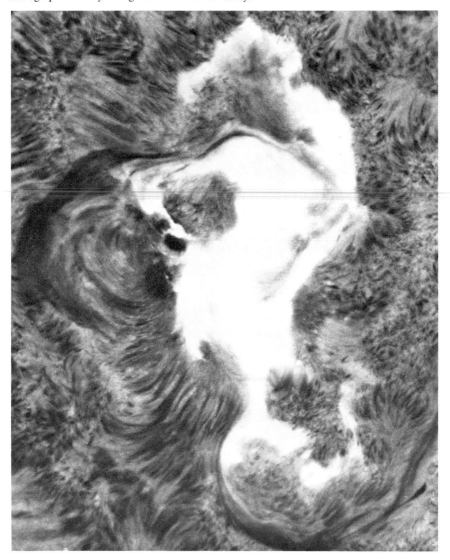

Figure 2-22. Soft X-ray images, showing time development of a flare, superimposed on the preflare magnetograph contours. Exposures for images A, B, C, D are of duration 1/16, 1/16, 1/4, and 16 sec respectively, thus compensating for the tremendous decrease in surface brightness after flare maximum. The emission in exposure D (postflare) is similar to the preflare emission in intensity and shape, except that the extended emission reaching to 100,000 km (2.5 arcmin) to the NW was created during the flare; it appears to be series of closed loops (unresolved in this figure) that expanded outward during the thermal phase of the flare. See Pallavicini et al. (1976) for details.

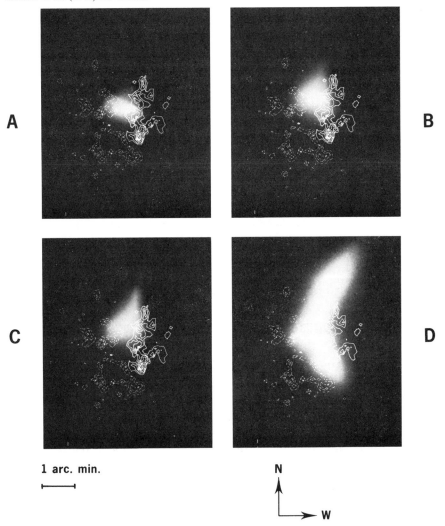

1 arc. min.

N
W

energy release? Does a second stage of acceleration occur associated with the blast wave? Does the postflare plasma cool primarily by radiation or conduction? Is there continued heating throughout the flare after the initial energy release?

Many of these questions will receive serious observational and theoretical study during the coming maximum of solar activity expected around 1979 and 1980.

Summary and Conclusions

In this brief overview of current solar physics, we have seen the sun as a mildly varying magnetic star. The magnetic fields are paltry compared to many other astrophysical sources. While the maximum surface field is several thousand gauss, the average surface field is only one gauss. Magnetic stars have average fields of 10^3 gauss, some white dwarfs have average fields of perhaps 10^7 gauss, and other collapsed objects such as neutron stars may have fields even far greater. Nevertheless the relatively weak solar fields dominate the structures and the mass and energy flow in the solar atmosphere, and give rise to the high-energy phenomena of solar activity. Their influence is also essential for most, if not all, of solar-terrestrial relations.

In spite of superb data and extensive theoretical work, however, we still know very little about solar magnetism. We do not know why sunspots exist, or what keeps them stable. We do not yet understand the generation or amplification of fields in the interior, or the 22-year magnetic cycle. We know very little of the complex processes of magnetic annihilation that must be occurring in solar flares.

Even putting magnetic fields aside for a moment, we are left with other fundamental unanswered questions, such as the structure of the solar interior with its puzzling lack of neutrino emission, the nature of solar convection, and the heating of the corona.

And yet, how much less we know of magnetic fields, stellar interiors, or stellar coronas elsewhere in the universe! Our frustration at not having answers is mitigated by knowing that we are asking difficult questions that could not easily be asked of any remoter body. And even as we probe these difficult and detailed astrophysical problems, we are aware that solar research, as it has many times in the past, will ultimately lead to real understanding of phenomena that we could otherwise only imagine in the more distant universe.

References

Babcock, H. W. 1961. The topology of the sun's magnetic field and the 22-year cycle, *Astrophys. J. 133*, 572–587.

Bahcall, J. N., and Davis, R. Jr. 1976. Solar neutrinos: A scientific puzzle, *Science 191*, 264–267.

Bahcall, J. N., Huebner, W. F., Magee, N. H. Jr., Merts, A. L., and Ulrich, R. K. 1973. Solar neutrinos. IV. Effect of radiative opacities on calculated neutrino fluxes, *Astrophys. J. 184*, 1–4.

Bray, R. J., and Loughhead, R. E. 1965. *Sunspots*, John Wiley and Sons, New York.

Brooks, J. R., Isaak, G. R., and van der Raay, H. B. 1976. Observations of free oscillations of the sun, *Nature 259*, 92–95.

Brueckner, G. E. 1976. ATM observations on the XUV emission from solar flares, *Phil. Trans. Roy. Soc. London*, in press.

Christensen-Dalsgard, J., and Gough, D. O. 1976. Towards a heliological inverse problem, *Nature 259*, 89–92.

Chupp, E. L., Forrest, D. J., and Suri, A. N. 1973. Solar gamma ray and neutron observations, in R. Ramaty and R. G. Stone, eds., *High Energy Phenomena on the Sun, Symposium Proceedings*, Goddard Spaceflight Center X-693-73-193, pp. 285–300.

Clayton, D. D., Newman, M. J., and Talbot, R. J. Jr. 1975. Solar models of low neutrino counting rate: The central black hole, *Astrophys. J. 201*, 489–493.

Cowling, T. G. 1953. Solar electrodynamics, in G. P. Kuiper, ed., *The Sun*, pp. 532–591, University of Chicago Press, Chicago.

Demarque, P., Mengel, J. G., and Sweigart, A. V. 1973. Rotating solar models with low neutrino flux, *Astrophys. J. 183*, 997–1004.

Deubner, F.-L. 1975. Observations of low wavenumber nonradial eigenmodes of the sun, *Astron. Astrophys. 44*, 371–375.

Dicke, R. H. 1974. The oblateness of the sun and relativity, *Science 184*, 419–429.

Dicke, R. H., and Goldenberg, H. M. 1967. Solar oblateness and general relativity, *Phys. Rev. Letters 18*, 313–316.

Dicke, R. H., and Goldenberg, H. M. 1974. The oblateness of the sun, *Astrophys. J. Suppl. Series 27*, 131–182.

Dilke, F. W. W., and Gough, D. O. 1972. The solar spoon, *Nature 240*, 262–264.

Dilke, F. W. W., and Gough, D. O. 1974. The stability of a solar model to nonradial oscillations, *Monthly Not. Roy. Astron. Soc. 169*, 429–445.

Dunn, R. B., and Zirker, J. B. 1973. The solar filigree, *Solar Phys. 33*, 281–304.

Eddy, J. A. 1976. The Maunder minimum, *Science*, in press.

Eddy, J. A., Gilman, P. A., and Trotter, D. E. 1976. Solar rotation during the Maunder minimum, preprint, to be published.

Fichtel, C. E., and McDonald, F. B. 1967. Energetic particles from the sun, *Annu. Rev. Astron. Astrophys. 5*, 351–398.

Foukal, P. V., Huber, M. C. E., Noyes, R. W., Reeves, E. M., Schmahl, E. J., Timothy, J. G., Vernazza, J. E., and Withbroe, G. L.   1974. Extreme ultraviolet observations of sunspots with the Harvard spectrometer on ATM, *Astrophys. J.* (*Letters*) *193*, L143–145.

Fowler, W.A.   1972. What cooks with solar neutrinos, *Nature 238*, 24–26.

Frazier, E. N., and Stenflo, J. O.   1972. On the small-scale structure of solar magnetic fields, *Solar Phys. 27*, 330–346.

Gabriel, A. H.   1975. A magnetic model of the solar transition region, *Phil. Trans. Roy. Soc. London A*, in press.

Gibson, E. G.   1973. *The Quiet Sun*, NASA SP-303, Washington, D. C.

Gilman, P.   1974. Solar rotation, *Annu. Rev. Astron. Astrophys. 12*, 47–70.

Golub, L., Krieger, A. S., Silk, J. K., Timothy, A. F., and Vaiana, G. S.   1974. Solar X-ray bright points, *Astrophys. J.* (*Letters*) *189*, L93–97.

Hill, H. A., and Stebbins, R. T.   1975. The intrinsic visual oblateness of the sun, *Astrophys. J. 200*, 471–483.

Hill, H. A., Stebbins, R. T., and Brown, T. J.   1975. *Proc. V. Intern. Conf. Atomic Masses and Fundamental Constants*, Paris, in press.

Hines, C. O.   1974. A possible mechanism for the production of sun-weather correlations, *J. Atmos. Sci. 31*, 589–591.

Hoyle, F.   1975. A solar model with low neutrino emission, *Astrophys. J.* (*Letters*) *197*, L127–131.

Kane, S. R.   1973. Characteristics of nonthermal electrons accelerated during the flash phase of small solar flares, in R. Ramaty and R. G. Stone, eds., *High Energy Phenomena on the Sun, Symposium Proceedings*, pp. 55–77, Goddard Space Flight Center X-693-73-193.

Kopp, R. A.   1972. Energy balance in the chromosphere-corona transition region, *Solar Phys. 27*, 373–393.

Krieger, A. S., Timothy, A. F., and Roelof, E. C.   1973. A coronal hole and its identification as the source of a high velocity solar wind stream, *Solar Phys. 29*, 505–525.

Kuperus, M., and Athay, R. G.   1967. On the origin of spicules in the chromosphere-corona transition region, *Solar Phys. 1*, 361–370.

Leighton, R. B.   1959. Observations of solar magnetic fields in plage regions, *Astrophys. J. 130*, 366–380.

Leighton, R. B.   1963. The solar granulation, *Annu. Rev. Astron. Astrophys. 1*, 19–40.

Leighton, R. B., Noyes, R. W., and Simon, G. W.   1962. Velocity fields in the solar atmosphere. I. Preliminary report, *Astrophys. J. 135*, 474–491.

Munro, R. H., and Withbroe, G. L.   1972. Properties of a coronal "hole" derived from extreme-ultraviolet observations, *Astrophys. J. 176*, 511–520.

Newkirk, G. Jr.   1968. Structure of the solar corona, *Annu. Rev. Astron. Astrophys. 5*, 213–266.

Noci, G.   1973. Energy budget in coronal holes, *Solar Phys. 28*, 403–407.

Noyes, R. W.   1971. Ultraviolet studies of the solar atmosphere, *Annu. Rev. Astron. Astrophys. 9*, 209–236.

Noyes, R. W., Withbroe, G. L., and Kirshner, R. P. 1970. Extreme ultraviolet observations of active regions in the chromosphere and the corona, *Solar Phys. 11*, 388–398.

Noyes, R. W., Foukal, P. V., Huber, M. C. E., Reeves, E. M., Schmahl, E. J., Timothy, J. G., Vernazza, J. E., and Withbroe, G. L. 1975. EUV observations of the active sun from the Harvard experiment on ATM, in S. R. Kane, ed., *Solar, Gamma-, X-, and EUV Radiation*, pp. 3–17, D. Reidel, Dordrecht-Holland.

Pallavicini, R., Vaiana, G. S., Kahler, S. W., and Krieger, A. S. 1976. Spatial structure and temporal development of a solar X-ray flare observed from Skylab, *Solar Phys.*, in press.

Parker, E. N. 1970. The origin of solar magnetic fields, *Annu. Rev. Astron. Astrophys. 8*, 1–30.

Parker, E. N. 1974. The nature of the sunspot phenomenon. I. Solutions of the heat transport equation, *Solar Phys. 36*, 249–274.

Pneuman, G. W. 1973. The solar wind and the temperature-density structure of the solar corona, *Solar Phys. 28*, 247–262.

Roxburgh, I. W. 1967. Implications of the oblateness of the sun, *Nature 213*, 1077–1078.

Rust, D. M., and Roy, J. R. 1971. Coronal magnetic fields above active regions, in R. Howard, ed., *Solar Magnetic Fields, IAU Symp. No. 43*, pp. 569–579, D. Reidel, Dordrecht-Holland.

Sagan, C., and Mullen, G. 1973. Earth and Mars: Evolution of atmospheres and surface temperatures, *Science 177*, 52–56.

Schwarzschild, M. 1958. *The Structure and Evolution of the Stars*, Princeton University Press, Princeton, N.J.

Severny, A. B., Kotow, V. A., and Tsap, T. T. 1976. Observations of solar pulsations, *Nature 259*, 87–89.

Sheeley, N. R. Jr., Bohlin, J. D., Brueckner, G. E., Purcell, J. D., Scherrer, V. E., and Tousey, R. 1975. The reconnection of magnetic field lines in the solar corona, *Astrophys. J. (Letters) 196*, L129–131.

Simon, G. W., and Leighton, R. B. 1964. Velocity fields in the solar atmosphere. III. Large-scale motions, the chromospheric network, and magnetic fields, *Astrophys. J. 140*, 1120–1147.

Spiegel, E. A. 1971. Convection in stars, *Annu. Rev. Astron. Astrophys. 9*, 323–352.

Spitzer, L. Jr. 1962. *Physics of Fully Ionized Gases*, John Wiley and Sons, New York.

Sturrock, P. A. 1973. Mass motion in solar flares, in R. Ramaty and R. G. Stone, eds., *High Energy Phenomena on the Sun, Symposium Proceedings*, pp. 3–11, Goddard Space Flight Center X-693-73-193.

Takakura, T. 1973. Theory of microwave and X-ray emission, in R. Ramaty and R. G. Stone, eds., *High Energy Phenomena on the Sun, Symposium Proceedings*, pp. 179–187, Goddard Space Flight Center X-693-73-193.

Tolstoy, I. 1963. The theory of waves in stratified fluids including the effects of gravity and rotation, *Rev. Mod. Phys. 35*, 207–230.

Ulrich, R. K.   1970. The five-minute oscillation of the solar surface, *Astrophys. J.* *162*, 993–1002.

Ulrich, R. K.   1975. Solar neutrinos and variations in the solar luminosity, *Science* *190*, 619–624.

Vaiana, G. S.   1976. The X-ray corona from Skylab, *Phil. Trans. Roy, Soc. London,* in press.

White, O. R., and Cha, M. Y.   1973. Analysis of the 5 min oscillatory photospheric motion. I. A problem in waveform classification, *Solar Phys. 31,* 23–53.

Wilcox, J. M., Scherrer, P. H., Svalgaard, L., Roberts, W. O., Olson, R. H., and Jenne, R. L.   1974. Influence of solar magnetic sector structure on terrestrial atmospheric vorticity, *J. Atmos. Sci. 31,* 581–588.

Withbroe, G. L.   1970a. Solar XUV limb-brightening observations. I. The Li-like ions, *Solar Phys. 11,* 42–58.

Withbroe, G. L.   1970b. Solar XUV limb-brightening observations. II. Lines formed in the chromospheric-coronal transition region, *Solar Phys. 11,* 208–221.

Wolff, C. L.   1972. The five-minute oscillations as nonradial pulsations of the entire sun, *Astrophys. J. (Letters), 177,* L87–91.

3

Star Formation and the Early Phases of Stellar Evolution

S. E. Strom

The study of star formation and the early phases of stellar evolution has received considerable attention over the past several years. Advances in infrared and radio technology have allowed astronomers to probe the physical conditions in the dark interstellar clouds, which are thought to be the birthplaces of stars, and to examine large samples of young stellar objects. Two- and three-dimensional hydrodynamic computer codes have been applied to the study of the early phases of the collapse of a gas cloud. These calculations, while still crude, have added considerably to our insight into the detailed processes that take place during these early phases. Furthermore, several recent studies have been aimed at understanding star formation on a galactic scale; as a result, we have achieved a somewhat better understanding of the physical mechanisms responsible for the initial collapse of dark-cloud complexes.

The relatively high level of theoretical and observational interest in problems related to star formation makes this area of research one of the most active in modern astrophysics.

A comprehensive review (references complete through October 1974) of observational studies aimed at understanding young stellar objects and dark-cloud complexes was published very recently by Strom, Strom, and Grasdalen (1975), and gives a more detailed discussion of the logical steps leading to the conclusions stated here. I will attempt to convey here some feeling for the current status of research in this field, and to identify problems that are likely to be of primary importance over the next several years as we try to gain a better understanding of the process of star formation and the early phases of stellar evolution.

The Current Theoretical Picture

At the current epoch in galactic history, star formation appears to take place in dark-cloud complexes (see Figure 3-1). These clouds have typical dimensions of ~ 10 pc, mean densities of about 5×10^3 atoms/cm^3, and temperatures of ~ 10 K. At some point in the lifetime of a typical dark cloud, the following sequence of events may transpire:

1. The internal (thermal plus turbulent) pressure of the cloud can become small compared to its self-gravity so that the cloud as a whole begins to collapse. The dust grains in dark clouds shield these clouds from radiative heating, and cloud magnetic fields may prevent cosmic rays from reaching and heating the interiors. Emission at millimeter wavelengths by simple molecules such as carbon monoxide may cool the interiors of these clouds, thus reducing the internal pressure and causing the cloud to collapse.

2. During collapse, the cloud "fragments" into a series of dense clumps, each of which collapses gravitationally. These dense fragments are thought to be the predecessors of individual stars or small complexes of stars.

3. A marginally stable dark cloud may be compressed by external pressure forces. If the cloud radius is decreased sufficiently, gravitational collapse and fragmentation may ensue, as described above.

The details of the cloud formation process, the thermal balance and velocity field in the cloud, and the mechanisms that determine the scale-length spectrum for the fragments are not well understood. The factors controlling the efficiency of star formation in the parent cloud complexes are also not understood. From a recent census of the total mass of dark-cloud material in our Galaxy, and from estimates of the number of stars per year formed in the Galaxy, it is clear that only a small fraction (10^{-3} to 10^{-2}) of the total mass of material in such clouds actually forms stars.

Any adequate representation of the cloud-collapse picture must address these problems. Moreover, it must explain why star formation seems confined to the spiral arms in our Galaxy, and how the dark-cloud material not involved in the star-forming process is dissipated on a time scale of order 10^7 years (the time it takes for young stars to become observable in spiral arms).

Despite our considerable ignorance of the factors that influence cloud collapse and fragmentation, progress has been made in understanding the collapse of individual cloud fragments of stellar mass.

Figure 3-1.   A photograph of the region near H-H 100 in the Corona Austrina dark cloud; top, north; left, east. A 098-02 plate and RG 610 filter define the effective bandpass. Note the nearly circular "white patch" in which H-H 100 is located. We believe that this region represents the high-density, placental cloud from which the H-H star was formed. Photograph courtesy of Dr. T. Gull.

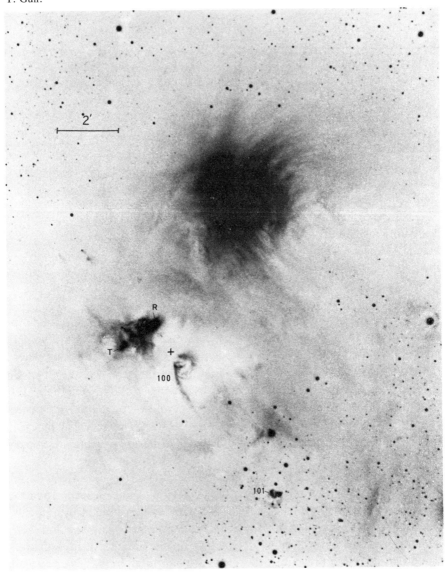

Over the last several years, a number of investigators have used hydrodynamic computer codes to study the collapse and evolution of homogeneous,
nonrotating, spherical clouds of star-like masses. There is considerable disagreement among the various investigators regarding the detailed evolution
of such a cloud (see Westbrook and Tarter, 1975). These disagreements result from several factors: the numerical treatments selected to study the collapse phases; uncertainties regarding appropriate opacities in the low-
temperature regions of the collapsing cloud; and differing approximations in
the treatment of time-dependent convective and radiative energy transport
in the cloud.

A few features of the collapse picture, however, are generally accepted in
all treatments. First, the collapse of the protostellar cloud is extremely non-
homologous; i.e., the cloud quickly develops a high-density core, which ac-
cretes infalling material from the outer parts of the cloud. During the early
phases of evolution, a protostar consists of a growing core of approximately
stellar mass surrounded by an optically thick cloud of circumstellar material.
Second, evolution of stars of all masses takes place on a hydrodynamic time
scale until a significant fraction of the envelope material is accreted or dissipated. At this stage, the stellar core begins to approach the main sequence of
the Hertzsprung-Russell (H-R) diagram along a quasi-static equilibrium
(QSE) track. The original version of the H-R diagram is shown in Figure 5-1,
Chapter 5.

The computations of Iben (1965), shown in Figure 3-2, provide a good representation of the QSE evolutionary tracks subsequent to the hydrodynamic
phases. Stars of high mass ($>3\ M_\odot$) begin their QSE tracks at a point where
radiative transport of energy in the interior dominates. These stars then
evolve along equilibrium radiative tracks. Low-mass stars begin their QSE
evolution along the so-called Hayashi track, where convection transports
energy throughout the outer envelope of the star. The Hayashi phase is
represented by the almost-vertical portion of each low-mass evolutionary
track in Figure 3-2.

The outstanding differences in the models involve the time scale during
which the infall of material controls the evolution of the protostar, and the
related problem of how long the star-like core is surrounded by optically
thick circumstellar material. Both radiation pressure from the luminous
stellar cores and mass outflow act to halt further infall. All investigators
agree that radiation pressure will eventually halt infall in massive pre-main

sequence (PMS) objects. In fact, Larson and Starrfield (1971) have argued that this effect may account for the currently observed upper mass limit of ~ 100 $M_\odot$ for individual stars. It is unclear at what point stellar winds become important. The development of a wind involves an understanding of the relationship between protostellar rotation and magnetic fields, but the

Figure 3-2. Evolutionary paths in the H-R diagram for models of young stars evolving toward the main sequence (Iben, 1965). These paths are theoretical results for models of mass $(M/M_0) = 0.5, 1, 1.5, 3, 5,$ and 9, as indicated. The small numbers along each path indicate the logarithm of the lifetime in years of the stars at that point. The shaded regions indicate the location of the high-mass Herbig emission-stars and the low-mass T Tauri stars.

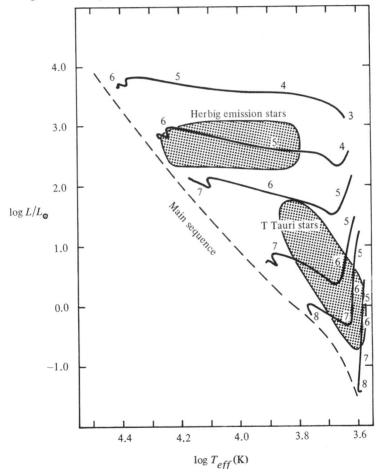

idealized spherical-collapse models available at present offer no theoretical guidelines to an understanding of this relationship.

Furthermore, there is little doubt that the role of rotation during collapse is crucial in determining whether a protostellar cloud forms a multiple star system, a star with planets, or a single star. Crude attempts have been made to estimate the effects of rotation on an initially spherical protostar cloud. However, there is no general agreement on the validity of the basic features of such estimates. Despite the uncertainties of our current theoretical picture of collapse and early PMS evolution, it is of great importance in guiding observational studies.

"Signposts" of Recent Star Formation

A discussion of young stellar objects is best guided by locating well-recognized "signposts" that point to recent star-forming activity.

OB Star Clusters and Associations

M. Walker (see, for example, Walker, 1956) conducted pioneering photometric and spectroscopic studies of the stellar population associated with these young groups of stars. The age of a group is estimated from the nuclear-burning lifetime of the most luminous stars in the group. As these luminous and massive stars exhaust their hydrogen fuel, they begin to depart from the upper end of the main sequence. From observations of these departures, ages in the range 10^6 to 10^7 years have been estimated for the young groups studied to date. A large number of stars of spectral type later than A lie above and to the right of the zero-age main sequence. These stars are apparently approaching the main sequence along equilibrium radiative tracks. Also discovered in Walker's surveys were a large number of late-type, emission-line, irregular variables of the T Tauri class. These stars were first found in dark-cloud complexes but were firmly identified as young stellar objects only after their discovery in the OB complexes. A few early-type, emission-line variables were also found by Walker.

Herbig Emission Stars

These stars are emission-line, irregular variables of early type, closely associated (as evidenced from the reflection nebulae they illuminate) with the dark-cloud material out of which they were presumably formed. They probably represent massive stars ($M \gtrsim 2\,M_\odot$) approaching the main sequence along equilibrium radiative tracks. Their ages appear to lie between 10^5 years and several times 10^6 years. The location of these stars and the T Tauri stars in the H-R diagram is indicated in Figure 3-2.

The Herbig-Haro Objects

These unusual objects are composed of single or multiple nebulous patches, some nearly stellar in appearance, found only in dark-cloud complexes and characterized by unusual emission spectra (Balmer lines, strong forbidden lines of oxygen and ionized sulfur, and little or no continuum emission). They are found only in regions that have characteristic ages $\gtrsim 10^6$ years.

The T Tauri Stars

These signposts of young stellar objects, characterized by irregular variability at optical wavelengths and by peculiar emission line spectra (described in more detail later in this chapter), are found in regions having a wide range of ages: near the OB complexes of age 10^6–10^7 years, near Herbig emission stars with ages 10^5–10^6 years, and in dark-cloud complexes where independent age estimates may be impossible since no other objects of recognized age-range are observed. Hence, the presence of T Tauri stars does not *necessarily* signify a region of unusual youth, although some stars of this class are undoubtedly very young. Criteria for selecting the youngest T Tauri stars are discussed later in this chapter.

Embedded Infrared Sources

At the earliest stages of stellar evolution, we expect that stars may be shrouded by envelopes of infalling circumstellar dust, by the [dense] condensations of gas and dust that represent the remains (which we call "placental clouds") of the original protostellar clouds, and by the lower-density dark-cloud material in the complex. Searches for optically obscured young stellar objects must therefore be carried out at infrared wavelengths, where dust opacity is considerably less than at visible wavelengths. Maps of dark-cloud complexes at wavelengths between 1 μm and 20 μm have been made by a number of observers. Maps at shorter wavelengths tend to select stars which are heavily obscured (visual extinction 10–40 mag) but have apparently normal characteristics (see Figure 3-3). For some cloud complexes, the number of infrared sources is sufficiently large to permit studies of their luminosity distribution. The derived luminosity functions suggest that we are observing the B- and A-type members of a cluster still shrouded by dark-cloud material. The age of some embedded clusters may be as high as 10^7 years.

Searches at $\lambda \gtrsim 5$ μm of dark-cloud complexes associated with H II regions have revealed young stellar objects of even greater luminosity. Some of these objects have infrared spectral-energy distributions that rise steeply from 2 to 20 μm. Because of the considerable uncertainty in the wavelength dependence of extinction in dense, dark-cloud regions, the nature of these luminous infrared sources is not clear. Some may be protostars character-ized by low temperatures and large radii. Many may be luminous pre-main-sequence stars of high mass obscured by up to several hundred magnitudes at visual wavelengths. Their high luminosity places them among the

Figure 3-3.   A plot of the sources observed in the course of the 2-μm surveys of the Ophiuchus dark cloud (Grasdalen et al., 1973; Vrba et al., 1975); top, north; left, east. The underlying photograph is the red plate of the Palomar Observatory Sky Survey. The bright star associated with the nebulosity is HD 147889 (Strom et al., 1975).

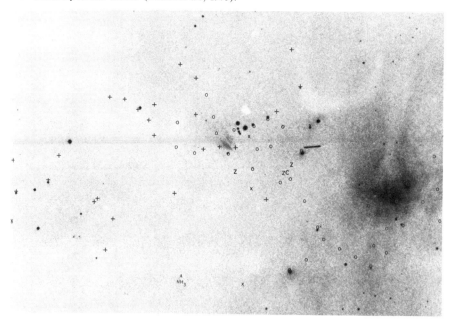

youngest known objects; 10^5 years is a reasonable estimate of their maximum age.

In addition to those luminous young stellar objects found in the course of mapping specific dark-cloud regions, the 10-μm and 20-μm sky surveys carried out by Walker and Price (1975) of the Air Force Cambridge Research Laboratories (AFCRL) appear to contain a significant number of such sources. Detailed photometric observations and moderate-resolution, spectroscopic observations currently underway will soon permit a more definitive statement regarding the evolutionary status of these luminous objects.

Molecular "Hotspots"

Surveys of dark clouds at the wavelengths of millimeter-wave, molecular-line transitions offer an extremely promising means of isolating regions of recent star formation; see Zuckerman and Palmer (1974) for a review. Carbon monoxide maps of dark clouds have revealed a number of "hotspots" of relatively high temperature (T $\sim$ 15–50 K as compared to the 5–10 K typical of dark clouds). Infrared maps of these hotspots typically reveal one or more bright infrared sources. These sources are probably directly responsible for heating the dust and gas in the nearby dark-cloud material to a higher kinetic temperature than that typical for the star-free regions of the cloud complex. Ages of objects associated with these hotspots may range from $\lesssim 10^5$ to 10^7 years.

Maps of dark-cloud complexes in those molecular line transitions, which are produced only under conditions of high density (e.g., CS, HCN), may select regions that contain recently formed stars. These dense regions probably represent the placental material associated with recently formed individual stars or small clusters of stars. Extremely young objects are usually found in the vicinity of such placental clouds; the ages of these objects are probably $\lesssim 10^5$ years.

H II Emission from Dark Clouds

Hot, luminous stars embedded in dark-cloud complexes of high density will produce small, circumstellar regions of ionized hydrogen (H II regions) which have high emission measures. Larger ionized-carbon regions are also expected. Emission from both C II and H II have been detected at radio wavelengths in the direction of a few dark-cloud complexes; for a recent discussion, see Brown and Zuckerman (1975). Some of the ionized regions appear to be directly related to bright sources detected during the mapping of the clouds. Others may be associated with much more luminous objects embedded deep within the cloud complex and consequently hidden from view at a wavelength as short as 2 μm. Radio continuum and recombination-line searches of dark clouds should lead to the discovery of heavily obscured luminous stars. Detailed studies of H II regions in dark clouds may give important clues to their initial evolution. However, without better models of the development of H II regions in a dense, dusty environment, it is impossible to estimate the probable age of these compact regions.

H_2O and OH Masers

Several bright infrared sources have been found associated with OH and H_2O masers located within dark clouds. In some cases, though not in all, the maser and the infrared source are spatially coincident with a compact H II region. In these cases, the maser is probably produced in a high-density ($n > 10^8$ cm^{-3}) placental cloud surrounding a newly formed, early-type star. Where no compact H II region is observed, the maser may act as a signpost, indicative either of a high-density protostellar condensation or of a placental cloud containing a luminous star that has not yet reached the main sequence. In either case, we are probably dealing with regions whose characteristic ages are not much greater than 10^5 years, since placental material has yet to be dissipated.

In some cases, the velocity of the maser may differ significantly from the dark-cloud velocity measured from mean CO emission. In these cases, the maser emission may arise in a shock region compressed to high density by the interaction of highly supersonic stellar winds and the ambient dark cloud material. Our theoretical understanding of the OH and H_2O maser emission processes (see Chapter 9) is far from complete. However, as the theory improves, observations of these sources should provide signposts as well as important information regarding the interaction of young stellar objects with placental or dark cloud material.

Properties of Young Stellar Objects

Despite the discovery of a number of unusual infrared sources associated with dark clouds, our basic understanding of the early stages of stellar evolution is derived primarily from objects that can be studied at optical wavelengths. As infrared spectroscopic techniques improve, it will be possible to study optically obscured objects in sufficient detail to elicit their basic physical properties. Furthermore, if higher spatial resolution can be attained at infrared wavelengths, it should be possible for selected objects, to distinguish between slightly reddened protostellar objects of low temperature, and highly reddened, pre-main-sequence or main-sequence objects of high temperature.

We shall restrict our discussion to those objects for which optical data are available: Herbig emission stars, the T Tauri stars, and the Herbig-Haro (H-H) objects.

The Herbig Emission Stars

Certain basic characteristics of this class of stars have been observed:

1. An absorption spectrum ranging from that of early B stars to that of late F stars.

2. Optical emission ranging in complexity from a simple spectrum that shows only $H\alpha$ emission, through the rich emission spectrum of V380 Orionis. In a few cases, Balmer continuum emission is also observed.

3. Direct association with dark-cloud material, as indicated by the illumination of nearby reflection nebulae.

4. P Cygni profiles (see Figure 3-4). In all but one case, the Herbig emission stars show Balmer emission lines with violet-displaced absorption features.

5. Irregular variability. At optical wavelengths, these objects vary with amplitudes ranging from a few tenths of a magnitude up to 4 mag.

6. Optical polarization. The electric vector directions differ from those of nearby objects and, furthermore, exhibit a definite rotation in position-angle with wavelength. Both observations suggest that the polarization is intrinsic to the Herbig emission stars.

7. Infrared excesses. The spectral-energy distributions of these stars at $\lambda \gtrsim 1$ μm depart significantly from those expected from reddened stars of early spectral type. In all cases, "excess" radiation is observed. In a few cases, there appears to be a definite evidence for a 10.2-μm "bump" often identified with emission from silicate material.

From estimates of surface gravity based on the Balmer line wings, the Herbig emission stars appear to lie in the domain of the H-R diagram occupied by high-mass stars approaching the main sequence along equilibrium radiative tracks. This fact, combined with their close association with darkcloud material, makes it almost certain that they are pre-main-sequence objects with ages in the range 10^5 to 10^6 years. Other characteristics of these stars seem best understood by assuming that they are surrounded by a hot ($T \sim 2 \times 10^4$ K), expanding, ionized region. The Balmer lines, Balmer continuum, and the infrared excess are produced by bound-bound and freebound transitions in this hot region. Optical variability may result from variations either in the intrinsic brightness or in the total amount of material in this envelope region. If we assume the envelope to be somewhat inhomogeneous, the optical polarization seems plausibly explained by electron

Figure 3-4.   An intensity tracing of a portion of the spectrum of P Cygni. Line profiles in this sample spectrum are combinations of emission and absorption features. The emission lines are produced in the star itself, and the absorption lines, displaced to the blue, are formed by attenuation of the star's spectrum by gases in front of the star. The absorption features are blue-shifted in this spectrum with respect to line center because the intervening gases are streaming away from the star. The most simple case of a P Cygni profile in this spectrum is the N II λ 3995 line at the left. In the center of this spectral region, we see an Hε profile in emission and absorption, but the profile is complicated by an additional absorption feature of interstellar calcium. The He I line, slightly to the blue of Hε, also shows P Cygni characteristics.

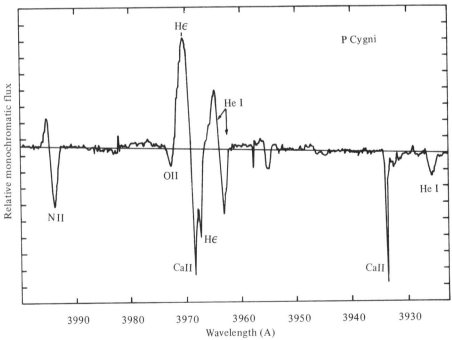

scattering in this region. The 10.2-μm bump may result from the heating of nearby, ambient, dark-cloud material, or possibly of remnant material located in a placental cloud surrounding the star. It is also possible that a dust envelope can be produced in the cooler region of the expanding envelope.

Important problems not yet solved with regard to these objects are as follows:

1. No detailed, self-consistent model of the emitting region has yet been made. Such a model would attempt a simultaneous explanation of all the emission processes. Observationally, the input required for such a model should include a detailed spectral-energy distribution for the "excess" radiation; calibrated spectrophotometry of the P Cygni profiles, with special attention to deriving accurate measures of the Balmer decrement; and spectrophotometry of Paschen and Brackett emission lines, which would provide important checks on the consistency of the model because these lines are likely to be optically thin, while the Balmer lines are probably optically thick.

2. As a corollary to problem 1, it is important to establish the mass-loss rate characteristic of these objects. The role of stellar winds in dissipating infalling envelope material will be better understood if one can make quantitative estimates of the wind characteristics. In addition, the relationship between these winds and stellar rotation may give some further clues to the origin of mass outflows in these stars.

3. Infrared spectroscopy of these stars, with particular attention to measuring characteristic emission-line strengths, may be of critical importance in attempts to classify similar, more heavily obscured objects not accessible to optical observations. Grasdalen (1974) at Kitt Peak National Observatory has already made considerable progress in this area.

4. The relative behavior of the various emission features has not been firmly established for this class of object. Simultaneous infrared and optical observations of these features will be essential in confirming the working hypotheses concerning the origin of optical variability and the sources of excess radiation.

The T Tauri Stars

Objects of this class exhibit certain characteristics:

1. An underlying absorption spectrum ranging between that of late F stars and that of middle M stars.

2. In some cases, broad absorption lines, which are usually interpreted as being broadened by stellar rotation.

3. Balmer-line emission.

4. Emission in the H and K lines of Ca II, as well as in lines of Fe I, Fe II, Ti II and [S II], in varying degrees. The strength and complexity of the emission-line spectra range widely. In a few cases, the emission-line spectra are similar to those observed for Herbig-Haro objects (see next section).

5. P Cygni profiles. With the exception of the class of T Tauri stars called "YY Ori stars" by Merle Walker (which at *some* times exhibit inverse P Cygni profiles), all T Tauri stars show violet-displaced absorption components, adjacent to Balmer emission lines or to Ca II emission lines or to both.

6. Excess Balmer-continuum and Paschen-continuum radiation. The superposition of Paschen-continuum radiation upon the photospheric absorption spectrum gives a washed-out or veiled appearance to the photospheric absorption lines.

7. Optical polarization. As for the Herbig emission stars, the optical polarization appears to be intrinsic. Further evidence for a stellar rather than an interstellar origin is the time variability in polarization observed for several T Tauri stars.

8. Irregular variability at optical wavelengths; the magnitude of such variation ranges from a few tenths of a magnitude to as much as 5 mag for members of the class.

A detailed study of these objects has recently been published by Rydgren, Strom, and Strom (1975). Their basic conclusions can be summarized as follows:

1. The emission features (Balmer continuum, Paschen continuum, infrared excess, Balmer lines) arise in a hot ($T \sim 2 \times 10^4$), dense ($n \sim 10^9$–10^{12} atoms/cm^3), circumstellar envelope.

2. The optical variability stems from changes in the emission measure, indicating changes in density or physical size of the envelope.

3. Mass outflows or stellar winds characterize most, if not all, of these objects.

4. Polarization results from electron scattering of photospheric radiation in the inhomogeneous envelope region.

5. The largest rates of mass outflow and strongest emission features are found in the youngest T Tauri stars, which are still confined within a dark

cloud or within placental material. The mass-outflow rates decrease as the stars approach the main sequence.

6. The large majority of T Tauri stars have relatively low mass ($M \lesssim 2$ $M_\odot$) and are approaching the main sequence along quasi-static equilibrium tracks (see Figure 3-2).

7. The mass outflows, coupled to the photospheric layers by strong magnetic fields, may represent the mechanism by which stars of near-solar mass slow their rotation to values consistent with the low radial velocities characteristic of main-sequence stars of this type.

A number of outstanding problems need further study:

1. Photosphere-envelope models should be constructed and compared in detail with observations. This would involve deriving quantitative estimates of the contribution of the envelope continuum-emission by measuring the line-center intensity in the photospheric spectrum. Thus, the contribution from the envelope can be isolated and compared more accurately with model computations.

2. Quantitative estimates should be made of the interaction between specific angular momentum and the rate of mass outflow. If accurate projected rotational velocities and estimates of mass outflow can be obtained from the analysis of high-resolution spectra, this would represent an important step towards providing an empirical basis for discussion of "spin-down" problems.

3. Estimates of magnetic field strengths are needed. Accurate high-dispersion studies of absorption lines sensitive to Zeeman splitting will be difficult to make, primarily because of the intrinsic broadening due to stellar rotation. However, even a few such measurements would yield important information concerning the behavior of the field as a function of time and of rotational velocity.

4. Indirect study of the circumstellar environment of the T Tauri stars is needed. These objects are among the most likely to be surrounded by protoplanetary material. Observation of such material would clearly be of great importance to understanding the development of the early solar system. Several T Tauri stars illuminate reflection nebulae (R Coronae Austrinae, T Tau, and the unusual object R Monocerotis). These systems may make possible the optical detection of thick clouds of orbiting circumstellar material. Synoptic studies of reflection nebulae might lead to the detection of T Tauri

stars, by the discovery of periodic changes in light and brightness across the face of the nebulae, which take place on a time scale faster than the apparent light-travel time between the variable regions.

5. Studies should be made of temperature structures in the upper photosphere. The combination of higher rotation and possibly greater magnetic-field strengths may produce greater heating in the chromosphere and upper photosphere. Such effects might be detected by comparing, in T Tauri stars and "normal" stars, the temperature-sensitive lines formed near the stellar boundary. High-resolution spectrograms at $\lambda \gtrsim 5000$ Å should provide a basis for such a study. Again, such studies might increase our understanding of the basic heating mechanisms that give rise to stellar chromospheres.

6. Optical polarization measures of different classes of emission lines are needed. Such observations might help sort out lines that are formed in different regions of the envelope. Comparison, say, of the polarization measured for hydrogen, helium, and forbidden lines might lead to a qualitative understanding of the differences in the regions of formation. L. V. Kuhi is already attempting such observations.

The Herbig-Haro Objects

The defining characteristics of the class are a patchy, sometimes semi-stellar, nebulous appearance; a low-excitation spectrum dominated by Balmer, [S II], and [O I] lines; very weak continuum emission; and the presence of such objects only in dark-cloud complexes.

A survey of objects of this class was recently completed by Strom, Gras-dalen, and Strom (1974). They observed *single* infrared sources associated with multiple-component Herbig-Haro (H-H) nebulosities; these sources are generally displaced from the optical objects. Strom et al. also reported polarization measurements which suggest that the individual nebulous patches comprising an H-H object are illuminated by a single source that coincides in location with the nearby infrared object.

Strom et al. concluded that H-H objects are reflection nebulae illuminated by "H-H stars" obscured at optical wavelengths by placental dark-cloud material. Spectroscopically, these stars are similar in character to, but more extreme than, some of the strongest emission-line T Tauri stars. Light reaches these optically observed nebulous patches through "holes" in the dark cloud. Radial velocity studies of the H-H nebulosity indicate that the H-H stars are losing mass at a higher rate than the T Tauri stars. The authors suggest that this strong stellar wind is in fact responsible for producing "holes" in the dark cloud, as the first observable manifestations of a mechanism that eventually destroys the placental cloud.

The location of H-H stars within placental clouds receives further support from maps of dark-cloud complexes made in CS and HCN molecular lines. Relatively large densities of 10^4–10^5 cm^{-3} are required to produce the observed millimeter-line emission from the molecules. In several cases, the H-H stars are located near the centers of high-density condensations.

Several kinds of observation would be crucial to a better understanding of the character of H-H objects. One would be the observation of optical polarization in the emission lines. If the direction of emission-line polarization is the same as observed in "broad-band" observations, then both the line-emission region and the continuum-producing region must be near the infrared H-H star. Another would be synoptic observations of the variability of H-H objects and H-H stars. Such observations, if carried out over sufficient time, should allow one to correlate variability of the star and the object, if the reflection hypothesis is correct. The differences in light-travel time from an H-H star to individual scattering patches may make this task difficult in practice. However, the *amplitude* of visual and infrared changes might provide a test of this hypothesis, since we assume that both infrared and visual radiation are produced by an emitting circumstellar envelope. Detailed studies of the spectral energy distribution of H-H objects would also be important. These should be compared with the distributions observed for the envelopes of extreme T Tauri stars. Such comparisons should permit detailed evaluation of the intrinsic relation, if any, between the two classes of objects.

A Scenario for Pre-Main-Sequence Evolution

The observations of Herbig emission stars, T Tauri stars, and Herbig-Haro objects provide an empirical basis for outlining the early stages of evolution.

After the initial collapse and hydrodynamic phases, a young stellar object is still embedded deep within its progenitor, i.e., within a large dark-cloud complex, and within its own smaller, denser, placental cloud of gas and dust. At an early age, perhaps as young as 10^4 years, the young stellar object develops a strong stellar wind. The winds at this stage are characterized by mass-outflow rates of 10^{-5} to 10^{-6} $M_\odot$/year and by outflow velocities of several hundred km/sec. (By comparison, the mass-outflow rate of the solar wind is 10^{-14} $M_\odot$/year.) The wind is probably sufficient to halt further accretion of gas and dust from the protostellar cloud. In the case of more massive stars, radiation pressure may be more important than the wind in halting accretion. On a time scale of 10^4 to 10^5 years, the stellar winds dissipate the placental dark-cloud material. During the first stages of dissipation, light from the newly formed star "leaks" out through small breaks in the placental cloud and illuminates patches of dark-cloud material. These illuminated patches are identified with the H-H objects. At this stage, emission from a circumstellar envelope dominates the photospheric contribution.

If the newly formed star lies close to the boundary of the dark-cloud complex, it will first be observed at an age of $\sim 10^5$ years as an extreme T Tauri star. In some cases, however, the young stellar object will remain embedded within the larger dark-cloud complex for times as long as $\sim 10^7$ years. The stars of higher mass may already have the spectral characteristics of the Herbig emission stars. However, Grasdalen (1973) argues that some high-mass stars observed early in their approach toward the main sequence are first visible as T Tauri stars. In his view, they undergo a rapid rise in luminosity (by a factor of ~ 100) and then join their equilibrium radiative tracks, becoming Herbig emission stars. He cites FU Orionis and V1057 Cygni as possible examples of this behavior.

As the star progresses toward the main sequence, the rate of mass outflow decreases and the envelope emission decreases as well. Possibly, during this stage, the low-mass ($M \lesssim 2 M_\odot$) stars lose a considerable amount of angular momentum as a consequence of the mass outflow. At ages $\sim 10^6$ years, most pre-main-sequence stars have lost their prominent, identifying spectral characteristics. Emission features are observable only by careful photometric study of Balmer lines and the Balmer continuum.

Because our detailed knowledge of early stellar evolution is restricted primarily to those objects for which optical data are available, we have yet to probe stages of stellar evolution earlier than several times 10^4 years. The most promising candidates, which deserve further scrutiny, are those luminous objects discovered in the AFCRL survey at 10 and 20 μm (Walker and Price, 1975) and in detailed 10-μm maps of dark-cloud complexes, and those objects presumably related to the continuum sources discovered at radio wavelengths in dark-cloud complexes. The next few years should witness significant progress in our understanding of the earlier phases of stellar evolution, as these objects receive closer observational scrutiny.

Star Formation in External Galaxies

The process of star formation seems intimately connected with the evolution of dark clouds. How did these complexes of gas and dust form? To what extent do external forces determine their creation? What determines the mass spectrum in dark-cloud complexes? How efficient is the star-formation process at present and in the past?

Perhaps the most profitable line of attack on these problems involves examining the global properties of systems that are actively forming stars at the current epoch.

Spiral Galaxies

The presence of OB stars and H II regions in the spiral arms of galaxies similar to our own provides ample evidence that stars have been formed during the last $\sim 10^7$ years in these systems. The fact that these newly formed stars trace a spiral pattern and are restricted to the arms suggests that a mechanism operating on a galactic scale triggers star formation in spiral galaxies. Of great significance is the location of lanes of dust directly adjacent to the spiral arms. It is reasonable to assume that the next generation of stars will be formed in these dust lanes and that the spiral pattern, in 10^7 years, will be outlined by these newly formed stars.

W. Roberts, F. Shu, C. C. Lin, and others (see Shu, 1973) have suggested that galaxy-wide star formation is triggered by the passage of disk gas through a gravitational disturbance in the disk, known as a density wave. When the gas encounters the spiral density-wave pattern, shock waves form, and the gas is compressed, and possibly forms dark-cloud complexes and, later, stars. The frequency of such star-forming events should depend on the frequency with which the disk gas circulates through the density-wave pattern. This frequency, in turn, depends on the rotation characteristics of the galaxy. In general, the star-formation frequency decreases radially outward in the galaxy so that fewer stars are formed per unit time and per unit volume in these regions. The efficiency of star formation per generation may depend on a number of factors, such as the amount and density of the unprocessed disk gas, and the metal-to-hydrogen ratio. It may also depend on the degree of compression in the shock region: the higher the compression, the higher the star-formation efficiency (since more clouds may be forced "over the brink" towards gravitational instability).

A partial observational test of these hypotheses was recently attempted by the author, in collaboration with K. M. Strom and E. Jensen. Because the ratio of element enrichment (as a by-product of nucleosynthesis in past generations of stars) and gas depletion (as a result of conversion of disk gas to stars) depends on the frequency of star-forming events, the metal-to-hydrogen ratio in the disk gas should vary directly with this frequency. Our results seem to confirm that the metal-to-hydrogen ratio in a spiral galaxy depends on the rate at which disk gas circulates through the spiral wave pattern. Further, this ratio seems to depend on the degree to which the gas is compressed in the shock region. The regions of greatest metal enhancement for a given star-forming frequency apparently suffer the greatest compression.

These studies encourage us to ask even more detailed questions concerning the factors that influence the efficiency of star formation. From maps of spiral-arm regions and spiral disks, it should be possible to determine how the efficiency of star formation at the present time depends on compression strength, metallicity, and gas density. The study of star formation in such external systems represents an extremely fruitful line of research because the influence of such a large range of physical characteristics seem amenable to fairly direct study.

The behavior of interstellar clouds in spiral-arm regions has been the subject of an extremely important recent study by Woodward (see Mouschovias, Shu, and Woodward, 1974). Using a complex hydrodynamic code, he has followed the time evolution of a spherical cloud embedded in a low-density, hot, "intercloud" medium. The intercloud medium moves relative to the cloud at a velocity that is determined by the gas flow in the arm region. The results of his analysis suggest that only a small fraction of the total mass of a typical cloud collapses to form stars, that this fraction depends on the relative cloud-intercloud velocity, and that the bulk of the dark-cloud material is "returned" to the general interstellar medium on a time scale of $\sim 10^7$ years. Furthermore, the qualitative morphological behavior predicted by Woodward's calculations seems in good agreement with the observed appearance of cloud complexes in our Galaxy. Unfortunately, the numerical complexity of the problem is sufficient to preclude computing more than a few trial cases. However, these calculations appear extremely promising because they offer a natural explanation both for the small fraction of cloud mass converted to stars, and for the mechanism by which cloud dissipation occurs. Further such studies are badly needed.

Irregular Galaxies

The presence of OB stars and H II regions in irregular systems such as the Magellanic Clouds offers ample evidence that star formation occurs at the present time in systems where star formation in spiral-arm shock regions seems improbable.

The work of W. L. W. Sargent, L. Searle, and others suggests, however, that there may be rather long quiescent periods during which such irregular systems are not actively forming stars. Triggering of star formation in the system may await supernova explosions, which force compression of surrounding galactic gas. Star formation in the supernova shock regions may follow a scenario qualitatively similar to that characteristic of spiral shocks.

Irregular galaxies seem to have lower metal-to-hydrogen ratios than those found in the solar neighborhood. Given a lower star-formation frequency, this may not be surprising. There is some indication that the ratio of high-mass stars to low-mass stars may be greater in such low-metallicity systems. This may be an indication of the role played by metals in controlling the number of stars formed with different masses. The lower cooling rates in metal-poor regions might prevent the collapse of a low-mass cloud (since the internal pressure may be too high). Further study of nearby irregular systems may hold important clues to the details of star formation under the low-metallicity conditions that may have prevailed in the early stages of spiral disk evolution.

Elliptical Galaxies

With few exceptions, the amount of gas remaining in these systems is not sufficient to support significant star formation at the present epoch. This seems due in part to the high efficiency of star-formation during the initial evolutionary phases of these systems. In fact, the morphology of elliptical galaxies cannot be understood unless these galaxies either collapsed from already-formed stars or clusters, or formed stars very efficiently during the time when intergalactic gas collapsed to form the systems. Environmental factors, such as galactic or intergalactic winds, may also play a role in precluding present-day star formation in elliptical galaxies. Such winds may prevent the accumulation of mass lost by old stars, which might otherwise form new stellar complexes in nuclear-bulge regions of elliptical systems.

Studies of the distribution of stars with different chemical compositions in elliptical galaxies should provide an important and relatively simple record of the past history of star formation and metal enrichment in such systems. This record may be the key to an understanding of the role of star formation in the early history of the universe.

At present, however, it is not at all clear how star formation can occur with the great efficiency required to account for the morphology of elliptical galaxies, particularly given the presumed low metallicity and consequent low cooling-rates in primordial material. The scenario for star formation during the early evolution of the universe was probably quite different from that which we believe applies in most galaxies that are actively forming stars today. Perhaps the study of low-metallicity blue-dwarf galaxies will provide some clues to help us understand the problem of star formation at the earliest epochs.

surface and is radiated away into space. This long-lived stage is called the main sequence, because most of the stars visible in space are in this stage of evolution and occupy a restricted region in the Hertzsprung-Russell diagram, which we shall discuss later.

The entire story of stellar evolution involves a continued release of gravitational potential energy through shrinkage of the interior, with pauses in the shrinkage when new nuclear fuels are ignited and can supply for a time the energy flow toward the surface of the star. Generally speaking, the ashes of one set of nuclear reactions become the fuel for the next set, which will burn at a still higher temperature. Not all parts of the star exhibit continued shrinkage; sometimes the stellar core will continue to shrink while the envelope expands, and vice versa, but we shall comment upon these complications later.

This process, of alternating shrinkage and nuclear fuel burning, cannot continue indefinitely. There are only a finite number of stages of nuclear burning possible before all of the available thermonuclear energy has been released from matter. There are also other forms of internal energy storage, quantum mechanical in nature, in which the energy is not available to flow to the surface and cannot be radiated away. With this form of energy is associated a "degeneracy pressure" which can stabilize some stars indefinitely against further shrinkage. If this stabilization does not become possible for a star, then, after the nuclear fuels have been completely exhausted, the only prospect in store for such a star is continued shrinkage until some catastrophe occurs. The ultimate aim of this chapter is to discuss the various endpoints of stellar evolution which can be reached in a diverse number of ways.

Equations of Stellar Structure

The virial theorem is a general relationship which must be satisfied by a stellar model, and it is very useful for giving us a general idea of what must go on in the course of stellar evolution, but in order to make detailed models we must consider the full set of equations which govern the physics of the stellar interior. The first of these equations is that of hydrostatic equilibrium. At any point in the interior of a star, the pressure must be high enough to hold up the weight of the overlying layers. Hence the difference of pressure between two adjacent points in the stellar interior will be given by the weight

References

Brown, R. L., and Zuckerman, B. 1975. Compact H II regions in the Ophiuchus and R Coronae Austrinae dark clouds, *Astrophys. J. (Letters) 202*, L125–128.

Grasdalen, G. L. 1973. V1057 Cygni and pre-main sequence evolution, *Astrophys. J. 182*, 781–808.

Grasdalen, G. L. 1974. An infrared study of NGC 2024, *Astrophys. J. 193*, 373–383.

Grasdalen, G. L., Strom, K. M., and Strom, S. E. 1973. A 2-micron map of the Ophiuchus dark-cloud region, *Astrophys. J. (Letters) 184*, L53–57.

Iben, I. 1965. Stellar evolution. I. The approach to the main sequence, *Astrophys. J. 141*, 993–1018.

Larson, R. B., and Starrfield, S. 1971. On the formation of massive stars and the upper limit of stellar masses, *Astron. Astrophys. 13*, 190–197.

Mouschovias, T. C., Shu, F. H., and Woodward, P. R. 1974. On the formation of interstellar cloud complexes, OB associations and giant H II regions, *Astron. Astrophys. 33*, 73–77.

Rydgren, A. E., Strom, S. E., and Strom, K. M. 1976. The nature of the objects of Joy: A study of the T-Tauri phenomenon, *Astrophys. J. Suppl.*, in press.

Shu, F. H. 1973. Spiral structure, dust clouds, and star formation, *Amer. Scientist 61*, 524–536.

Strom, S. E., Grasdalen, G. L., and Strom, K. M. 1974. Infrared and optical observations of Herbig-Haro objects, *Astrophys. J. 191*, 111–142.

Strom, S. E., Strom, K. M., and Grasdalen, G. L. 1975. Young stellar objects and dark interstellar clouds, *Annu. Rev. Astron. Astrophys. 13*, 187–216.

Vrba, F. J., Strom, K. M., Strom, S. E., and Grasdalen, G. L. 1975. Further study of the stellar cluster embedded in the Ophiuchus dark cloud complex, *Astrophys. J. 197*, 77–84.

Walker, M. F. 1956. Studies of extremely young clusters. I. NGC 2264, *Astrophys. J. Suppl. 2*, 365–387.

Walker, R. G., and Price, S. D. 1975. The AFCRL infrared sky survey, vol. I, Air Force Cambridge Res. Lab. Tech. Rept. TR-75-0373.

Westbrook, C. K., and Tarter, C. B. 1975. On protostellar evolution, *Astrophys. J. 200*, 48–60.

Zuckerman, B., and Palmer, P. 1974. Radio radiation from interstellar molecules, *Annu. Rev. Astron. Astrophys. 12*, 278–313.

4

Endpoints of
Stellar Evolution

A. G. W. Cameron

Understanding the evolution of the stars is one of the oldest goals of astronomy. However, only with the development of the science of astrophysics during the present century has it been possible to make substantial progress toward that goal. Indeed, a large number of the new discoveries in physics which have been made during the last few decades have had important implications for the theory of stellar interiors.

To the extent that one can ignore the rotation of the stars and regard them as spherically symmetric objects (generally a good assumption), the physics of their hot gaseous interiors is relatively simple. This simplicity may be misleading, for the failure to detect the energetic weakly-interacting neutral particles called neutrinos emitted from the solar interior is one of the major puzzles of modern astrophysics, and it warns us to be cautious in assuming that we have a good understanding of the interiors of the sun and other stars.

However, because the stars are objects in hydrostatic equilibrium, and change their dimensions very slowly throughout the greater part of their evolutionary lifetimes, we can use a simple principle to understand the general trend of their evolution. This is the virial theorem. Let U be the gravitational potential energy of a stellar configuration. Let K be the internal kinetic energy of motion. This kinetic energy may take any of a number of forms: the kinetic energy of the individual thermal motions of the particles in the interior, the kinetic energy of bulk motions in the interior such as those due to convective turbulence, or the kinetic energy of larger organized motions such as rotation of the configuration. Then the virial theorem states that

$$2K + U = 0. \tag{4-1}$$

Thus the virial theorem states that when a bound configuration is formed,

half of the released gravitational potential energy is stored as interna energy, and the other half is lost from the system. Thus, for a planet in a cular orbit about a star, the kinetic energy of its motion is half of the gra tional potential energy which has been released in binding the star and planet together. In a nonrotating star the internal energy is half of the leased gravitational potential energy, but it is stored in the form of h which provides the necessary pressure to hold up the outer layers of this against the force of gravity which tries to press them into the center.

In principle, all the main features of stellar evolution can be understoo terms of the virial theorem. A star is formed when a cloud of gas collar from the interstellar medium toward high density. At first, internal hea readily radiated away from the collapsing gas because of the relatively g transparency of the gas in the infrared. As the gas becomes denser, its op ity increases, and the energy released by the collapse can be stored in interior. The gas settles down to form a star at a radius at which the sto internal energy is half the released gravitational potential energy.

In ordinary stars, energy is lost through electromagnetic radiation fr the surface. A gradient in the temperature will be set up in order to acco plish this; the center of the star will be relatively hot and the surface lay relatively cool, so that energy can flow from the center toward the surfa by any of several mechanisms, which will be discussed below. In the a sence of a temperature gradient, there would be no preferred direction f the flow of energy. As energy is radiated away from the surface, the st shrinks just enough to provide not only the energy that has been radiate away, but also an equal amount of energy which is added to that internal stored. Thus the interior of the star shrinks at a rate which is governed by th flow of energy to the surface and its loss into space.

This shrinkage leads to a continual heating of the stellar interior. The tim will come, when the temperature at the center has risen to the range 1 t 2×10^7 K, that a new source of energy will appear in the interior: hydroge thermonuclear reactions. The amount of energy which can be released through conversion of hydrogen to helium is very large compared to the amount of gravitational potential energy that is released in a star up to the point at which hydrogen burning commences. Therefore, according to the virial theorem, the star will settle into a long-lived state in which the internal thermal energy remains half of the released gravitational potential energy, but where the energy generated in the central regions flows to the

of the material per unit surface area in the shell lying between those two points. Expressing this more quantitatively, we have

$$\frac{dP}{dr} = -\frac{GM(r)\rho}{r^2},\qquad(4\text{-}2)$$

where P is the pressure at the radial distance r, G is the gravitational constant, ρ is the density in gm/cm³, and $M(r)$ is the mass in grams contained inside the radial distance r. This mass is a function of the radial distance r, and is given by the simple equation

$$\frac{dM(r)}{dr} = 4\pi r^2\rho.\qquad(4\text{-}3)$$

If we knew how the temperature varied with pressure and density throughout the stellar interior, then the above two equations would be sufficient to allow us to calculate the structure of the star. However, except in very special circumstances, we do not know what the temperature distribution is in the interior, in advance of the construction of the model, but we must determine what the temperature distribution is at the same time that we determine the distribution of the other quantities.

We have already mentioned that a star will establish a temperature gradient between the center and the surface which will be responsible for transporting energy, generated in the interior, to the surface where it can be radiated away. Let us digress for a moment to discuss the various ways in which this energy transport can occur.

One method of energy transport, familiar to anyone who stirs a cup of coffee with a metal spoon, is thermal conduction. This mechanism of energy transport is not important in most stellar interiors, but does become important under conditions in which the electrons are degenerate, which we shall discuss in a number of contexts. Energy transport by thermal conduction is usually very efficient when the electrons form a degenerate gas.

Another form of energy transport is radiative transfer. Electromagnetic radiation is emitted by any of a number of microscopic processes, travels a short distance through the interior of the star, and is reabsorbed by other processes. The rate of emission of electromagnetic energy depends upon the temperature; under conditions in which we can speak about "black-body radiation," the rate of emission of energy varies as the fourth power of the temperature. Because of the greater efficiency for generating energy in regions of higher temperature, the photons which are created there tend to diffuse away and to be absorbed in regions of lower temperature, thus transporting energy from high-temperature regions toward low-temperature ones. This is the principal mechanism of energy transport in many parts of the interiors of normal stars.

The mechanisms of energy emission and absorption in stellar interiors mostly involve electronic processes. Electrons can absorb or emit energy by jumping between the discrete quantum states of atoms. However, in a stellar interior the temperatures are so high that most of the time the common atoms are stripped of their electrons. Electromagnetic energy will be emitted when a free electron is captured by an ion, jumping down from the continuum of free electronic states to one of the bound states. Such an electron will shortly absorb a photon and jump from one of the bound states into the free continuum. Electrons can also both absorb and emit energy by jumping between the states of the free continuum, as long as they are reasonably close to an ion when they do so, so that it is easy for the ion to participate in the simultaneous conservation of energy and the momentum which must occur during such transitions. In addition to these processes, and additional processes which occur mainly at lower temperatures, electromagnetic energy is also readily scattered by the free electrons (Thomson or Compton scattering).

A third mechanism of energy transport involves convection. In this there is a large-scale mass motion of material in the interior of the star, so that gas which is hotter than the average material in its neighborhood can rise to a high level and radiate its excess energy to the material at its new position, and gas which is unusually cool can fall to a lower level and absorb energy from its new position. This is a very efficient mechanism for energy transport.

A fourth mechanism of energy transport involves mechanical waves running through the interior of the star. Generally speaking, such wave motion can be generated only in a convective region. The efficiency for generating such waves is usually quite small, so that such wave motions are not an important mechanism for transporting energy through stellar interiors, in general. However, in ordinary stars like the sun, generation of waves by convection taking place near the surface plays a dominant role in heating the outer parts of the atmosphere, raising them to a temperature much above the effective temperature at which the star radiates its energy into space. We shall not be further concerned with this form of energy transport (see Chapter 2 in this volume).

If energy transport takes place by radiative transfer, then the temperature gradient in the interior is proportional to the opacity of the material and to the luminous energy flux; the precise relationship is

$$\frac{dT}{dr} = -\frac{3}{4ac}\frac{\kappa\rho}{T^3}\frac{L(r)}{4\pi r^2},$$ (4-4)

where T is the temperature, a is the radiation density constant, c is the speed of light, κ is the opacity of the material in cm²/gm, $L(r)$ is the luminosity, and $L(r)/4\pi r^2$ is the luminous flux carried through the interior of the star by the radiative transfer. In any small volume in the interior, the luminous flux which flows into the volume from the bottom will be augmented by any energy generated within the volume by thermonuclear reactions or by gravitational potential-energy release. Hence the equation describing the variation of the luminous flux within the interior is

$$\frac{dL(r)}{dr} = 4\pi r^2 \rho \epsilon,$$ (4-5)

where ϵ is the energy generated by nuclear reactions or by the gravitational potential energy release. It should be noticed that ϵ can be a negative quantity, in the event that the part of the star under consideration is increasing its radial distance, or if, for example, there is neutrino-antineutrino pair emission from the small volume so that energy is removed from the luminous flux being transported by radiative transfer and escapes directly from the interior of the star.

Radiative transfer is a dominant form of energy transport only if the temperature gradient is not too large. If the temperature gradient becomes too large, then convection will take over the bulk of the energy transport. The critical temperature gradient at which convection begins is the *adiabatic temperature gradient*. This means the gradient in which a blob of gas, moving from high towards low pressure, is always in pressure equilibrium with its surroundings, *while neither receiving energy from its surroundings nor losing energy to its surroundings*. Let us suppose that the temperature gradient is steeper than this, i.e., the temperature falls faster with increasing radial distance than would the adiabatic temperature gradient. Give our hypothetical blob of gas an upward displacement, but allow it to remain in pressure equilibrium with the surroundings. According to the perfect-gas law, the pressure is proportional to the product of the density and the temperature. Because of the high temperature gradient in the surroundings, the blob would experience a restoring force that would push it back toward its surroundings, hence at a lower density. This means that the blob of gas is buoyant, and will rise even farther. On the other hand, if the temperature gradient had been smaller in absolute value than the adiabatic value, then the raised blob of gas would have found itself at a lower temperature than its surroundings, the density would be higher than that in the surroundings, and the blob would experience a restoring force that would push it back toward its original level. Thus the temperature gradient must slightly exceed the adiabatic value in order that energy can be transported by convection. The rate of energy transport is proportional to the difference between the actual temperature gradient and the adiabatic value. But the efficiency of such convective energy transport is so high that the actual temperature gradient will be only microscopically greater than the adiabatic value. For this reason, it is generally sufficient in stellar interiors to approximate the actual temperature gradient by the adiabatic value when convection occurs. This gives for the temperature gradient the following expression:

$$\frac{dT}{dr} = \left(1 - \frac{1}{\gamma}\right) \frac{T}{P} \frac{dP}{dr}, \tag{4-6}$$

where γ is the ratio of specific heats which is used in calculating adiabatic transformations. This equation is obtained by differentiation of the adiabatic relationship $T = $ constant $P^{(\gamma-1)/\gamma}$.

When one is constructing a model of a stellar interior, it is customary to compute the temperature gradients from both equations (4-4) and (4-6), and to use the lower in absolute value of the two values thus obtained. This ensures that a region will be taken to be convective if a higher-than-adiabatic temperature gradient would otherwise be needed to transport the luminous energy flux.

These differential equations governing the interior structure of a star are nonlinear, and in general must be solved by numerical integration. They must also be supplemented by equations for the physical quantities which are contained within the differential equations. Thus, for example, we need an equation of state to express a relation between the pressure P, and temperature T, and the density ρ. We need expressions for the rates at which nuclear reactions contribute to the energy generation function ϵ. We need to know how the opacity κ varies as a function of temperature and density. And we also need to know how the ratio of specific heats γ varies with the same quantities. The inclusion of accurate expressions for these quantities simply adds to the nonlinearity of the differential equations which must be integrated.

Because of the extensive numerical computations required, the development of high-speed computers has played a very significant role in the understanding of stellar evolution. Some of the detailed problems in the more advanced stages of stellar evolution continue to challenge the largest computers available today.

Nuclear Energy Generation

For several decades we have been seeking new forms of energy through rearrangement of the structure of nuclei. The greatest amount of nuclear binding energy per nucleon occurs for nuclei in the vicinity of iron. Thus we can release nuclear energy by causing very large nuclei to divide into smaller pieces, or by causing very light nuclei to fuse into larger pieces. Nuclear fission has been used for energy generation for some time, but nuclear fusion has so far been achieved only explosively. The problem is that it takes high temperatures and pressures to produce thermonuclear reactions. These are achieved easily in stellar interiors, where gravitation provides containment, but only with difficulty in the laboratory.

In order to achieve nuclear fusion, it is necessary to bring two nuclei into physical contact at their surfaces. But two such nuclei contain positive charges, the protons in their interiors, and the repulsion between the positive charges on two nuclei exerts a very strong force which repels them from each other. It requires a lot of work, or energy, to bring two such nuclei into surface contact in the presence of such repulsive forces. The normal thermal kinetic energy that an ion may have is almost never sufficient to achieve this.

Fortunately, once two ions are in contact at their surface, the very strong nuclear attractive forces take over, and such nuclei can easily be pulled together into one large nucleus. Therefore it is only in the range of distances in which the nuclei are close together, but not yet in contact, where the strong repulsive forces are felt. This is the region of the Coulomb potential barrier.

One of the fundamental properties of quantum mechanical behavior is that particles can penetrate through such potential barriers, even though they may not have sufficient energy to come together over the top of the barriers. If two such particles collide at very low relative energy, then the probability that they will penetrate their mutual potential barrier is extremely small, though finite. As the relative energy is raised, the probability of barrier penetration rapidly increases, although it remains very small. Two such ions may bombard one another an enormous number of times at half the energy needed to surmount the mutual potential barrier, and eventually they will succeed in penetrating through the barrier by quantum mechanical tunneling. Because the probability of the barrier penetration is such a sensitive function of the bombarding energy, the rates of such thermonuclear reactions are extremely sensitive to the temperature of the gas in which they occur.

At any given moment in a stellar interior, the thermonuclear reactions which go the fastest are generally those which involve the lowest ionic charges in the nuclear collisions. This means that many of the stages of nuclear energy generation involve collisions between nuclei having the smallest available charge. This is complicated for hydrogen-burning reactions, because the only way that collisions between two protons can lead to a heavier nucleus, a deuteron, is that the two-proton system must undergo a beta decay during the process of the collision itself. This is the slowest known type of nuclear reaction, and hence a number of nuclear reactions which involve collisions between protons and ions of higher charge can take place at rates comparable to the very slow nuclear reaction between two protons. This means that the complete sequence of nuclear reactions which can be involved in hydrogen burning is extremely complex, and we shall not go into them here.

The later stages of nuclear energy generation are also quite complex, because the basic collisions may involve collisions between nuclei having six or more protons, but these collisions can produce protons, neutrons, and alpha particles, which are readily absorbed on nuclei of quite high atomic number at the very high temperatures involved, and hence many different products are possible. Still further complexities occur at very high temperatures because some nuclear reactions can proceed through absorption by nuclei of very high energy photons, which cause the ejection of protons, neutrons, and alpha particles, with the subsequent absorption of these light particles on other nuclei present.

The principal stages of nuclear energy generation are listed in Table 4-1. It may be seen that the products, or ashes, of one set of thermonuclear reactions frequently become the fuel in the next set. The hydrogen-burning reactions require a temperature of at least 10^7 K, at which they will occur in stars of relatively low mass, and up to 3×10^7 K or even higher in the most massive stars. From that range we have to jump to temperatures around 2×10^8 K for helium reactions to take place. The temperature continues to jump by significant factors between the various stages of energy generation, until temperatures of the order of 3×10^9 K are required for silicon burning. Many stars are unable to achieve such high central temperatures, and come to the endpoint of their stellar evolution without having advanced beyond helium burning. Only stars substantially less massive than the sun cannot advance beyond hydrogen burning, but the universe is much too young for them to have left the main sequence.

The final energy-generation stage of silicon burning forms nuclei in the vicinity of iron. We have already noted that this represents the release of all of the nuclear energy which is available. Nevertheless, as we have already discussed, the interior of a star which has managed to pass through the process of silicon burning must relentlessly contract toward higher temperatures and densities. The flux of high-energy photons bombarding the iron nuclei becomes enormous. Under these conditions, the iron nuclei will eventually undergo a reverse transformation, in which they are broken down into helium.

Table 4-1 Stages of nuclear energy generation.

Name of process	Fuel	Products	Approximate temperature (K)
Hydrogen-burning	Hydrogen	Helium	$1-3 \times 10^7$
Helium-burning	Helium	Carbon	2×10^8
		Oxygen	
Carbon-burning	Carbon	Oxygen	
		Neon	8×10^8
		Sodium	
		Magnesium	
Neon-burning	Neon	Oxygen	1.5×10^9
		Magnesium	
Oxygen-burning	Oxygen	Magnesium	2×10^9
		to sulfur	
Silicon-burning	Magnesium	Elements	3×10^9
	to sulfur	near iron	

This is an energy-absorbing process. If this should happen in the interior of a star, the only source of energy available for it is the gravitational potential energy which can be released by shrinkage of the core. Such shrinkage would then happen catastrophically, and this is one of the paths which may lead to the endpoints of the evolution of some stars.

The above process might be considered a form of negative nuclear-energy generation. Another process which could also be considered a form of negative energy generation is the formation of neutrino-antineutrino pairs in the deep interior of a star. There are many different processes which can be involved here, but the net effect of these is to convert a quantum of energy into a neutrino and an antineutrino, which have such a small interaction with the material at the center of the star that they can readily escape, carrying energy off into space directly away from the interior. The rates of these neutrino-antineutrino energy-loss processes are also sensitive functions of the temperature, increasing rapidly as the temperature is raised, although not as rapidly as the energy generation processes increase with the temperature. This distinction is important, because it means that if the temperature in the interior of the star is increased, nuclear energy burning will sooner or later occur at a rapid enough rate to offset any possible rates of loss of energy by the emission of the neutrinos and the antineutrinos. It may be recalled that the virial theorem requires that for every unit of energy that is lost, either by radiation from the surface or by neutrinos or antineutrinos from the deep interior, twice the amount of gravitational potential energy must be released by shrinkage of the stellar core. Thus, since the rates of energy loss by emission of neutrinos and antineutrinos can become truly enormous in the later stages of stellar evolution when the interior has reached a very high temperature, a star will pass very rapidly through the later stages of its nuclear energy generation and resulting evolution.

Equations of State

An equation of state is a relation between the pressure, temperature, and density of a gas. In most ordinary stars the only equation of state needed for the interior is the perfect-gas law. However, more complicated equations of state are needed in the advanced stages of stellar evolution where the electrons are often in a state to be described shortly, called degenerate, and the equation of state is consequently more complicated.

The perfect-gas law is

$$P = \frac{R\rho T}{\mu},\tag{4-7}$$

where R is the gas constant and μ is the mean molecular weight of the particles in the gas. The astrophysicist finds a few complications in using this equation, however, owing to the fact that most of the atoms can be fully recombined near the surface of a star, but they become progressively fully ionized toward the deeper interior. As more and more electrons are stripped off the ions, the mean molecular weight in the interior decreases, since the same amount of mass is divided among more particles on the average. Thus the mean molecular weight is variable in the stellar interior and this must be taken into account in constructing the models.

For use under conditions of high density, it is convenient to divide the pressure of a gas into two contributions:

$$P = P_i + P_e\tag{4-8}$$

where P_i is the pressure contribution to the ions, and P_e is the pressure contribution due to the electrons.

Let us consider a blob of gas which we will squeeze toward high densities, meanwhile allowing plenty of time for energy to leak out of the blob, so that the temperature will stay constant. We assume that we start with fully ionized material. Initially both the ions and the electrons will have a Maxwell-Boltzmann distribution of their energies. However, when we reach very high densities, the energy distributions of the electrons must be altered for quantum mechanical reasons.

There is a fundamental principle of quantum mechanics, called the Pauli principle, which states that not more than two electrons can be placed in any single quantum state, and those two electrons must have opposite directions of their intrinsic spins. The number of quantum states available to the electrons in our blob is truly enormous, but the number of states in a unit energy interval is progressively reduced as the blob shrinks and its volume decreases. Eventually a condition is reached in which there are fewer quantum states available for the electrons at lower energies than the Maxwell-Boltzmann distribution of energies predicts that the electrons should have. Under these circumstances some of the electrons must be pushed toward higher energy. As the process continues, the shrinkage of the blob pushes more and more electrons toward higher energy, until nearly all of the quantum states are filled up to an energy level called the Fermi level. Under these circumstances the electrons form a degenerate gas, and the pressure which they exert because of their rapid motion is very much higher than a thermal gas of electrons would exert had they continued to have a Maxwell-Boltzmann distribution of energy. The great bulk of the energy possessed by the electrons under these circumstances is not thermal energy in the ordinary sense, for this energy cannot be radiated away from the star if the temperature should be allowed to drop. All that would then happen would be that the electrons would fill all of the available quantum states up to the Fermi energy, and at absolute zero temperature there would then be no electrons beyond the Fermi energy. Such a gas would be completely degenerate. It would continue to exert a strong pressure. It also has a very high thermal conductivity. Because in general there are many more electrons than ions, and because the average electron in a degenerate gas exerts more pressure than the average ion does, the electrons will be mostly responsible for the pressure in the stellar interior under these circumstances.

It also makes a difference whether or not the electrons become relativistic. At low temperatures and low densities the pressure of the gas varies as the 5/3 power of the density under adiabatic conditions. But when the density becomes very high (much in excess of 10^6 gm/cm^3), the Fermi energy of the electrons becomes so high that most of the electrons have velocities close to that of light, behave relativistically, and the pressure varies more nearly as the 4/3 power of the density under adiabatic conditions. Thus as the density at the center of the star progressively increases, the pressure increases more slowly in the relativistic regime than in the nonrelativistic regime. This places a limit upon the amount of mass which can be supported solely by electron degeneracy pressure.

Let the density in the blob continue to increase much further. The Fermi energy of the electrons becomes truly high, many millions of electron volts or even many tens of millions of electron volts. Under these circumstances, the energetic electrons react with protons in any nuclei that are present, converting those protons into neutrons, with the production of neutrinos. Thus, at very high densities, matter will be progressively converted from ordinary nuclei into a gas predominantly composed of neutrons. At even higher densities, at the order of 10^{15} gm/cm^3, these neutrons will also form a degenerate gas, because there will be too few quantum levels available for them to maintain a Maxwell-Boltzmann distribution of energies. Thus at very high densities the gas will have a strong pressure contribution from the degenerate neutrons. This is very difficult to calculate with great accuracy, because the pressure is modified by the nuclear forces between the neutrons, and we still understand these forces only imperfectly. But meanwhile, before the degeneracy pressure of the neutrons becomes strong, we will have eliminated most of the electrons from the system, thus reducing very greatly the contribution from the electron degeneracy pressure. No objects stable against gravitational collapse can exist in nature if they have their central densities in this intermediate region, in which the electron degeneracy pressure has been weakened but the neutron degeneracy pressure has not yet become strong.

If we should increase the density of the blob of material still further, a great variety of complex events occurs. Reactions among the neutrons, protons, and electrons produce a great variety of other particles which add to the populations of particles in the interior. These include mu and pi mesons and a great variety of hyperons, which can be regarded as excited states of the neutron and proton. We need not delve into this complexity, because it makes very little difference to stellar structure, since the pressures exerted by the exotic mixture of nucleons, hyperons, and lighter particles continue to increase with increasing density.

The various types of stars which can exist in nature reflect the great differences exhibited by the equation of state in going through these various density regimes.

Evolution of Lower-Mass Stars

By lower-mass stars we shall mean stars with masses comparable to that of the sun, or larger by a small factor. Stars significantly less massive than the sun cannot have passed beyond the hydrogen-burning stage of their evolution during the lifetime of the universe, and hence the advanced stages of their evolution lie in the distant furture. For the moment we shall also consider only those stars which are single in space, not members of binary systems.

Figure 4-1 shows a schematic diagram of the evolution of a star of one solar mass. This is drawn in the form of a Hertzsprung-Russell diagram in which the surface temperature of a star is plotted along the abscissa, with the temperature increasing toward the left, and the luminosity of the star increases upwards along the ordinate. The luminosity is usually expressed in stellar magnitudes, which are distributed on an inverse logarithmic scale.

From the upper left of the diagram down toward the lower right is drawn a line called the main sequence in the Hertzsprung-Russell diagram. Most stars observed in space belong to the main sequence on such a diagram; they are in their hydrogen-burning phase of evolution. The more massive stars are on the upper left of the main sequence; the less massive ones are on the

Figure 4-1.   Schematic Hertzsprung-Russell diagram showing evolutionary sequence of a star containing one solar mass on the main sequence.

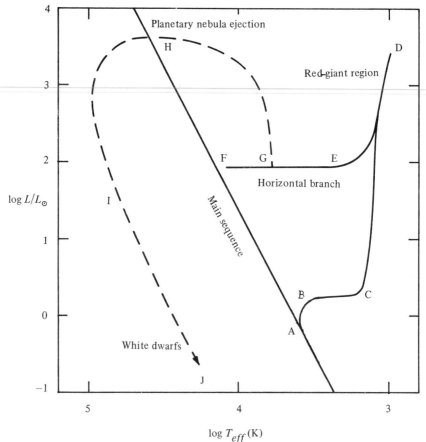

lower right. We choose to follow the evolution of a star of one solar mass; its position on the main sequence is marked at the point A.

A star such as the sun will spend about 10^{10} years on or near the main sequence, while it burns hydrogen into helium in approximately the central 10% of its mass. During this time it rises from A to B in Figure 4-1. In the process of hydrogen burning, four hydrogen atoms combine to form one helium atom. Thus, at the beginning of the process we have four protons and four electrons, and at the end of the process we have two protons, two neutrons, and two electrons. Thus two electrons have been lost in the process. In the deep interior of the star the principal process acting to confine the photons toward the center is the electron scattering process; the loss of electrons thus makes it easier for the energy to diffuse from the center of the star toward the surface, and this accounts for the small rise in luminosity in the star which occurs as the hydrogen becomes exhausted in the central core of the star. After the exhaustion of the central hydrogen fuel, the core recommences its contraction in order that the released gravitational potential energy can maintain the flow of energy toward the surface of the star. This compresses and heats the hydrogen in the region surrounding the helium core, so that soon hydrogen burning in a shell surrounding the core will take over the whole burden of energy generation on the star. The next stage of evolution of the star then has energy generated in a hydrogen-burning shell source, with the helium that is produced being added onto the mass of the core. The core continues its slow shrinkage, but at a rate which is sufficient to allow the energy readily to leak out from the core. Thus, for a considerable time, the temperature in the core never rises much above the temperature in the surrounding hydrogen-burning shell source. This continuing increase of density without a corresponding increase in temperature leads the central region of the star into the electron degenerate condition. There then occurs a high thermal conductivity in the core, which assists the outward conduction of energy and maintains the central region of the star at a uniform temperature.

As the mass of the helium core increases and the core shrinks toward very high density, the hydrogen envelope of the star expands and the star moves from B to C in the Hertzsprung-Russell diagram. This expansion can also be qualitatively understood on the basis of a virial-theorem argument. If the star had continued to shrink homologously, with the relative density distribution in the interior being approximately preserved, then the temperature at the center of the star would have progressively increased along with the density. Instead, the temperature has been constrained not to rise in this manner. Therefore the central region of the star is storing less than the normal share of energy which we would expect it to have on the basis of the virial-theorem argument that half of the released gravitational potential energy should be stored as internal energy. If the core stores less than its share of the internal energy, then the outer part of the star, the envelope, must store more than its share of internal energy. In order to do this, the envelope swells up and occupies a very large radius. However, the luminosity of the star does not change during this interval, so that the same amount of energy must be emitted per second from a much larger surface than was the case before. A smaller surface temperature suffices to do this, and this accounts for the movement to the right in the Hertzsprung-Russell diagram.

This movement to the right does not continue indefinitely. At the point C, the star develops a convection zone extending from near the surface deep into the extended envelope. We have already noted that the convective process is extremely efficient in transporting energy. Thus, when the outer convection zone reaches deep into the envelope, energy can be brought much more rapidly to the surface and radiated away into space. The nuclear energy generation adjusts its rate accordingly. During this time, as the luminosity of the star rises, it moves from point C to point D in the Hertzsprung-Russell diagram. We call such a star a red giant.

It is an observed property of red-giant stars that they lose mass rapidly into space, often as rapidly as about 10^{-6} solar masses per year. The process of mass loss is probably similar to that of the solar wind, but on a much larger scale. Several tenths of a solar mass may be lost in this way while the star is in the vicinity of point D.

Near point D, the high luminosity of the star places a large demand upon the energy generation process, which must be greatly increased in its rate. Thus the hydrogen-burning shell source processes hydrogen fuel more rapidly into helium, and adds helium to the core at an increased rate. The increased rate of mass addition to the core causes the core to shrink more rapidly, so that the energy is less easily transported away from the surface layers of the electron degenerate region, and the temperature at the center of the core begins to rise significantly above the temperature in the hydrogen-burning shell source. When this temperature reaches about 10^8 K, helium thermonuclear reactions commence. However, it is a property of an electron-degenerate gas that the pressure is very insensitive to the temperature. Therefore we can have a thermal runaway in the electron degenerate core, in which the energy generated by the helium thermonuclear reactions is stored in the core, raising the temperature, and greatly increasing the rate at which the helium reactions proceed. This process continues until the average energy of the thermal motion of the electrons becomes comparable to the Fermi energy of the electrons, so that the electrons assume a Maxwell-Boltzmann distribution of energy corresponding to a very high temperature, and hence are no longer degenerate. But now the pressure has suddenly risen in the core, and this increased pressure will cause the core to expand, producing a new stellar configuration in which helium thermonuclear reactions generate energy at the center of the core, and the hydrogen-burning shell source has been switched off, since the temperature at the base of the hydrogen envelope decreases. Because the core of the star has much more nearly its normal share of the internal energy, the envelope of the star shrinks to a more normal value, the surface temperature increases, and the star moves to the left in the Hertzsprung-Russell diagram. As the envelope shrinks, there is no longer a deep outer convection zone, and hence energy is less efficiently transported to the surface and the luminosity of the star falls. The result is that the star has moved from D to somewhere along the line marked EF in Figure 4-1, a portion of the Hertzsprung-Russell diagram called the horizontal branch. Just where it will lie along this line is uncertain, since this depends upon the amount of mass which has been lost in stellar winds during the red-giant portion of the star's evolution, and apparently this can differ somewhat from one star to another. The stars in this region of the Hertzsprung-Russell diagram tend to pulsate, producing light variations that characterize them as variable stars.

The subsequent motion of the star in the Hertzprung-Russell diagram is not known from theoretical calculations with very much precision. It is evident that the star will burn out its helium to carbon and oxygen within the core, but the core will then recontract into the electron degenerate regime. The star will establish a double shell source to provide the energy, one being the former hydrogen-burning shell source and the other being a helium-burning shell source immediately surrounding the core. Because much more energy is released in the hydrogen burning process than in the helium burning process, normally the hydrogen-burning shell source would provide much more of the star's luminosity than the helium-burning shell source.

However, this is a very unstable situation, and frequently the helium-burning shell source flares up and temporarily takes over most of the burden of the nuclear energy generation in the stellar interior. It is these complications which have made it difficult to follow the evolution of the star with precision.

However, some general aspects of the stellar behavior are clear. When the mass of the material overlying the nuclear-burning shell source becomes relatively small, it will no longer be possible to maintain the high temperature required to burn the helium. Therefore the helium-burning shell source will die out, and the hydrogen-burning shell source will proceed at a very rapid rate because the overlying material does not provide much opacity to impede the flow of radiation out of the stellar interior. We may think of the star as being schematically at about the point G in Figure 4-1, with a tendency to evolve toward higher luminosity and smaller envelope radii, leading toward the point H in Figure 4-1.

The very high luminous flux of radiation working its way through the outer envelope exerts a strong outward radiation pressure on the envelope. By now the envelope probably contains about 0.1 solar mass of hydrogen. As the luminosity increases, the radiation pressure alone becomes more and more effective in holding up the outer layers of the star, and eventually the time comes when this radiative pressure will exceed the gravitational force binding the envelope to the core. When this happens, the outer envelope is lifted off the star, and ejected into space with an outward velocity of a few tens of kilometers per second. Although the general nature of this process seems clear, not everyone is agreed upon the details, and it is not clear whether the envelope is ejected in many small spurts or all at once.

The envelopes ejected in this way are called planetary nebulae. There are a large number of them in the Galaxy, consistent with all stars of about solar mass eventually producing a planetary nebula, and they are quite beautiful. A picture of one is shown in Figure 4-2. The gas in the planetary nebula is excited and glows, as a result of energy which is radiated by the compact stellar core which has been left behind.

This compact core is composed mostly of electron degenerate matter, but it is also very hot. Most of the internal pressure is due to the degenerate electrons, and hence the core will be unable to contract substantially further in the subsequent course of its evolution. That evolution is very simple indeed. The large amount of heat which has been stored in the interior gradually leaks into space, and the star cools to become a white dwarf. The approximate track in the Hertzsprung-Russell diagram followed by the star as it does this is shown by the sequence HIJ in Figure 4-1. The evolution downwards in the diagram is rapid at first, but gradually slows down, and several billion years can be spent in the vicinity of J. Very large numbers of these white-dwarf stars have been observed in the Galaxy.

It seems probable that a very similar course of evolution will be followed by stars having masses up to about four times that of the sun. There is a possibility, which is not yet understood theoretically, that stars of mass greater than four solar masses may also undergo an additional stage of carbon burning before ejecting their outer envelopes and forming white-dwarf stars. However, it is also possible that these stars may undergo supernova explosions, which we shall consider in the following section. As the main-sequence mass of the star increases, the mass of the final remnant left as a

Figure 4-2. Photograph of the planetary nebula NGC 7293. The white dwarf is clearly visible in the center of the nebula. Hale Observatory photograph.

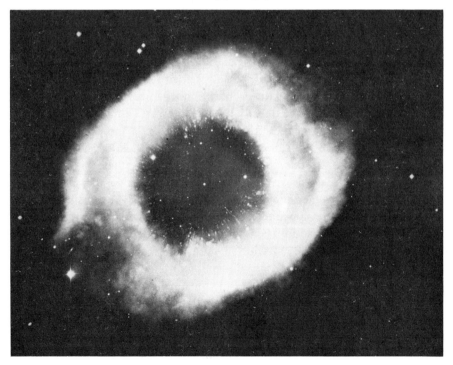

white-dwarf star also increases. It seems that stars with main-sequence masses of about one solar mass leave a white dwarf having a mass about 0.5 times that of the sun. The maximum mass of a stable white-dwarf star is about 1.4 times that of the sun. Beyond that point, the degenerate electrons in the stellar interior are so relativistic that they are unable to support additional mass, so that if such mass were to be added to the surface of a white-dwarf star near its maximum limit, the white dwarf would be forced to collapse toward a higher density configuration.

Thus the white dwarf star represents one possible endpoint of stellar evolution.

Evolution of Massive Stars

Let us now consider the evolution of a star of perhaps ten or a few tens of solar masses. This evolution is schematically sketched in Figure 4-3. As before, there is a line extending from the upper left of this Hertzsprung-Russell diagram toward the lower right, which is the main sequence. We start with a star located much farther toward the top of the main sequence than was the case in Figure 4-1. Let the star be located at the point A.

When the star has exhausted its central core of hydrogen, the core will shrink, and hydrogen burning will be established in a shell surrounding the core, just as in the case of the lower mass stars. However, it is a characteristic of massive stars that the density in their interiors is much less than that in a star like the sun, and therefore the contraction of the core during the course of hydrogen shell burning leads to the onset of helium burning before the electron degenerate regime can be reached. Nevertheless, the helium core contracts relatively slowly and hence does not rise as much in temperature as it would if the star were contracting homologously (i.e., all radial distances shrinking by the same factor), and hence the envelope of the star expands, also as in the case for the less massive stars. This takes a star from the point A to the point B, at which helium burning begins in the center.

When helium burning takes over the main burden of energy generation in the star, the core expands somewhat, and the envelope contracts somewhat, so that the evolutionary track swings back to point C. After the completion of helium burning in the core, a double shell source structure is again established, a hydrogen-burning shell source and a helium-burning shell source, and the star continues its expansion into the red-giant region. After the star reaches the point D, a deep outer convection zone is formed in the envelope, and the luminosity of the star can begin to increase, also as in the case of the less massive stars.

The subsequent behavior of the star is not reliably determined and is only schematically sketched in Figure 4-3. Some general features are evident. At some point, carbon burning (see Table 4-1) will commence at the center, causing a core expansion, envelope contraction, and a swing of the star to the left in the Hertzsprung-Russell diagram. After the exhaustion of the carbon fuel at the center, we revert to nuclear-burning shell sources, this

Figure 4-3.   Schematic Hertzsprung-Russell diagram showing evolutionary sequence of a massive star leading toward a supernova explosion.

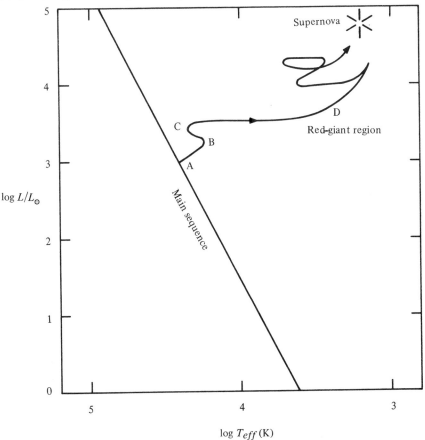

time three of them, and swing back toward the right in the Hertzsprung-Russell diagram. This dance to and fro in the Hertzsprung-Russell diagram may occur each time a new energy source is switched on at the center. But eventually the star runs out of available energy sources, before reaching silicon burning if its mass is not too great, or after having undergone silicon burning in the case of the most massive stars.

In each case the action of the nuclear shell sources surrounding the core is to keep adding material to the core, which eventually must become an electron degenerate configuration. But the mass of this core cannot exceed the maximum stable mass of a white-dwarf star. When the mass of the core becomes large enough, electrons combine with protons in nuclei to form neutrons, emitting neutrinos, thus leading to the removal of pressure near the center of the star, and the collapse of the core. This is the start of a supernova explosion.

Supernova explosions are immensely complicated events. There may or may not be some remaining nuclear fuel available in the collapsing core which can be triggered to burn on a very rapid time scale during the collapse of the core. The core is certainly surrounded by large amounts of nuclear fuel, representing the ashes of various advanced stages of stellar evolution, and this fuel can burn very rapidly either as a result of a rapid collapse of the interior portions of the star, or as a result of the passage of a strong supernova shock wave through the material in the later course of the supernova explosion. The energy released by this rapid nuclear burning raises the internal pressure, and helps to throw off the stellar envelope, but it may not be sufficient to do this entirely by itself.

When the core commences its collapse, it is composed of electron degenerate material. In the course of the collapse most of the electrons combine with protons in the nuclei to make neutrons, so that the pressure in the core cannot rise very rapidly during the collapse. However, when the density approaches 10^{15} gm/cm^3, the neutrons suddenly exert a large degeneracy pressure, which will halt the collapse of the central portions of the core. There will then be a central neutron core upon which continues to rain down the outer portions of the original electron degenerate material. This process releases a tremendous amount of heat in the center of the core, owing to the very large amount of released gravitational potential energy resulting from the collapse. Thus the temperature can rise to very high values within the neutron core, and huge numbers of neutrinos and antineutrinos can be created there. Ordinarily, neutrinos and antineutrinos readily escape from stellar interiors without making collisions, but under these circumstances, the density is so high that the neutrinos and antineutrinos do not escape easily from the core, but must diffuse their way out of the core after a large number of scatterings from the particles that are present.

There is a good chance that this outward diffusion of neutrinos and antineutrinos plays a crucial role in the supernova explosion. The outgoing flux of neutrinos exerts a very large pressure on the descending material as a result of scattering from it; one of the principal problems in the current theoretical investigations of supernova explosions is to determine whether this pressure exerted by the neutrinos and antineutrinos plays the principal role in the ejection of the outer layers of the star, or whether the excess pressure generated by nuclear reactions in the material surrounding the core can do this.

Supernova explosions are truly majestic spectacles. Many solar masses of material can be thrown off, and the ejected material can radiate tremendous amounts of energy into the surrounding space during the ejection process itself. For a brief period, the light emitted by the supernova can become comparable to the total light emitted by all of the other stars in its galaxy. Supernovas may seem to be rare events, but their numbers are consistent with the notion that all the more massive stars undergo this spectacular final stage of stellar evolution.

Meanwhile, what has been left behind is a stellar remnant containing a degenerate gas of neutrons. The physics of the supernova explosion is not well enough known to allow us to predict the mass that will be left in the neutron star with any degree of confidence. Nor are we absolutely sure of the upper limit to the mass of a stable neutron star. If we should keep adding mass to the surface of the neutron star, the pressure in the interior would have to increase continually in order to support the additional mass and, sooner or later, general relativistic effects will render the star unstable. As a general rule we may state that any configuration will be crushed by gravitation after its mean density becomes high enough.

The maximum mass of a stable neutron star depends upon the character of the nuclear forces between the neutrons and the other particles, protons, hyperons, and mesons, which may be present in the interior. These forces are still imperfectly understood, and hence the upper limit to the stable mass cannot be calculated with precision. However, it appears to lie near two solar masses. Since this limit is greater than the mass of the electron degenerate core which participated in the collapse, there is an excellent probability that the outer less dense envelope of the star will all be ejected, and that the stellar remnant will lie in the range of stable neutron-star masses. The existence of pulsars, which have been identified as rotating neutron stars, argues in favor of this conjecture (see Chapter 5 in this volume).

In the present status of stellar evolutionary theory, it thus appears that all of the more massive single stars will undergo supernova explosions and form neutron-star remnants. If the amount of material left in the remnant could have exceeded about two solar masses, then the neutron stars would have been crushed by gravitation, and would have formed eternally collapsing black holes instead. If this happens at the endpoint of the evolution of any of the single massive stars, the mechanism by which it should occur is not understood.

Evolution of Binary Systems

Stellar evolution can become even more complicated when the stars participating in it are the members of close binary systems. Probably more than half of the stars in our Galaxy are members of binary systems. Let us commence our consideration of such a system with two stars of somewhat unequal mass upon the main sequence.

The more massive of the two stars will have the shorter main-sequence evolution time. Following the exhaustion of hydrogen in the core, the hydrogen-burning shell source will commence operation, and the envelope will start to expand toward the red-giant configuration.

However, as the envelope expands, its outer layers will eventually come under the gravitational control of the binary companion star. These layers will then go into orbit about the binary companion, forming a gaseous disk which will be dissipated and accreted onto the surface of the binary companion. Theoretical evolutionary studies have shown that this process can go all the way to completion; the entire hydrogen envelope of the originally more massive star can be transferred onto the originally less massive companion in this fashion. What is left behind is a helium white dwarf. Its mass is that of a stellar core and can be quite small.

The binary companion is now likely to have considerably more mass than did the original primary of the system. It will finish its main-sequence evolution in a relatively short time. When hydrogen is exhausted in its core, its envelope will also expand toward a red-giant configuration. This is likely to lead to a reverse mass transfer onto the white-dwarf star.

The subsequent evolution of systems like this is obviously very complex and has not been studied in great detail. We may have several episodes in which most of the mass is now upon one of the stars, now upon the other. We may have times in which most of the mass forms a common envelope around two electron degenerate cores. These electron degenerate cores cannot become very massive without undergoing a stage of central helium burning. Meanwhile, all of the mass transfers onto electron degenerate cores that are occuring may lead to some interesting X-ray emission phenomena from such compact binary systems. Occasionally, nuclear runaways in the thin nondegenerate envelope may throw off a small amount of mass in nova explosions.

There are obviously a number of different endpoints of the evolutionary process which can occur in such systems, depending upon the original masses and separations of the stellar components. Enough mass may be lost from the system so that two white-dwarf stars are left in orbit about each other. Or one of the electron degenerate cores may grow large enough to undergo collapse. The result would be a supernova explosion in the binary system.

In a supernova explosion a considerable amount of mass may be ejected. If this is less than half of the total mass of the system, then the stars will remain bound in the binary system, but will acquire quite eccentric orbits. This may leave a neutron star and a more ordinary binary companion still undergoing advanced stages of stellar evolution. Or, if the collapse which triggered the supernova explosion occurred at the time that both electron degenerate configurations were located in a common envelope, we would be left with a neutron star and a white dwarf in a mutual eccentric orbit. A recently discovered pulsar in a binary system seems to be of this type (Hulse and Taylor, 1975).

Still more complicated phenomena must occur if the binary system originally contained very massive stars. The stellar remnants left following mass transfer in these cases are likely to be too hot and of too low density to form electron degenerate configurations. On the other hand, with repeated mass transfers within such a system, details of the internal evolution through many stages of nuclear burning are likely to differ markedly from those for single stars. The outcome of this evolution can only be a subject of speculation. Some very massive compact X-ray binary systems exist in which one of the companions appears to have become a *black hole, a general relativistic configuration which is in a state of indefinite collapse after having crushed matter beyond neutron-star densities*. Perhaps the evolutionary conditions in the binary system with multiple exchanges of material between the components can lead to a core massive enough to collapse to the black-hole configuration. Perhaps, on the other hand, the internal evolution is more like that of the single stars which we have described, so that we have a supernova explosion in the system, leaving behind a neutron star. But if the neutron star remained bound to the system, and its companion remained a very massive star, then mass exchange onto the neutron star is likely to have built up the mass of the neutron star until gravitational crushing took place and a black hole was formed. Subsequent mass exchange onto the black hole would simply form a gaseous disk, which would be dissipated and continue to add mass onto the black hole. This form of mass exchange would be quite irreversible.

Thus we have seen that a wide variety of phenomena can be expected to occur in a close binary system. We may get all three of the possible end-points of stellar evolution in such a system: white dwarfs, neutron stars, and black holes.

References

Cited in Text

Hulse, R. A., and Taylor, J. H.   1975. Discovery of a pulsar in a binary system. *Astrophys. J. (Letters) 195*, L51–53.

General References

Recent Review Articles

Iben, I.   1967. Stellar evolution within and off the main sequence, *Annu. Rev. Astron. Astrophys. 5*, 571–626.

Weidemann, V.   1968. White dwarfs, *Annu. Rev. Astron. Astrophys. 6*, 351–372.

Cameron, A. G. W.   1970. Neutron stars, *Annu. Rev. Astron. Astrophys. 8*, 179–208.

Hewish, A.   1970. Pulsars, *Annu. Rev. Astron. Astrophys. 8*, 265–296.

Ostriker, J. P.   1971. Recent developments in the theory of degenerate dwarfs, *Annu. Rev. Astron. Astrophys. 9*, 353–360.

Woltjer, L.   1972. Supernova remnants, *Annu. Rev. Astron. Astrophys. 9*, 353–360.

Iben, I.   1974. Post main sequence evolution of single stars, *Annu. Rev. Astron. Astrophys. 12*, 215–256.

Tayler, R. J.   1974. Ed., *Late Stages of Stellar Evolution,* IAU Symposium No. 66, Reidel, Dordrecht-Holland; Boston, Mass.

Standard Textbooks

Chandrasekhar, S.   1939. *An Introduction to the Study of Stellar Structure,* University of Chicago Press, Chicago.

Aller, L. H.   1954. *Astrophysics: Nuclear Transformations and Nebulae,* Ronald, New York.

Schwarzschild, M.   1958. *The Structure and Evolution of the Stars,* Princeton University Press, Princeton, N.J.

Aller, L. H., and McLaughlin, D. B.   1965. Eds., *Stars and Stellar Systems,* vol. 8, *Stellar Structure,* University of Chicago Press, Chicago.

Chiu, H. Y.   1968. *Stellar Physics,* Blaisdell, Waltham, Mass.

Clayton, D. D.   1968. *Principles of Stellar Evolution and Nucleosynthesis,* McGraw-Hill, New York.

More Advanced Books

Harrison, K. B., Thorne, K. S., Wakano, M., and Wheeler, J. A.   1965. *Gravitation Theory and Gravitational Collapse,* University of Chicago Press, Chicago.

Zeldovich, Ya. B., and Novikov, I. D.   1971. *Relativistic Astrophysics,* University of Chicago Press, Chicago.

Weinberg, S.   1972. *Gravitation and Cosmology,* Wiley, New York.

Misner, C. W., Thorne, K. S., and Wheeler, J. A.   1973. *Gravitation,* Freeman, San Francisco.

5

Neutron Stars, Black Holes, and Supernovae

Herbert Gursky

There is a certain human fascination in spectacular events, and those which occur in the sky have always been part of recorded history. Often there is a simple physical cause of these events: an eclipse is just the chance alignment of one object with another. In other cases, their true nature is more bizarre than anyone could have first imagined. Certain of the ''new'' stars, which frequently shone in daylight, are now known to be the supernovae. The magnitude of the event, not recognized until early in this century, is such that at its brightest, almost as much light is emitted as comes from all the stars in the parent galaxy! One could only guess the origin of this enormous energy—the collapse or explosion of an entire star. At about the same time as the supernovae were discovered, the existence of a new class of stars was revealed: the white dwarfs, which have about the mass of the sun, but only the size of the earth. The study of these objects and the elucidation of their true nature took place in the period 1915–1940 and led to the realization of the possible existence of a family of ''collapsed'' stars. The white dwarfs are the most benign members; more compact configurations are neutron stars and black holes.

During the 1930s, the opinion arose that there was a connection between supernovae and collapsed stars. A star somehow exploded or collapsed, possibly at the end of its life, leaving behind one of the collapsed stars as a remnant. In the past decade, through a remarkable combination of luck, new observational techniques, and theoretical insight, this simple idea has been demonstrated.

In this chapter, I will discuss these two phenomena, supernovae and the collapsed stars, primarily from an historical and observational point of view. Of the collapsed stars, I will concentrate on neutron stars and black holes.

White dwarfs provide an important beginning for understanding the material in this chapter, but it is now believed that they form quietly and may, in certain instances, be the stage in the life of a star immediately preceding the supernova stage. On the other hand, as will be described later, there is a positive observational and theoretical basis for believing that neutron stars and black holes follow as a result of supernovae.

In the next section, we will trace the historical development of the theory of collapsed stars. The understanding of the subject did not spring up in a simple or logical fashion. The story comprises an important element in what is as dramatic a revolution in physics and astronomy as the one that took place in the century spanning Tycho to Newton.

Historical Background and Theory of Collapsed Stars

Before the beginning of this century, astronomers began the study of the physical characteristics of individual stars and constructed models of their internal structure in terms of large gaseous bodies in which gravitational forces held matter together against the ordinary pressures of gas and radiation. This view has proven to be essentially correct, at least for the early portion of a star's life, and, with the addition of energy generation by nuclear reactions, is the basis for present theories of stellar structure and evolution.

However, some of the earliest observations indicated that conditions existed which could not be accounted for by this classical approach. One of the first versions of the Hertzsprung-Russell diagram as prepared by Russell (1914) is shown in Figure 5-1. The main sequence is clearly outlined; however, also present, in the lower left is the single star, 40 Eridani B, which had an unaccountably low luminosity for its color.

The astounding characteristic of stars of this kind, which soon came to be called white dwarfs, was their density, greater by many orders of magnitude than that of other stars or of any known material, at least 10^5 g/cm^3. The observational data were irrefutable. The basic result, the very high internal density, can be obtained straightforwardly from the following considerations. The luminosity L of a star is given by

$$L = 4\pi R^2 \sigma T_{\text{eff}}^4 = 4\pi D^2 F,$$

where R is the radius of the star, T_{eff} is its effective temperature (which is defined by this equation), σ is the Stefan-Boltzman constant (5.669×10^{-5} erg

Figure 5-1. The original version of the Hertzsprung-Russell diagram as prepared by Russell, based on stars in the solar neighborhood. Figures of this kind are the principal observational tool for studying the evolution of stars. The main sequence, where stars reside for most of their lives, is delineated by the two heavy lines. The white dwarf, Eridanus B, at + 11 mag and spectral class A, is clearly anomolous with respect to normal stars. Adapted from Russell (1914).

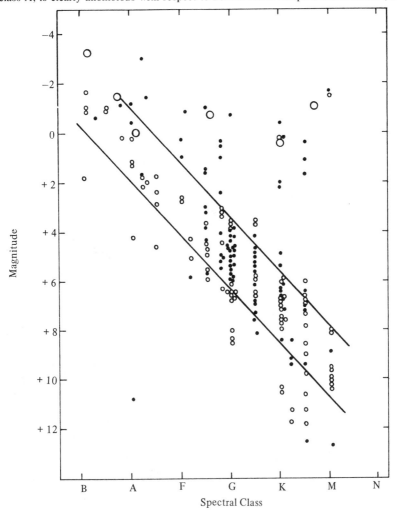

sec^{-1} cm^{-2} K^{-4}), D is the distance to the star in cm, and F is the total flux measured at the earth in erg sec^{-1} cm^{-2}. The density ρ is then,

$$\rho = \frac{M}{(4/3)\pi R^3} = \frac{3M\sigma^{3/2}T_{eff}^6}{4\pi D^3 F^{3/2}}.$$

The flux F can be obtained by converting the visual magnitude, V, to a bolometric magnitude M_{bol} and using the fact that $M_{bol} = 0$ is equivalent to a total flux of 2.5×10^{-5} erg sec^{-1} cm^{-2}. For Sirius B, shown in Figure 5-2 as an example, $V = 8.7$, $T \approx 10^4$ K (spectral class A), $D = 2.7$ pc $= 8.3 \times 10^{18}$ cm and $M \approx 1$ $M_\odot$ (Allen, 1973, p. 235). For such a star, the bolometric correction ($M_{bol} - V$) is about -0.3; thus, $M_{bol} \approx 8.4$ and $F = 1.1 \times 10^{-8}$ ergs sec^{-1} cm^{-2} since $M_{bol} = -2.5 \log (F/2.5 \times 10^{-5})$. Finally, $\rho \approx 3 \times 10^5$ gm/cm³!

This is only a crude result because the relation between V and M_{bol} is only

Figure 5-2. Photograph of Sirius. Sirius B is the faint star just to the right of Sirius A, which is greatly overexposed. Photograph from the Harvard College Observatory plate collection.

very approximate; see Oke and Shipman (1971) for a modern treatment of the relation between radiated power and visual measurements. However, it is accurate enough to illustrate the point that the internal density of white dwarfs is qualitatively greater than that of normal stars.

The nature of these objects was inexplicable until more progress was made in quantum mechanics. Classically, such high densities implied the existence of very high internal temperatures and these stars should have been much brighter than actually observed. During the early 1920s, Bose, Einstein, Fermi, and Dirac worked out the fundamental statistical properties of matter based on the new quantum mechanics and the application of the Pauli exclusion principle. In the same year that Fermi developed his quantum statistics, Fowler (1926) proposed that these new concepts should be applied in treating the high density found in white dwarfs. Fowler's ideas were applied and developed further by Milne, Stoner, and Chandrasekhar, and by about 1930 the theory of the fundamental properties of white dwarfs had been developed. Thus, within a decade, astronomers and physicists had discovered and explained a new state of matter that could be understood only in terms of twentieth-century concepts. In white dwarfs, the electrons, already stripped from the atoms, are as densely packed as permitted by the Pauli exclusion principle, a condition known as "degeneracy." Under these conditions the electrons act as if they are in one enormous atom; they occupy a spread of energy up to a maximum (the Fermi energy) with all possible energy levels filled. Thus, as in an atom, the electrons cannot radiate since they find their way to lower energy blocked by other electrons. The matter is effectively very cold in spite of its high internal energy. With the discovery of the neutron, it became apparent that one could also have a star built of a degenerate neutron gas. However, the density of such a system would be orders of magnitude greater than is the case for a white dwarf. Such a configuration was explicitly described by Gamow (1937).

At about the same time, a remarkable theoretical discovery was made. Chandrasekhar (1931, 1932) and Landau (1932) found that as a consequence of special relativity the rate of change of pressure with density would decrease at very high densities and a maximum stable mass existed for stars that have exhausted their internal sources of energy. Landau stated that the addition of matter over this critical value would lead to the collapse of the star without limit because of the ever-increasing gravitational forces. For a personal account of the steps that led to this fundamental discovery, see Chandrasekhar (1969, 1972).

The Basic Theory of Collapsed Stars

   We can here summarize the main points of the reasoning that led to a mass limit for cold stars and to a prediction of their eventual collapse. After having exhausted most of its nuclear fuel, a star is expected to be composed either of nuclei embedded in a gas of electrons, as in the case of a white dwarf, $\rho \sim 10^5-10^8$ gm/cm^{-3}, or of a gas of neutrons, protons, and electrons in equilibrium, which is a neutron star with $\rho \sim 10^{14}-10^{16}$ gm/cm^3. In both cases, the material of the star may be described as a degenerate Fermi gas held together by the gravitational force of the system. The pressure required to balance gravity results from the zero-point energy, allowing for the Pauli exclusion principle, rather than from the kinetic energy in an ideal gas. In the case of the white dwarfs, the main contribution to the pressure comes from the gas of degenerate electrons, whereas the nuclei contribute mainly to the density; in the case of neutron stars, both pressure and density are generated by the gas of neutrons.

   The main formulae describing these states can be obtained from the theory of Fermi-Dirac statistics (cf. Landau and Lifshitz, 1958). However, the basic physics can be formulated in the following qualitative arguments as presented by Hawking and Ellis (1973). We demonstrate below that for a nonrelativistic degenerate gas, equilibrium can be established, whereas for a relativistic gas the weaker dependence of pressure on density prevents any equilibrium mass.

   We consider a cold star composed of particles of mass m and number density n. The particles comprising the gas are fermions, either electrons or protons and, by the Pauli exclusion principle, each particle will occupy a volume $1/n$. Then by the uncertainty principle ($\Delta x\, \Delta p \approx \hbar$), the momentum of the particle will be of order

$$p_x \approx \hbar n^{1/3}.$$

If the gas is nonrelativistic the velocity of the particles will be $\sim p_x/m = \hbar n^{1/3}/m$; if the gas is relativistic the velocity will be close to c, the velocity of light.

   Now the pressure, as it is for a simple gas, is the momentum transfer per unit area, or

$$P = (\text{momentum}) \times (\text{velocity}) \times (\text{number density})$$
$$= \hbar n^{1/3}(\hbar n^{1/3}/m)n = \hbar^2 n^{5/3}/m.$$

   In ordinary matter, composed of electrons and nuclei, m must be taken to be m_e, the mass of the electrons, which because of their so much smaller mass have the larger velocity and, thus, make the most significant contribution to the pressure. When the gas becomes relativistic, the pressure relation becomes

$$P \approx \hbar n^{1/3}(c)n = \hbar c n^{4/3}.$$

We can also crudely derive these mass limits. The gravitational force (Newtonian) on a typical unit volume in a star of mass M and radius R is given by

$$F = \frac{GM}{R^2} \Delta m = \frac{GM}{R^2} n m_n,$$

where m_n refers to the mass of nuclei, since the electrons contribute only a negligible mass and G is the gravitational constant. This force must be balanced by a pressure gradient P/R; thus,

$$\frac{P}{R} = \frac{GMnm_n}{R^2}$$

or

$$P = \frac{GMnm_n}{R} = GM^{2/3}(nm_n)^{4/3},$$

where we have substituted for R its value obtained from the mass density relation $nm_n = M/R^3$. In the nonrelativistic regime, equilibrium is achieved when

$$GM^{2/3}(nm_n)^{4/3} = \hbar^2 n^{5/3}/m_e.$$

From this equation the mass is given by

$$M = \hbar^3 n^{1/2} m_e^{-3/2} m_n^{-2} G^{-3/2}.$$

In this case, for any specified density n, a value of the mass M can always be found. In the relativistic regime,

$$GM^{2/3}(nm_n)^{4/3} = \hbar c n^{4/3}.$$

The density n cancels out of this equation and one obtains a star of mass M:

$$M = \left(\frac{\hbar c}{G}\right)^{3/2} m_n^{-2} \approx 1.5 \, M_\odot.$$

Since this quantity is independent of density, it means that this same mass is obtained independent of radius. This is the limiting mass; more massive stars cannot be supported by electron degeneracy pressure no matter how small they are. This was the discovery of Chandrasekhar and Landau: that the pressure dependence on density changed in going from nonrelativistic to relativistic conditions and, as a consequence, there arose a finite limit to the mass of a star.

The above equations describe a white dwarf. If the object is a neutron star, the nonrelativistic momentum becomes $\hbar n^{5/3}/m_n$ since no free electrons are present; however, in the crude form presented here, the limiting mass is the same.

Detailed computations of the configurations of equilibrium for white dwarfs were presented by Chandrasekhar (1935) and those for neutron stars by Oppenheimer and Volkoff (1938). The value of the critical mass for white dwarfs was found to be $M_{crit} \sim 1.39\ M_\odot$, the one for neutron stars $M_{crit} \sim 0.7\ M_\odot$.

Then Oppenheimer and Snyder (1939) focused attention on the process of gravitational collapse itself. For the first time, it became evident that this phenomenon is of basic importance for our understanding of the nature of space and time. They showed that the phenomenon of gravitational collapse with its processes of time dilatation, light deflection, and gravitational redshifts, is a unique instance where a fully relativistic theory of gravitation can be seen at work.

The work of the 1930s on the equilibrium of cold stars is summarized in Figure 5-3, which is a plot of the equilibrium mass of cold stars. The values of the mass are obtained by starting with a star with an assumed central density ρ, and integrating an equilibrium equation outward, defining a new density at each point. The mass is simply

$$M = \int_0^r \rho 4\pi r^2 \rho(r)\ dr.$$

The equilibrium equation results from equating the downward gravitational forces and the internal pressure. The equations must be relativistically correct. The pressure relation comes from the equation of state and begins with the equation of a nonrelativistic degenerate electron gas and ends with a relativistic degenerate neutron gas (cf. Rees, Ruffini, and Wheeler, 1974, p. 14).

There is no "classical" equivalent of this figure. Ordinary matter in quantities of stellar size cannot be supported unless it is very hot. As the matter cools, if its mass is less than about 0.1 $M_\odot$, it will become a planet like Jupiter (or the earth, depending on its chemical composition). At higher masses it will become a white dwarf according to the curve of Figure 5-3. Because of the high energy content of degenerate material, there is not a great deal of difference between "hot" white dwarfs and "cold" white dwarfs.

The significance of the curve in Figure 5-3 is as follows. The portion $\rho = 10^5 - 10^8$ gm/cm^3 is the domain of the white dwarfs. Between $\rho = 10^8 - 10^9$, the mass reaches the Chandrasekhar mass limit. Beyond this point, the slope of the curve becomes negative and there are no stable configurations until the slope becomes positive again at $\rho \sim 10^{13}$ gm/cm^3. The instability can be understood as follows. Suppose we somehow construct a star of about 0.6 $M_\odot$ and 10^{12} gm/cm^2. Now we perturb the radius to make it smaller by a small amount. The central density tries to rise, but then the star finds itself too massive for its new central density, causing yet an additional increase in density, and on and on. The star cannot come to rest until it reaches the positive slope at a density of $\sim 10^{15}$ gm/cm^3, which is the domain

Figure 5-3. Equilibrium mass for cold stars. The solid portion of the curve is the region where stable stars reside. On the dashed portions, stars are unstable against collapse. The portion of the curve labelled Oppenheimer-Volkoff was derived ignoring interactions between neutrons. The higher curve is a more recent derivation which makes use of such interactions. Adapted from Rees, Ruffini, and Wheeler (1974).

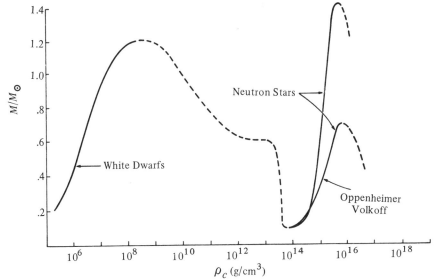

of the neutron stars. Our star has "collapsed" from its original density of 10^{12} to a density of 10^{15} and shrunk in radius by a factor of ten. A similar mass limit exists for the neutron stars, although its exact value is not as well known as that of the white dwarf. However, once the neutron star mass limit is exceeded we know of no way to stop the ensuing collapse. What happens during this final collapse? We can get a clue by calculating the potential energy of a single proton on the surface of stars of decreasing size but of 1 $M_\odot$, thus of increasing density, as given by

$$\phi = \frac{GM_\odot m_p}{R}$$

$$\approx GM_\odot^{2/3}\rho^{1/3}m_p,$$

as shown in Table 5-1. The significance of the figures given there is that the rest energy of a proton is 938 MeV; thus, at $\rho \approx 10^{18}$, the surface potential of a proton exceeds its rest energy; i.e., when

$$\frac{GM_\odot m_p}{R} > m_p c^2$$

or

$$\frac{GM_\odot}{Rc^2} > 1.$$

But a much more dramatic effect is found when one considers the nature of space outside of a uniform, massive star. The basic equation for a line element in space, in its differential form in general relativity, is

$$ds^2 = -\left(1 - \frac{2GM}{c^2 r}\right)^{-1} dr^2 - r^2(d\theta^2 + \sin\theta \, d\phi^2) + \left(1 - \frac{2GM}{c^2 r}\right) c^2 \, dt^2.$$

Table 5-1   Potential energy of a proton on a stellar surface.

$\rho(\text{gm/cm}^3)$	$\phi(MeV)$
10^6	0.11
10^9	1.1
10^{12}	11
10^{15}	110
10^{18}	1100

When $r = 2GM/c^2$ the radial term diverges, and for r less than this value these terms reverse sign. This value of r is the Schwarzschild radius. As an object approaches this radius, any light it emits is increasingly diminished by the increasing Doppler shift as the object gains velocity, by the gravitational redshift, and by the gravitational light deflection, which reduces the solid angle through which light can escape. When the object crosses the Schwarzschild surface, it disappears entirely, having lost its ability to transmit light to any external observer. For this reason this surface is said to be a closed trapped surface. Having entered it an object cannot reemerge. All that can be sensed of its presence is its gravitational field.

As we noted, with an object of 1 $M_\odot$, this condition of "blackness" arises when ρ is greatly in excess of nuclear densities; however, the effect is very general and not dependent at all on the particular density. We list in Table 5-2 the Schwarzschild radius R_S for a variety of "bodies" of a given mass and average density. The first entry in the table is the solar-mass black hole we are discussing here. The second is an object of galactic mass, but confined to a region only a fraction of its normal size ($\sim 10^4$ pc). Presumably this is the ultimate fate of our own Galaxy, but clearly when it does occur, the density is so low (about 0.01 that of air) that the laws of matter on a local scale are completely unaffected, and one will pass into the black hole without incident. The final entry could describe the universe in which we reside; in other words, we may at this moment be living in an enormous black hole.

The purpose of this exercise is simply to demonstrate that the condition of "blackness" of a black hole is neither surprising nor spectacular. What is dramatic, but only when we get to objects of ~ 1 $M_\odot$ is that a black hole can be accompanied by such enormous gravitational forces that the usual nuclear forces, the strongest material forces of which we have knowledge, are overcome.

Table 5-2   Characteristics of black holes.

$M/M_\odot$	$\rho(\text{gm/cm}^3)$	$R_S(=2\ GM/c^2)$
1	2×10^{16}	3 km
10^{11}	2×10^{-6}	0.03 pc
10^{22}	2×10^{-28}	3000 Mpc

What happens beyond this point is pure speculation. Unless some new laws of physics intervene, the matter will shrink down to a singular point. Unfortunately, even if this "point" was close by, we could never "see" it. According to our present understanding of physics, any radiation produced by the singularity would never emerge, any probes we would send out could come only as close as several kilometers before being swallowed up within the Schwarzschild radius. See Penrose (1969); Rees, Ruffini, and Wheeler (1974); and Ruffini (1975, p. 59) for more complete descriptions of the phenomenology of black holes.

All this was known thirty-five years ago; the great advances since then have come in the observational domain.

The Discovery of Neutron Stars and (Perhaps) Black Holes

The important observations began early in this century with the discovery of the supernovae. The modern work dates from the discovery in 1885 by E. Hartwig of a nova in the central region of the Andromeda Nebula, M 31. Its visual brightness at maximum was about +5.4 mag; with the modern distance to M31 of about 600 kpc, the peak luminosity of this object was about 10^{43} erg/sec, which is the luminosity of about 10^{10} stars! However, it was not until the 1920s that this result was generally accepted. It took until then to establish the size of our Galaxy and the fact that nebulae like M 31 were distant versions of the assemblage of stars in which the sun resides, as was postulated by Kant and others around 1800 (cf. Bok and Bok, 1974).

Baade and Zwicky (1934) suggested that the supernova process was the result of the transition from a normal star to a neutron star, the essential point being that the energy release in such a process is comparable to the change in gravitational potential energy of a star which collapses from its 'normal' size of 10^6 km down to the size of a neutron star of the order of 10 km.

In the 1950s, work on stellar nucleosynthesis (cf. Burbidge et al., 1957) led to physically realistic models of stars in the stage before supernova explosions. The supernova process was seen as the result of a catastrophic change of state occurring in the core of a highly evolved star, one possibility being the transformation of an iron core to a helium core as a first step in this process. Cameron (1958) suggested that the degenerate iron core would collapse to a neutron core through inverse beta decay; namely, nuclear reactions of the form $A^z + e \rightarrow A^{z-1} + \nu$. In this model, the resulting implosion and ensuing explosion would blow off the outer envelope of the star and leave behind the core as a neutron star. Thus, the 20-year old idea of Baade and Zwicky received at least a general qualitative explanation.

Meanwhile, the conditions of matter under extremely high pressures and the conditions following collapse were re-examined by a number of scientists. Different models lead to different ranges of masses, radii and density distributions for the equilibrium configurations of neutron stars. This work was accelerated when the first X-ray sources were discovered by Giacconi et al. (1962). It was believed that neutron stars could radiate large fluxes of X-rays, possibly as a consequence of very high surface temperature. However, subsequent observational effort yielded no positive evidence for the existence of neutron stars.

It can be argued that neutron stars and black holes will be invisible for all practical purposes. Stars are visible because the internal energy generated by nuclear burning radiates from their surface. The sun, an average star, will be visible to a distance of ~ 2500 pc in a modest telescope, about a quarter of the way across the Galaxy. The white dwarfs, in which nuclear burning is very slight, can be seen to about 100 pc with large telescopes. However, in the case of neutron stars, in which no nuclear burning is possible, the surface radiation that might render them visible must be much less than that of white dwarfs, and black holes by their very nature should be invisible! How then might we ''see'' them? One possible idea pointing to their observation was that they might be powerful radiation sources because of accretion of matter—either from the interstellar medium or from a neighboring star. This idea was first advanced to account for the powerful radiogalaxies and quasars, by both Salpeter (1964) and Zeldovich (1964). The source of energy, instead of nuclear burning, is the matter falling through the intense gravitational field of the compact star. Another suggestion was that, to detect a black hole, one should exploit its invisibility and search for its presence as an unseen, massive companion in a binary system (Guseinov and Zel'dovich, 1966).

Great impetus was given to the subject when Scorpius X-1, the brightest X-ray source in the sky, was identified optically as a faint, blue, starlike object (Sandage et al., 1966). The star has properties similar to the old novae, which led to conjectures that the X-rays were being produced by accretion onto a white dwarf or a collapsed object in a binary system. Shklovskii (1968) gave a detailed picture of Scorpius X-1 as a close binary system in which one of the members is a neutron star and the other member a cool dwarf star. "A stream of gas coming out of the second component is permanently incident on the neutron star · · · . This accreting material, enormously compressed, should then emit X-rays · · · . The optical object accompanying the X-ray source might be a cool dwarf star with half of its surface heated by a strong flux of hard X-rays from the source." For a review of the many discussions on the binary nature of X-ray sources that emerged at the time of the optical discovery of Sco X-1, see Burbidge (1972).

Suddenly, however, the observational picture changed dramatically. In 1967, a new type of radio telescope intended for the study of interplanetary scintillation came into operation in Cambridge, England, under the direction of A. Hewish. For this purpose, the telescope was designed with two features not particularly common to such installations—sensitivity at low frequency, and high time resolution. Within a month, a radio source was detected whose emission was not steady, but pulsed with a precise period of 1.33 seconds. This object, known as CP 1919, was the first pulsar to be discovered. In their initial paper, Hewish et al. (1967) suggested that this phenomenon was associated with a compact star, possibly the radial pulsations of a white dwarf or a neutron star; see also Wade (1975) for a personal account of this discovery.

The idea that these objects were rotating, magnetized neutron stars was advanced, among others, by Gold (1968) and by Pacini (1968); within a short time, observational evidence was found to confirm this view. Brief, sporadic pulses were detected from the direction of the Crab Nebula by Staelin and Reifenstein (1968), and a pulsar with a period of ~33 msec was identified as their source by Comella et al. (1969). The lengthening of the pulse that had been suggested by Gold was also soon discovered, with a fractional change of ~1:240 per year (Richards and Comella, 1969).

As shown by Gold (1969), these observations provided very compelling evidence that one was dealing with a rotating neutron star. In the first place, among all known stellar configurations only a neutron star could be stable when rotating (or vibrating) with such a high frequency. Secondly, the pulse period was observed only to decrease (slow down), as expected from a rotating object that was losing rotational energy. A vibrating object would tend to increase its frequency. Finally, the slow-down rate was consistent with a loss of energy of $\sim 10^{38}$ erg/sec if one assumed the object to be a neutron star of ~ 1 solar mass. This amount of energy is comparable to the radiated energy from the entire Crab Nebula.

Thus, with this single object, a variety of astronomical problems were clarified. First, it provided strong evidence for the existence of neutron stars; second, it provided a simple mechanism for the observed energy generation in the Crab Nebula; and third, it provided confirmation of the ideas of nucleosynthesis that tied together the formation of neutron stars with supernova explosions.

In late 1970, the observational situation in X-ray astronomy improved considerably with the launch of NASA's satellite Uhuru, a project conceived and directed by R. Giacconi. Whereas almost all previous results in X-ray astronomy were derived from very brief sounding-rocket or balloon flights, Uhuru provided continuous observations of X-ray sources with good sensitivity, time resolution, and angular resolution.

One of the first objects to be studied by Uhuru was Cygnus X-1. This source apparently was seen during the discovery observations in 1962 (Gursky et al., 1963) and was the first source found to be variable (Bowyer et al., 1965). In 1966, following the discovery of the optical counterpart of Sco X-1, a rocket survey of the Cygnus region, made with the specific intent of locating sources with sufficient precision to allow a search for optical and radio candidates, was apparently successful in the case of Cyg X-2, but not successful in the case of Cyg X-1 (Giacconi et al., 1967).

The Uhuru data on Cyg X-1 revealed that its X-ray emission was highly variable on a very short time scale. In many instances, the emission appeared to be organized in single pulsations or bursts with a suggestion of regularity (Oda et al., 1971). The conclusion was that there seemed to be a periodicity comparable to or less than the Uhuru time resolution of 0.1 sec; this in turn required an object smaller than a white dwarf. Since there was no evidence of a supernova explosion, it was argued that this was not a neutron star, but rather a black hole. These arguments are now known to be weak; nevertheless, the following chain of reasoning and subsequent observations, which rely heavily on optical observations (Webster and Murdin, 1972; Bolton, 1972), lead to the conclusion that Cyg X-1 is best understood as a black hole.

1. The object is in a binary system.

2. The short time scale variability for the X-ray emission requires a compact source.

3. The energetics of the source is simply explained by the release of gravitational energy into kinetic and thermal energy during the process of mass accretion onto a compact star.

4. Analysis of the binary elements leads to a lower limit of 5 $M_\odot$ for Cyg X-1, well above the absolute upper limit for the maximum mass of a neutron star or a white dwarf.

There are several other X-ray sources with characteristics similar to those of Cyg X-1 although with less stringent limits on the masses. Also, the X-ray emissions of two X-ray binaries, Hercules X-1 and Centaurus X-3, are found to be regularly pulsing with periods of 1.2 and 4.8 sec, respectively, which are obviously reminiscent of the radio pulsars. Centaurus X-3 was the first X-ray source shown definitely to be a binary system, based on the observation of the Doppler shift of the 4.8-sec pulse period and of regular eclipses (Schreier et al., 1972).

In the remaining sections of this chapter, we will describe in more detail the observation of supernovae, pulsars, and binary X-ray stars, and the unified picture which emerges of the origin and nature of the collapsed stars.

Supernovae

Neutron stars and black holes apparently begin their lives with a supernova explosion. We are not so much interested here in why a supernova occurs, but rather in the question of total energy release, which tells us something about how neutron stars and black holes originate; and in the question of the kinds of stars that become supernovae, which relates neutron stars and black holes to the more normal stars and completes the story of stellar evolution.

Characteristics of Supernovae

Given the basic understanding that a supernova is the sudden flaring of a star by a factor $\sim 10^{10}$, it was immediately realized that a search of the nearest hundreds of galaxies would yield a few examples every year for astronomers to study and understand. F. Zwicky was the first to organize a search on a large scale, and with W. Baade produced dramatic results; some dozen supernovae were found in three years of work. It was also obvious that supernovae must have occurred in our own Galaxy, and were the origin of Tycho's and Kepler's "new" stars and of the new star that produced the Crab Nebula. The Crab supernova was witnessed by Chinese astronomers in 1054 A.D. (cf. Mayall and Oort, 1942). When Zwicky began his search in 1936, only 20 supernovae had been observed worldwide since 1885. At the present time, nearly 400 have been reported worldwide. Of these, 250 have been discovered at Palomar where Zwicky began his work (Zwicky, 1974, p. 6). In addition, we have evidence of several hundred ancient supernova explosions which took place in our own Galaxy, indicated by the radio emission, which persists for many thousands of years after the initial event.

The supernovae yield several different kinds of information. First is the "light curve," which is simply the variation of the brightness with time; second is the spectrum of the radiation, which also varies significantly with time. Both will be discussed here. The rate at which supernovae occur in various galaxy types and their location within the galaxy will be discussed in the next section.

One outstanding characteristic of supernovae is that they are of two types, as first discovered by Minkowski (1941). The light curves of type I supernovae appear to represent a homogeneous class. As illustrated in Figure 5-4, there is an initial rapid rise to a maximum and a rapid decline, all occurring within about a month, followed by a slow decline, in which the light intensity decreases exponentially with a decay time of between 40 and 50 days.

After correcting for distance and the effects of extinction, Minkowski (1964) estimated the absolute magnitude of supernovae of type I to be about -19, with a total spread of no more than about 1 magnitude. The 1885 supernova in M 31, seen at $+5.4$ mag was at an absolute magnitude of -18.6 without the extinction correction.

The type II light curves, also shown in Figure 5-4, are not nearly so distinctive. In general, the early phase (~ 1 month) develops more slowly than those of type I, but at later times they decay more rapidly.

Whereas the light curve of the supernovae of type I almost begs for a simple explanation, that is not true for their spectra. Minkowski, discussing the spectrum of SN 1937c wrote that "the appearance of the spectrum leaves little doubt that [it] is composed essentially of wide and partially overlapping emission bands." However, it is equally possible that the spectrum might

Figure 5-4. Light curves of type I (upper box) and type II supernovae. Adapted from Minkowski (1964).

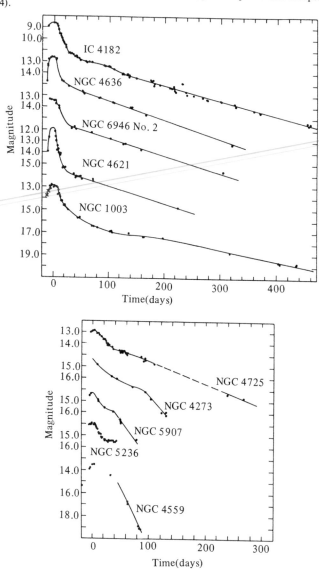

consist of broad absorption lines cutting into an underlying continuum! The interpretation of these spectra is made difficult by our inability to identify these features with the radiation from specific atomic species; the latest view is that one is seeing black-body radiation ($T \approx 10,000$ K) modified by broad emission and absorption lines.

The spectra of type II supernovae are similar in their complexity to those of type I. However, certain features of the type II spectrum are identifiable; in particular, hydrogen is seen in emission. Nevertheless, the spectral data from supernovae have provided a few key pieces of information which yield basic characteristics of the explosion. The identification of spectral lines allows a direct determination of expansion velocities. In the case of type II, this velocity is in the range 10^4 km/sec, based on the observation of the so-called P Cygni profiles (combination of emission lines and blue-shifted absorption lines) and line broadening. The same technique yields higher expansion velocities for the type I supernovae—2×10^4 km/sec—but with less certainty since the lines are not identified. The spectra also reveal an underlying continuum of radiation which can be interpreted as originating from a black body radiating at a temperature of 5000–10,000 K (Kirshner et al., 1973).

What can we learn from these meager facts? First, by reason of their great energy release the supernovae stand apart from every other galactic phenomenon. The only similar event of which we have knowledge, the ordinary novae, are fainter by at least 10^4. The differences in brightness among various supernovae are much smaller than this factor.

The total radiated power can be estimated from the light curve. At peak intensity, the absolute magnitude of a type I supernova is about -19. Assuming a temperature of 10^4 K, we obtain a bolometric magnitude of about -19.4. Since absolute magnitude refers to the apparent magnitude if the object is at 10 pc, this works out to a peak luminosity of $\sim 10^{43}$ erg/sec. The half duration of the initial peak of the luminosity is about 10 days; thus, the total radiated power is $\sim 10^{49}$ erg. The supernovae of type II are between 2 and 4 times less powerful than this.

A second measure of the energetics is the kinetic content of the expanding envelope; namely, $1/2\ mv^2$, where m is the mass of the envelope and v is its expansion velocity, found to be between 1 and 2×10^4 km/sec.

It is not so easy to determine the mass. If, simple-mindedly, we assert that $M \approx M_\odot$ then the kinetic energy becomes $\sim 10^{51}$ erg.

There is one argument that yields at least a minimum mass. We can determine the area of the radiating surface by the same technique we used for determining the radius of Sirius B, using the luminosity of 10^{43} erg/sec and a surface temperature of 10^4 K. This number is about 2×10^{31} cm². In order for significant continuum radiation to be emitted, this surface must be at least partially opaque to Compton scattering. This implies a column density of about 5 gm/cm²; thus, the minimum mass is $\sim 10^{32}$ gm or about 0.1 $M_\odot$. Thus, the *minimum* kinetic energy release is 10^{50} erg.

This number is enormous. It is comparable to the total energy radiated by the sun during the entire 10^{10} years of its existence. Clearly something is happening which involves the fundamental energy-producing mechanisms of the entire star. Another way to consider the energy release is the energy per nucleon, which is about 1 MeV. By comparison, when hydrogen "burns" to helium in a star, the energy release is 6 Mev/nucleon. The heavier elements do not burn so efficiently, but the burning of oxygen and carbon can in fact yield energy releases between 0.3 and 0.7 MeV/nucleon. Thus, considering the uncertainties involved, nuclear burning is a possible source of the energy release comprising the supernova.

The other possibility is the energy release during gravitational collapse. In the preceding section we noted that there exists a finite mass limit for white dwarfs, which if exceeded would cause the star to collapse. The matter in a white dwarf may collapse to a neutron star, in which case there is a release of energy corresponding to the (negative) increase of gravitational energy of the new stellar configuration by an amount $\sim GM^2/R$. Taking R for a neutron star of ~ 10 km, this quantity is about 3×10^{53} erg, which is more than ample to provide the observed release of kinetic energy.

We are entering here into one of the most complex domains in astrophysics. It is beyond the scope of this chapter to discuss the phenomena of the supernova explosion, but we can at least present the following scenario. A star, possibly only the core, finds itself too massive to be maintained by internal pressure. The star starts to shrink and the internal pressure increases, which causes certain nuclear reactions to proceed more rapidly. If inverse beta decay occurs, stealing both energy and phase space, the collapse proceeds even faster. If thermonuclear reactions occur involving C^{12} or O^{16}, energy may be released explosively. It is likely that both processes occur; if the former dominates, the star will collapse to either a neutron star or a black hole; if the latter, the star may simply disperse all its matter as a cloud. Another complicating factor is how to stop the collapse once it has passed beyond a certain point, since the kinetic energy of collapse can be so much greater than the energy release in subsequent nuclear reactions. Here the key point is apparently the creation of large fluxes of neutrinos. When the collapsing gas becomes very hot, and the density is very great, neutrino production dominates over γ-ray production. These neutrinos may in fact blow off the outer envelope of the star in spite of their infinitesimal interaction rate with matter.

The Stellar Origin of Supernovae

As a corollary to trying to understand the physical nature of an astronomical phenomenon, one must try to understand its parentage; specifically, what kinds of stars produce supernovae, and when in their lifetimes does such an event occur. We have no direct evidence on this question, since there is no case where the star preceding the supernova was identified and studied. We do have a certain amount of indirect evidence, based on the study of the types of galaxies in which supernovae occur and of where within the galaxy they occur. This evidence yields the kind of information we want because of the tendency of certain stars to occur predominantely in certain galaxies or in certain parts of galaxies. The first realization of this phenomenon came from Baade, who categorized stars into population types (cf. Blauuw, 1965). Stars of population I are the youngest, most massive stars, occurring principally in the arms of spiral galaxies (indeed, these young, very luminous stars, give the spiral arms their distinctive appearance). Stars of population II are the oldest stars and are the main constituent of globular clusters and elliptical galaxies. The best information on the distribution of supernovae in galaxies is given in Table 5-3, adapted from a study of Tamman (1974) of 75 supernovae occurring in a sample of 408 nearby galaxies.

The following important conclusion can be derived from the numbers in Table 5-3. Type II supernovae occur almost exclusively in spiral galaxies such as our own. This leads to the conclusion that the type II supernovae occur in young population I stars, since it is these stars which are unique to the spiral galaxies. This result confirms an apparent observational fact, that supernovae in spiral galaxies tend to occur in the arms (cf. Moore, 1974), which can also be used to estimate that the mass of the progenitor is between 5 and 10 $M_\odot$. This mass is derived from the known lifetime of massive stars.

Table 5-3 Frequency of supernovae according to galaxy type.[a]

	Galaxy type		
	Elliptical	Spiral	Irregular
Total galaxy mass (in $10^{11} M_\odot$)	720	130	40
Corrected N_{SN}	17	100	5
SN per $10^{11} M_\odot$ each 100 yr	0.07	2	4
Numbers of SN observed (type I/type II)	6/1	17/44	7/0

[a] I have not listed uncertainties, which are quite large. The galaxy mass and the number of supernova events (N_{SN}) are derived quantities (N_{SN} starts with the observed number of supernovae, but must be corrected upward to account for visibility in the galaxy). Thus the final supernova rates are uncertain by at least 50%. The last line lists the actual numbers of supernovae, by type, observed in the three galaxy types.

These stars are assumed to be born in the spiral arms, and with time they tend to drift out. Thus, only the more massive stars would be expected to complete their life cycles and still be found in the arms.

Beyond this, the interpretation of these data becomes a bit muddy. The ratio of type I/type II supernovae in elliptical galaxies is apparently reversed in spiral galaxies, which has been used as a basis for arguing that the type I supernovae occur principally in old and low-mass stars ($\sim 1\ M_\odot$). However, the rates of type I production in ellipticals and in spirals are so different that this conclusion is hard to support; namely, only a small number of young stars in elliptical galaxies is required to give rise to the observed supernovae. Furthermore, the failure to find type II supernovae in the irregular galaxies (like the Magellanic Clouds) is very puzzling, since these galaxies are believed to contain significant numbers of young, population I stars.

The spectroscopic data muddy the picture further. The failure to find hydrogen in the spectra of type I supernovae is generally believed to represent a real deficiency of the element in the ejected gas, an indication of an old star which has burned all its original hydrogen. However, "old" in this sense means that the star is highly evolved, and does not necessarily imply a long period of time.

Summary

The supernovae represent a phenomenon involving an entire star in which at least 10^{50} erg of energy is released in a very short time. We can understand this theoretically as the collapse of a star, with an ensuing explosion, as may occur near the end of a star's life. There are at least two distinct types of supernovae, which differ in their radiated light, energetics, and stellar progenitors. The evidence is that the type II supernovae originate in massive population I stars, but we cannot decide, from the available data, on the origin of the type I supernovae.

Before leaving this subject, I must mention several other important facets of supernovae phenomena.

1. The supernova explosion returns to the interstellar medium about 1 $M_\odot$ of material greatly enriched with heavy elements. This material, especially the heavy elements, probably plays a crucial role in the continuous birth of stars, not to mention the planet on which we reside.

2. The energy release, in the form of kinetic energy, radiation, magnetic fields, and cosmic rays is probably an important factor in determining the appearance of the interstellar medium.

There are also more speculative possibilities. For example, at least several times in the approximately 3 billion years that life has existed on earth, a supernova explosion has occurred in the immediate vicinity of the sun (within 5 to 10 pc). The effect would be to increase significantly the cosmic radiation on earth for a period of 10^4–10^5 years. The resulting increase in genetic mutation rate could have been a powerful stimulant in altering the evolution of life forms.

Pulsars and the Discovery of Neutron Stars

The Crab Nebula, illustrated in Figure 5-5, is the only historical supernova in our Galaxy which is now prominent optically; in fact, it is the first entry (M 1) in Messier's eighteenth-century catalogue of nebulosities. It is bright throughout the electromagnetic spectrum and, for this reason, it has always occupied a central role in the study of supernova phenomena. As an example, this object in the 1950s provided the key evidence that synchrotron radiation (the radiation produced by relativistic electrons as they gyrate in a magnetic field) was an important mechanism by which radio emission is produced by a great variety of astronomical sources.

Baade, in the 1940s, studied the central parts of the nebula in an attempt to find the "remnant" star left over from the supernova explosion (Baade, 1942). He did, in fact, identify a candidate, a 17th-magnitude star that possessed a featureless spectrum, and later was believed to be a white dwarf

Figure 5-5. Photograph of the Crab Nebula. Photograph courtesy of the Mount Wilson and Palomar observatories. Copyright by the California Institute of Technology and the Carnegie Institution.

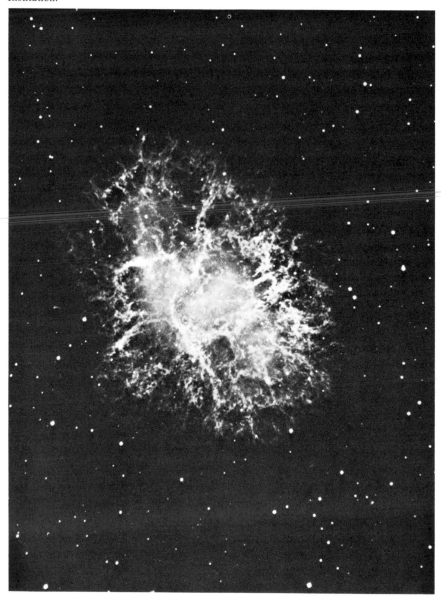

(Minkowski, 1942). The matter was not pursued at that time since this particular star could not be positively related to the nebula. Baade was at least ten
years ahead of the times and simply did not have the appropriate tools to
study this particular star. A more recent picture taken with a similar telescope, but using photomultipliers and a phased oscilloscope instead of photographic film, is shown in Figure 5-6. The light from the star pulses regularly with a period of 0.033 sec. Baade was also unlucky; if the period had
been several times greater, he might have seen the star flicker perceptibly.
We now know from a variety of evidence that this star, a pulsar, is indeed
the remnant of the explosion. This object, plus the general characteristics of
pulsars, is the subject of this section.

Characteristics of the Pulsars

The first pulsar, now known as CP 1919, was discovered in July 1967. A
portion of the original records recording this remarkable object is shown in

Figure 5-6.   The Crab Nebula pulsar as seen in an oscilloscope. The oscilloscope sweep is synchronized to the pulsar period of 0.03 sec. The main peak appears twice. The smaller peak
between the two is the interpulse. Photograph courtesy of Professor E. Groth, Princeton University.

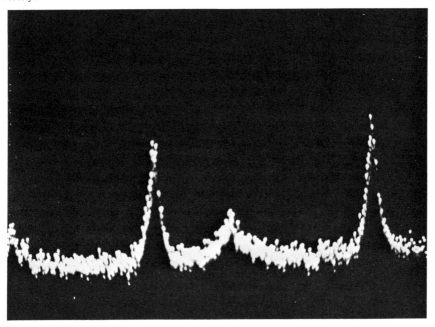

Figure 5-7. The discovery paper by Hewish et al. (1967) delineated the central features of these objects; namely, the organization of the radio emission into pulses, and the constancy of the period. In the case of CP 1919, the period in 1967 was found to be better than a few parts per million, the pulse duration is about a quarter of the 1.3-sec period. They also observed a variation of arrival time with frequency (frequency drift or dispersion), which they correctly attributed to the frequency dependence of the index of refraction of the interstellar medium due to its electron content, as well as to significant pulse-to-pulse variability. By now we know of more than 100 pulsars in the Galaxy. The properties described above have been studied in great detail, as have the systematic and random variations in the pulse shape and pulse period, the spectrum and polarization of the emission, the distribution around the sky, and possible associations with known objects (cf. Groth, 1975).

By far the most fruitful data, for our understanding the pulsars, are the pulse period and its variations.

Because the radio emission is "pulsed" and single pulses are readily distinguishable, at least in many instances, the determination of the period is

Figure 5-7. Original data records of the pulsar CP 1919 obtained by Hewish et al. (1967); a, the record of the source when it is strong. A pulse appears each second and particularly prominent pulses appear at 39, 40, and 55 sec; b, pulses obtained from two independent radio receivers at slightly different frequencies. Adapted from Hewish et al. (1967).

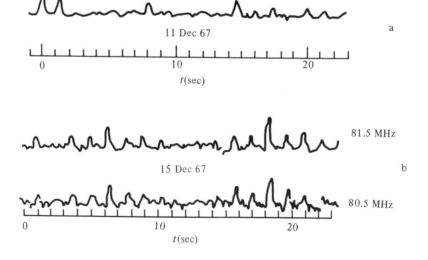

simply a matter of counting the number of pulses received during a specified time interval. (The reader can convince himself that the technique will work even if there are significant time intervals when no pulses are recorded.) Two factors complicate the analysis; one is that the earth's rotation and its motion around the sun add an appreciable frequency shift. A second is that the magnitude of the correction for these effects depends on the declination (angular distance from the equator) of the sources.

The first positive indication of variations in period was found by Davies, Hunt and Smith (1969). They studied four pulsars, including CP 1919, and found changes of between $1:10^6$ and $2:10^7$ parts per year. This measurement eliminated the possibility that these objects were in binary systems unless the binary period was at least several tens of years. This follows directly from Kepler's laws. If the objects are assumed to be of $\sim 1\ M_\odot$, with an orbital period of one year, the orbital velocity would be ~ 30 km/sec, whereas the observed period change of $1:10^6$ if caused by a Doppler shift corresponds to a velocity change of only 0.3 km/sec within the year. However, as noted by the authors, "it may be significant that the periods of all four pulsars seem to be changing in the same sense. The increases in period seem to indicate a real physical change."

By now, period variations have been measured in a large number of pulsars. Remarkably, as shown in Figure 5-8, all measurements reveal that the period is changing in the same sense; that is, the period increases with time. Furthermore, there appears to be a simple relation between the increase in period and the period; namely, the pulsars of longer period undergo smaller changes of period. This remarkable result, plus the values of the periods themselves, is the essential point in understanding the nature of the pulsars. First, the very small values of the period, some less than one second, indicate that we are dealing with small, compact stars, smaller even than a white dwarf. The reason is that the object must be "doing" something—vibrating or rotating—in order to yield the pulsed emission. A naive argument, that the object must be smaller than the distance light can travel in one pulse period, yields sizes of $\sim 10^5$ km, much smaller than that of a normal star, but larger than that of a white dwarf. More realistically, dynamical calculations of white dwarfs reveal fundamental vibrational periods of many seconds and comparable maximum rotation periods. This argument alone forces one to consider neutron stars as more likely candidates for the pulsars than white dwarfs.

The fact that periods always increase with time means that one is dealing with rotation and not vibration, assuming, as seems reasonable, that the pulsars lose energy (through radiation) with time. A vibration period tends to decrease or remain constant as energy is lost, as is the case for a pendulum. The realization that the pulsars rotate also clarifies the nature of radio pulses; namely, that we are seeing a beam of emission, similar to that from a lighthouse, as it sweeps across the earth.

Figure 5-8. Plot of pulsar period (in sec) versus the rate of change of pulsar period (d/dt). Adapted from Groth (1975).

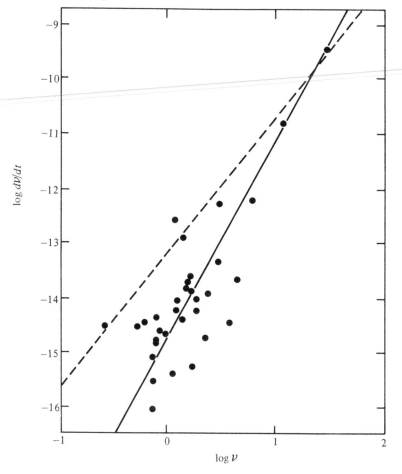

The fact that there appears to be a simple relation between the change in the period and the duration of the period indicates that one is dealing with a single class of objects with similar properties. Adopting as a working hypothesis that the pulsars are rotating, neutron stars, we obtain a clear interpretation: the longer-period objects are the older ones. In fact, it is common practice to express the pulsar slowdown as the period divided by the period change per year, which is related to the time, in years, that the pulsar required to slow down to its present period from some initial high rate. On this basis, some of the pulsars may be $\sim 10^7$ years old.

Another kind of period variation, "glitches," are seen in several pulsars. One year of data from NP 0532 in the Crab Nebula is shown in Figure 5-9, which is a plot of period residual (difference between a true arrival time and a best-fit arrival time). On 29 September 1969, a discontinuity of about 2×10^{-3} sec is present, which is the glitch. There is a real change of about $1:10^8$ of the pulse period from that before and after the discontinuity. Within less than a month the pulsar period had returned to its initial value. Another kind of period (or phase) variation seen in this figure is a "wandering" of the period around a central value.

Figure 5-9.   Change of period of the Crab Nebula pulsar over a two-year interval. What is actually plotted is the difference in arrival time (phase) between the predicted and the actual observed arrival times. A "glitch" occurs on 29 September 1969. Data obtained at Princeton were of the optical emission from the pulsar. The Arecibo data are of the radio emission. Adapted from Boynton et al. (1972).

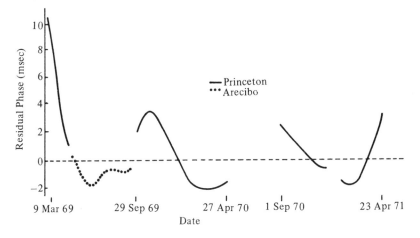

In all the glitches seen to date, the pulsar sped up. This can be explained qualitatively in the following way. As the pulsar (a rotating neutron star) ages, it rotates more slowly and possibly cools. Both changes make the star tend to shrink, which induces stresses in the crust. At a certain point the stresses exceed the yield strength of the crust and the star shrinks abruptly (perhaps producing the stellar analogue of an earthquake!). Since angular momentum must be conserved, the period decreases slightly.

This is only part of the story. The period will eventually return to its former value, implying that the crust is not tightly coupled to the (fluid?) interior, which carries most of the angular momentum. In this view only the rotation rate of the crust will change. Immediately after the glitch, the crust is spinning slightly faster than the core and friction between the two returns the crust almost to its initial period. See Pines (1975) for a discussion of the interior of neutron stars.

The shape of the pulses varies widely among different pulsars. All retain the same qualitative behavior shown in Figure 5-7; namely, a discrete, well-defined pulse is present and there is little, if any, steady radio emission. Two other phenomena relating to pulse shape are worth noting; one is the "interpulse," present in certain pulsars, which is a second, smaller pulse within a single period. This interpulse typically occurs close to halfway between the main pulses. Second are the marching subpulses. Frequently the pulse has the appearance of a series of subpulses; however, the subpulses do not remain at fixed phase with respect to the average pulse envelope. They tend to shift systematically from one pulse to the next.

Another outstanding characteristic of the pulsars is the high degree of polarization of the emission. It can approach 100% and often changes systematically across the pulse. It has not yet been possible to develop a definite model that will explain these various pulse characteristics. The combination of the observed power, polarization, and pulse complexity tells us that the radiation is produced by the coherent motion of electron streams (coherence simply means that the electrons are moving together in bunches). The radiation process is not very different from what takes place in a radio antenna. Instead of a metallic conductor, the pulsars require a strong magnetic field to guide and structure the electric currents and yield the observed polarization and pulse structure. Ironically, the detailed reasons as to how the radio beam is produced may not be important astronomically.

The Crab Nebula Pulsar, NP 0532

The story of the discovery of this pulsar has already been given. In the same sense that the study of the Crab Nebula was important twenty years ago and more for our understanding of a variety of supernova phenomena, the Crab Nebula pulsar has been a key object in the past decade in helping us to understand the compact stars and other phenomena.

We argued above that the pulsars were rotating neutron stars. NP 0532, by itself, provides the most direct demonstration of this possibility. The pulse period, if it corresponds to the rotation of the star, requires an extraordinarily high density. For a gravitationally bound object to be stable against rotational breakup, the gravitational acceleration at the surface must exceed the centripetal acceleration; otherwise, the surface material cannot remain bound to the star. This requires

$$\frac{GM}{R^2} > \omega^2 R$$

or

$$\frac{M}{R^3} \approx \rho > \frac{\omega^2}{G}$$

where ω is the angular velocity and G is the constant of gravitation. M/R^3 is the density of the star (approximately). Thus, given the rotation period, we determine a limit on ρ. Using the period of 0.033 sec ($= 200$ radians/sec) yields $\rho > 6 \times 10^{12}$ gm/cm^3. This is an enormous figure, many orders of magnitude greater than the density of a white dwarf, and close to the density of nuclear matter—10^{14} gm/cm^3—expected in a neutron star.

Another argument for this picture, although not so direct, is as follows. The period of NP 0532 is observed to slow down as do many other pulsars, by an amount 1.3×10^{-5} sec/year, or 4×10^{-13} fractions of a period per period. This implies a loss of energy, which can be derived as follows. The kinetic energy of a rotating body is given by

$$E = (1/2)I\omega^2$$

and the dissipated power when ω changes is

$$P = dE/dt = I\omega \, (d\omega/dt).$$

If for the moment of inertia $I(\approx MR^2)$ we take the expected parameters of a neutron star ($M = 1\,M_\odot$, $R = 10$ km), this relation yields $P = 3 \times 10^{38}$ erg/sec. This number exceeds significantly the radiated power of the pulsar but is comparable to the total radiated power of the nebula. Thus, NP 0532 as a rotating neutron star is the driver, somehow, for the entire nebula and solves a longstanding mystery. When it was first understood that the synchrotron process was responsible for the nebular radiation it was also realized that there needed to be a "source" of energy continuously accelerating or injecting electrons; otherwise these electrons would decay in energy within ~ 50 years. However, the 20–30 year span over which the Crab has been studied in modern times reveals no significant changes in its luminosity. The present view is that the pulsar somehow robs its rotational energy to accelerate electrons, which are ejected into the nebula and provide the observed synchrotron emission.

It is possible that this same "accelerator" is responsible for the galactic cosmic rays. NP 0532 emits γ-rays of 10^{11} eV (Grindlay, 1972), which probably results from the Compton scattering of primary photons from the pulsar on very high-energy electrons ($\sim 10^{14}$ eV). This is the first evidence, direct or indirect, for the existence of such energetic particles in the Galaxy, except for those cosmic ray particles which reach the earth.

The Crab Nebula pulsar is unique in many respects. It is the fastest and most powerful (intrinsic) pulsar. It is the only pulsar seen at optical, X-ray and γ-ray energies as well as in the radio region. The pulse shape, shown in Figure 5-10, is duplicated at all wavelengths. In fact, with such data one can derive a very precise limit for any possible variation of the speed of light with frequency, from the fact that the variation in arrival time is less than 10^{-3} sec across this enormous frequency range.

The Relation of Pulsars to Supernovae

The association of the pulsar NP 0532 with the Crab Nebula resolves at least one aspect of supernovae; namely, that supernova explosions, in certain instances, produce a neutron star remnant. The evidence must be regarded as conclusive, not only because of the physical position of the pulsar within the nebula, but also because the "lifetime" of the pulsar, as derived from the period slowdown, is commensurate with the known age of the nebula, nearly a thousand years.

At least one other association of pulsar and supernova remnant is known, PSR 0833 − 45 and the Vela supernova remnant (Kristian, 1970). The Vela remnant in many respects is much more dramatic than the Crab, but it is in the southern hemisphere, which has limited its observability, and is much older, which has eliminated the possibility of historical records. The pulsar period is 0.089 sec—only the NP 0532 pulses more rapidly—and its age, as derived from slowdown, is ~ 10^4 years, which is commensurate with the age as derived from the expansion of the remnant. This pulsar is about 0.6° from the geometric center of the nebula, whereas the nebula is about 5° across. There is one other possible pulsar-SNR association, namely, PSR 0611 + 22 and IC 443. In this case the pulsar is actually *outside* the observable remnant; thus, the "association" may be a chance coincidence of position. Not one of the remaining hundred or so known supernova remnants is found with an accompanying pulsar. There are several possible explanations for such a negative result; one is that the neutron star is not left behind in all supernova

Figure 5-10.   Mean pulse shape of the Crab Nebula pulsar at various wavelengths. The broadening seen at low-frequency wavelength is caused by frequency drift in the interstellar medium. Adapted from Hewish (1970).

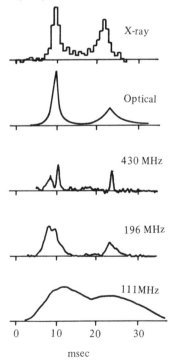

events; another is that the radio "beam" from the pulsar has the shape of a narrow cone, and we see only the small fraction in which the cone sweeps across the earth. (Alternatively, the beam could be in the shape of a sheet, in which case the earth would intercept all pulsar beams.) However, there is a strong observational bias which makes these data hard to interpret; namely, that the remnants are often very distant from us (thousands of parsecs) compared to the detection limit for pulsars (hundreds of parsecs). The limited sensitivity of present pulsar surveys and the greater frequency drift make the more distant pulsars difficult to observe.

The reverse association has also not been made; i.e., in no case has the discovery of a pulsar led to the discovery of a hitherto unknown supernova remnant. However, the absence of such other evidence does not prove that the remaining pulsars are not related to supernovae. The uniqueness of this class of objects and certain shared features argue in fact that all pulsars have such an origin. However, the possibility that all supernovae produce pulsars cannot be demonstrated at this time.

We discussed in the previous section the fact that supernovae are of two types, at least. Unfortunately, we do not have positive evidence for the relation between these types and pulsar remnants. One possible clue comes from the distribution of pulsars in the Galaxy, shown in Figure 5-11, which

Figure 5-11. Galactic distribution of pulsars. The coordinate system is one in which the equator is the Milky Way and the point 0 is the direction of the galactic center. Adapted from Groth (1975).

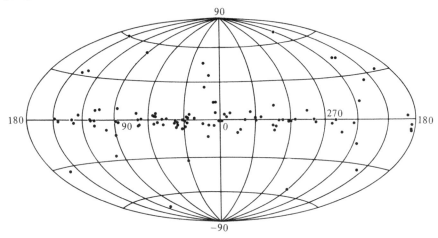

shows the pulsars concentrated along the Milky Way, as expected for galactic objects. We noted earlier that the type II supernovae seem to be associated with young (population I) stars, which lie predominantly in the spiral arms. Equivalently, one expects these young stars to lie very close to the galactic plane; however, this does not appear to be the case. Thus, this distribution tends to favor the idea that pulsars originate from supernovae formed from older stars—the type I supernovae. Here again observational biases may dominate the data; in particular, because of frequency drift, pulsars close to the plane of the Galaxy, where the interstellar density is high, are more difficult to observe than those out of the plane. Also there may be a real physical cause for dispersion of the pulsars out of the plane. Some pulsars have exceptionally high velociites with respect to the sun (several hundred kilometers per second). This is not unexpected and could result from a small asymmetry in the supernova explosion, for example. If widespread, such large space velocities could make a distribution typical of young stars mimic that of old stars. Thus, the question as to which type supernova leads to pulsars must be left open.

Another distributional property of the pulsars is that only one is found in a binary system. Many arguments have been advanced as to why this must be so—that the supernova explosion disrupts the binary, and the material from a stellar companion quenches the radio emission. However, one must take seriously the possibility that pulsars are produced predominantely in single stars. The one binary pulsar, discovered by Hulse and Taylor (1975) is an exceptional object with a pulse period of 59 msec. It has an orbital period of 7.7 hr and its orbit is highly eccentric. This object may be the first discovered example of an entirely different class of pulsars.

Summary

The two outstanding conclusions derived from the study of pulsars is that they are neutron stars and that they are produced in supernova explosions. Beyond this, it has not been possible to obtain clear information. We have the beginnings of information on the structure of neutron stars, but we must still rely heavily on theory to tell us what to expect observationally. Also, definitive evidence on which types of supernovae lead to pulsars is still lacking.

However, this is still a young discipline. Given the very recent and exciting discovery of a binary pulsar, it is clear that these objects will yield some of their secrets to the more concerted observational efforts now being made.

The Binary X-ray Stars

X-ray astronomy by now is known to comprise a number of diverse elements. For example, it appears that all supernova remnants and rich clusters of galaxies may be strong X-ray sources. What concerns us here are the eighty or so discrete sources clustered along the Milky Way which radiate in the keV energy range. These are galactic objects and are not found in association with other known classes of objects. They are among the most luminous objects in our Galaxy, and most of this power must be emitted in the form of keV photons. The X-ray emission is highly variable on almost every time scale on which they have been observed. In contrast to the pulsars, there is no simple "signature," except their great power, which describes this class of objects.

The essential clue to understanding the nature of these objects is the finding that at least some of the sources occur in close binary systems. This, along with the short time variability, the great power, and other bits of information have allowed the development of a credible model that accounts for many of the observations and can be supported from theoretical and stellar evolutionary points of view. This model comprises a compact star in close orbit around a normal star. Matter coming from the normal star is heated as it falls through the gravitational field of the compact star. The efficiency of this process for making X-rays is prodigious; the order of 10% of the rest energy of the infalling matter is radiated, mostly in the form of X-rays, if the compact star is a neutron star or black hole.

Characteristics of the X-ray Sources

Figure 5-12 shows the distribution of the galactic X-ray sources as found by the X-ray satellite Uhuru (Giacconi et al., 1974). The sources show a distribution characteristic of galactic objects, a concentration near the plane of the Galaxy, particularly at low galactic longitudes. The power observed at the earth ranges from 10^{-7} erg sec^{-1} cm^{-2} for the strongest to 10^{-10} erg sec^{-1} cm^{-2} for the weakest in the energy range 1–10 keV. To convert this to intrinsic power we must know something about their distance; however, these data by themselves do not yield an unambiguous answer. The average latitude of the X-ray sources is about 5°. If the X-ray sources are very young, they would lie within about 100 pc of the galactic plane and their distance would be about 1 kpc; if, however, the objects are very old, their average distance from the plane could be ~ 1 kpc and their distance from us would be ~ 10 kpc.

There are two other pieces of information which indicate that the average distance of these sources is closer to 10 than to 1 kpc. One is the discovery of discrete sources in the Magellanic Clouds which, except for being weak, appear to be the same as the galactic sources. We see these at a power ~ 10^{-10} erg sec^{-1} cm^2; at the 65-kpc distance of the Magellanic Clouds, their intrinsic power is close to 10^{38} erg/sec. Since many of the X-ray sources found between $\pm 40°$ longitude must be radiating at similar power levels, they must lie at distances of 5–10 kpc.

Another argument is the relative absence of a "halo" of diffuse radiation at low galactic longitudes. If the sources were as close as 1 kpc, the more distant, unresolved sources would be expected to reveal themselves as such a halo. Its absence implies again distances of 5–10 kpc.

With such distances, the intrinsic power of the X-ray sources must lie in the range 10^{36}–10^{38} erg/sec. A power of 10^{38} erg/sec is of particular significance since it corresponds to the "Eddington limit" of luminosity. At higher luminosities, for a 1 $M_\odot$ object, the outward force due to radiation pressure exceeds the surface gravity and material will be ejected from the surface. It is

Figure 5-12. Galactic distribution of X-ray sources as seen by the Uhuru satellite. Adapted from Gursky and Schreier (1975).

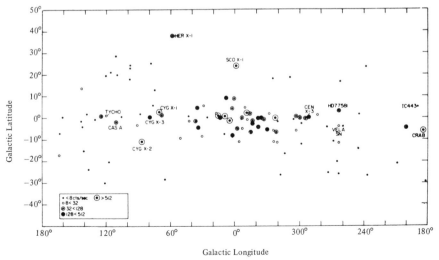

for this reason that the most luminous stars in the Galaxy (the supergiants) radiate at this level. If the star tries to radiate more power, the excess energy will depart in the form of kinetic energy of evaporating gas (stellar winds).

Another important fact that emerges at this point is the very small number of X-ray sources at these great luminosities—a few hundred in the galaxy, or only $1:10^9$ stars. This can indicate either the rarity of the conditions leading to the X-ray sources or a very short phase of X-ray production. In any event the small number of X-ray sources and their great power must be accounted for.

None of these X-ray sources radiates steadily in the sense that stars do, but the variability that occurs shows no simple pattern as is the case for the pulsars. Two of the X-ray sources, Centaurus X-3 and Hercules X-1, do indeed pulse (Figure 5-13), with periods of 4.8 and 1.2 sec, respectively. The similarity of this pulsing to the pulsars leads to the presumption that these are also rotating neutron stars, especially for the more rapidly pulsing Her X-1.

More dramatic variability is found in Cygnus X-1, as shown in Figure 5-14. Here the variability occurs on a time scale of milliseconds; however, the variability is perfectly random, as nearly as can be determined. Another very different kind of variability is shown in Figure 5-15, which shows both the optical and the X-ray variability of Scorpius X-1. This X-ray source exhibits two "states"; one, when the X-ray (and optical) emission is variable on a time scale of minutes, the other when the X-ray intensity is lower by a factor of two and less variable.

Figure 5-13.   X-ray pulses from Centaurus X-3 as seen by the Uhuru satellite. The gradual rise and decline of the pulse amplitude is caused by the collimators as the source is transited. Each bin is 0.096 sec and the pulse period is 4.8 sec. Adapted from Schreier et al. (1971).

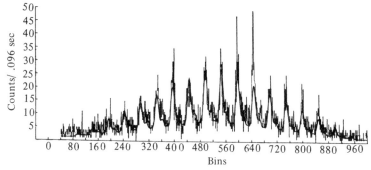

It is almost the case that each X-ray source exhibits its own brand of variability. However, one kind of variability occurs in a number of X-ray sources, which is common in optical astronomy. It was discovered first in Centaurus X-3 (Schreier et al., 1972) that the X-rays regularly "disappeared" for brief periods. It was soon realized that this was an eclipse, the disappearance being caused by the periodic interposition of a large body. Positive evidence for this conclusion is shown in Figure 5-16. The pulse period from Cen X-3 is found to vary sinusoidally with a period of 2.1 days, the same period as found for the eclipse of the X-ray intensity. The interpretation of these data is very clear. The variation of pulse period must be caused by the Doppler shift as the X-ray source moves in an almost perfectly circular orbit around the central object. In support of this view is the fact that the center of the

Figure 5-14. Rapid time variability of Cygnus X-1. The inset shows the burst seen at ~318 sec. The cross-hatched data in the inset are statistically significant and illustrate time variability on a scale of 1 msec. Data were obtained from proportional counters while the X-ray sources were being viewed during a rocket flight from White Sands, N. M. Adapted from figure provided by Dr. E. Boldt, Goddard Space Flight Center.

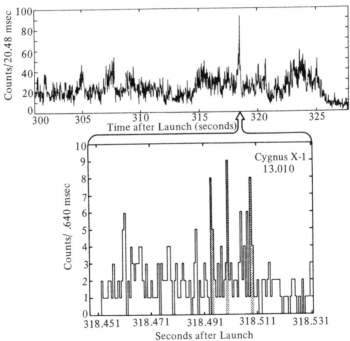

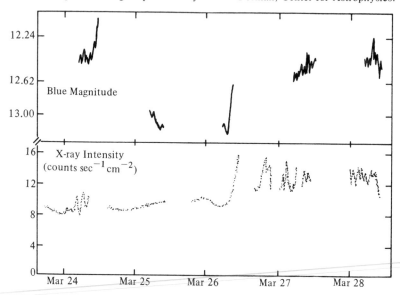

Figure 5-15. Simultaneous X-ray and optical observations of Sco X-1 between 24 and 28 March 1971. The X-ray data were accumulated by Uhuru and are the lower density points. Both X-ray and optical data show significant time variability, particularly in the time interval after March 26.3. Adapted from figure provided by Dr. W. Forman, Center for Astrophysics.

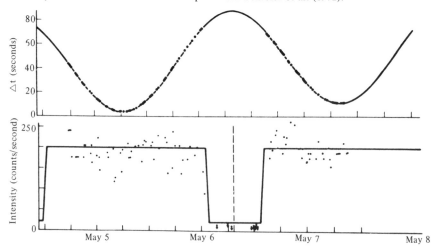

Figure 5-16. Doppler shift and eclipse data for Centaurus X-3. The lower curve is the average intensity between 4 and 7 May 1971. For a 0.2-day interval on May 6 the X-ray intensity fell to a low value consistent with zero. During the same time interval the 4.8-sec pulse period showed a sinusoidal variation in period (actually measured is the difference between the predicted and observed arrival times of the pulse, assuming a fixed period). The phasing between the two curves is just what is predicted if the time of low intensity is an eclipse and the pulse period variation is caused by motion in a circular orbit. Adapted from Schreier et al. (1972).

eclipse occurs when the Doppler shift changes from negative to positive; that is, the X-ray source changes from moving away from us to moving toward us, just what is expected as the source is eclipsed behind the central object. The other pulsing X-ray source, Her X-1, also exhibits both the periodic Doppler shift and synchronized eclipse. Four other nonpulsing X-ray sources show eclipses.

These data are complemented and extended by optical observations. In contrast with the pulsars (of which only one, NP 0532, is seen optically), nine X-ray sources are so observed. It is generally believed that we fail to find others simply because the optical counterparts of the X-ray sources are not particularly distinctive, and positional uncertainties—typically as large as arc minutes—prevent unique selection of the correct candidate in most cases.

The nine optically identified X-ray sources are listed in Table 5-4. In seven of these objects the optical data reveal direct evidence of the binary system, in the periodic variation of the stellar spectral lines (spectroscopic binaries). In Cyg X-3, the optical emission is not observed because of obscuration, but a strong infrared source is seen, which varies in phase with the X-ray source. In Cyg X-2, an ordinary G-type star is seen, although there is no other direct evidence for a second object except for the X-ray emission. A strong case can be made that most of the galactic X-ray sources are binaries, not just these nine. One simple point is that the probabil-

Table 5-4   Characteristics of identified X-ray sources.

Catalogue number	Name	Binary period (days)	Optical magnitude	Companion spectral type	Remarks
3U 1956 + 35	Cyg X-1	5.6	9	0.97Ib	
1118 − 60	Cen X-3	2.1	13	B0Ib-III	4.8-sec X-ray pulse period
0900 − 40	Vela XR-1	8.9	6	B0.5Ib	284-sec X-ray pulse period
1700 − 37		3.4	6	07f	
0115 − 37	SMC X-1	3.9	13	B0Ib	
1617 − 15	Sco X-1	0.8	>13	?	
1653 + 35	Her X-1	1.7	15	late A	HZ-Her 1.2-sec pulse period
2030 + 40	Cyg X-3	0.2	?	?	Heavily obscured, seen only at wavelengths $\lambda \geq 1 \mu m$
2142 + 38	Cyg X-2	?	14	G	

ity of an eclipse occurring is inversely proportional to the square of the separation of the two stars. Thus, it is likely that we are seeing the closest binaries—for others, the combination of orbit separation, inclination, and poor observability make the chance of seeing an eclipse very low.

The Binary Model of X-ray Sources

When two stars form a close pair, several factors make their appearance and history qualitatively different from that of single stars. One simple point is that only for stars in binary systems can we determine the masses. Of greater significance is the fact that there can be mass transfer from one star to the other, especially in the close binaries, and this can have two dramatic effects. For one, the matter falling through the gravitational field of a compact star can be a substantial energy source, much greater than even nuclear burning. For the other, the transfer of material can alter the structure of both stars and their subsequent evolution. All these elements are of importance for the X-ray sources.

The binary model is illustrated in Figure 5-17, which shows a normal star and a compact companion. Clearly, in such a system there is a point on the line joining the two stars where the gravitational potentials are equal; a bit of matter at the point could just as well fall to one star as to the other. When

Figure 5-17. Schematic drawing of binary system with mass transfer. The solid line is the critical Roche surface. The star is shown filling the Roche surface and is considerably distorted from circularity. Materials flow to the compact star through the inner Lagrange point (L_1) and either falls inward, forming a disk around the second star, or leaves the system through L_2. The dashed lines show the orbits of each star about the center of mass of the system. Adapted from Gursky and Schreier (1975).

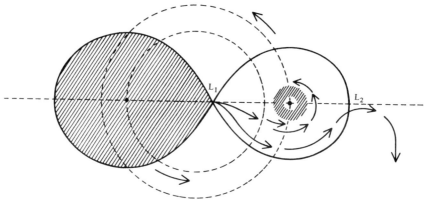

account is taken of the centripetal acceleration, a figure-eight equipotential surface, called the Roche surface, is defined, and matter finding itself along that surface can find its way to either star. The significance of this surface is that at some stage of its life one of the stars may begin to expand as part of its normal evolution off the main sequence and may fill its Roche surface. Any further expansion then stops, and material is transferred through the common point of the Roche surface (the inner Lagrange point) and into the gravitational influence of the companion star. In this case a gas "stream" may form; numerous examples of such mass transfer are known, particularly among the close binary pairs of white dwarfs.

There is direct evidence for the existence of this kind of gas stream in the X-ray sources. Note that the geometry (specifically the conservation of angular momentum) constrains the stream to a certain quadrant with respect to the compact star; in particular, the stream might be expected to show up against the compact star just before it is eclipsed by its companion. This, in fact, is observed in several of the eclipsing X-ray sources.

The binary model resolves the question of the energy source of the X-ray emission. A proton falling onto a white dwarf will release about 100 keV of kinetic energy; onto a neutron star the energy release will be 100 MeV (a black hole, as we discuss below, is more complex). The yearly mass-transfer rates required to yield an X-ray luminosity of 10^{36} and 10^{38} erg/sec observed from the X-ray sources ranges from 10^{-5}–10^{-10} $M_\odot$/year. It is known from other considerations that in fact such transfer rates can occur.

Can we decide at this point whether the compact star is a neutron star (black hole) or a white dwarf? There is at least one simple argument that indicates the neutron star to be the more likely candidate. We consider that the kinetic energy falling onto the compact star simply heats the surface and is re-radiated as black-body radiation. The surface temperature will then be

$$T = (L/4\pi R^2 \sigma)^{1/4}$$

where L is the luminosity (10^{36}–10^{38} erg/sec). Using $R = 10$ km for a neutron star yields temperatures in excess of 10^7 K, for which the peak of the emission would be keV X-rays; for a white dwarf with $R = 10^4$ km the temperatures would not reach 10^6 K, and the emitted power will be UV radiation. This argument does not actually require that the emission be black-body radiation from the stellar surface, only that the emitting region be comparable in size to the parent star and that the opacity be very high.

The actual chain of events leading to X-ray emission from neutron stars and black holes is likely to be exceedingly complex. Material falling toward the surface of a star will become compressed and heated. Shock waves may form. The final temperature of the region will be determined by the density near the star's surface and the radiating area. Since the infalling material most likely has excess angular momentum, a disk will form (cf. Pringle and Rees, 1972). A neutron star may have a strong magnetic field, which allows the material to flow to the surface only in the polar regions. In the case of a black hole, the compression and heating of the gas outside the Schwarzschild radius is the primary source of radiation. In addition, if the black hole is rotating, additional energy may be extracted at the expense of the rotational energy when the particles spiral in beyond the Schwarzschild radius.

There is no single object which provides the kind of direct evidence for the nature of the X-ray sources provided by NP 0532 and the Crab Nebula for the pulsars. However, for the two objects, Cyg X-1 and Her X-1, one can make reasonable direct arguments that we are seeing, respectively, a black hole and a neutron star.

Cygnus X-1 is not observed to eclipse in X-rays; rather, the binary nature is revealed through the study of the optical counterpart, which is an early supergiant (Webster and Murdin, 1972; Bolton, 1972). The spectral lines show a sinusoidal Doppler shift characteristic of a binary system with a period of 5.6 days. The half amplitude of the velocity equivalent to the Doppler shift is about 70 km/sec. For binary systems, the so-called mass function can be derived from Kepler's laws:

$$f(M) = \frac{M_u^3 \sin^3 i}{(M_s + M_u)^2} = \frac{4\pi^2 P}{G}.$$

Here M_u is the mass of the unseen object, in this case the X-ray source, M_s is the mass of the seen object, P is the period, V is the observed velocity, and i is the inclination angle (measured between the line of sight and the normal to the orbit plane). For the parameters listed above, $f(M) = 0.2\ M_\odot$. To go further we must assume that the visible star is what it seems to be, an early-type supergiant, which has a mass of order 20 $M_\odot$. In this case, we can derive a *minimum* mass for the X-ray source, assuming $i = 90°$, which is about 5 $M_\odot$. The best estimate of the mass, based on a realistic inclination angle (obviously $i < 90°$ since the X-ray source would otherwise be eclipsed), is actually between 10–15 $M_\odot$. The significance of such a large mass is that it is substantially larger than the maximum mass permitted for either a neutron star or a white dwarf. Thus, since the high time variability requires a compact star, we must have here a completely collapsed configuration; namely, a black hole.

Clearly, the evidence for Cyg X-1 as a black hole is not nearly so complete as what we presented for identifying Sirius B as a white dwarf or NP 0532 as a neutron star, since we rely on two assumptions: first, that the observed star is ~ 20 $M_\odot$ and second, that the X-ray emission occurs at the unseen star. We are reasonably certain of the first of these. There are certain stars of lower mass (and lower luminosity) that can mimic the supergiants; however, the distance has been measured to Cyg X-1 (Margon, Bowyer, and Stone, 1973; Bregman et al., 1973) to be about 2 kpc, which is consistent with the star's being a supergiant. Also, there are four other X-ray sources seen with a very similar companion star, which indicates the existence of a class of objects of this kind. Regarding the second assumption, a number of alternative hypotheses have been advanced (triple star systems, for example) but none has been found viable and we are left with the black hole idea as the most direct interpretation of the observations.

As noted, Her X-1 is a pulsing X-ray source with a pulse period of 1.2 sec. We can measure the orbital velocity of both the X-ray source and the optical object, which gives us a better idea of the mass of the X-ray source—about 1 $M_\odot$. Since the pulse period is shorter than can be accommodated by a white dwarf, we must accept the idea that this object, like the pulsars, is a rotating neutron star.

This object adds another piece of information in favor of the idea that mass accretion is occurring at the rate needed to supply the observed energy. The pulse period, as shown in Figure 5-18, both increases and decreases by a few parts per 10^7 per year (pulsar periods always show a long-term increase). An increase in period is naturally explained by matter falling onto the neutron star, carrying with it the angular momentum it possessed in its last orbit. The fact that there is both an increase and a decrease in period indicates that there is a delicate balance between the speedup due to accretion and slowdown due to dissipation. Centaurus X-3 shows similar behavior. Only in these two sources, because they are pulsing, can the periods be measured with the accuracy required to reveal this effect.

Have We Discovered A Black Hole?

In the natural sciences, discovery involves two independent elements; one is the observation of some natural phenomenon and the other is the theoretical understanding of that phenomenon. Although the two are generally treated independently in the history books, it is more correct to consider them together. Thus the white dwarfs were not "discovered" until R. H. Fowler elucidated their nature in 1925, ten years after the accumulation of facts that established their basic characteristics. Discovery is frequently complicated in astronomy by a lack of the complete and rigorous information that is typically present in other sciences. We cannot subject astronomical systems to the kind of detailed, probing experimentation possible in the laboratory; rather we must accept the few bits of data nature throws our way.

It is generally agreed that the paper by Oppenheimer and Snyder (1939) constitutes the discovery paper for the theory of black holes. Observationally, we have in the case of Cyg X-1 a compact star whose mass is significantly greater than can be accommodated as a white dwarf or a neutron star. Thus, it is likely to be a black hole. However, this statement is simply a reiteration of the theory of mass limits of compact stars. Thus, although the present data do not yet constitute the observational discovery of a black hole, they certainly are the most important step yet taken in the direction in

Figure 5-18. The pulse period of Hercules X-1 between December 1971 and March 1973. Adapted from Gursky and Schreier (1975).

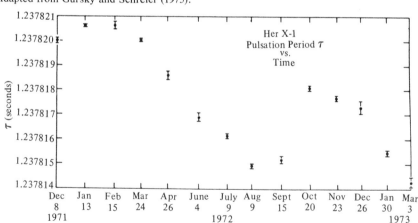

finding these elusive objects. They remove our fear that perhaps stars have always found a way to shed enough mass during their lives to come within the various mass limits. More important, these data point to where in the Galaxy we may find other candidates for black holes.

The essence of a black hole is the existence of the closed trapped surface at the Schwarzschild radius; the observation of such a surface would establish without any doubt the existence of a black hole. One might consider watching matter disappear as it approached the hole; however, the only candidate objects are extremely luminous as a result of the release of gravitational energy. Thus the prospects for the success of this simple method of observation are not good. However, there are numerous other effects such as gravitational focusing, redshifts, and the emission of gravitational waves which may be unique to black holes.

Origin of the Binary X-ray Sources

Solid information exists about the origin of binary X-ray sources, the most important being the existence of five sources—Cyg X-1, 3U 0900 − 40, 3U 1700 − 37, Cen X-3, SMC X-1 (see Table 5-4)—in which the optical companion is an O- or B-type supergiant. This kind of star is massive and young (massive stars burn their nuclear fuel very rapidly and do not survive for more than $10^6 \times 10^7$ years). Hence, these stars are very rare; there are no more than several thousand in the entire Galaxy.

The occurrence of these five systems is extraordinary. The rarity of both X-ray sources and supergiants means that the coincidence cannot be due to chance. Consider the numbers in the following manner. We see four such combinations in the Galaxy (the fifth is the small Magellanic Cloud) out to a distance of about 2 kpc, which means there are perhaps 50–100 of this class in the entire Galaxy. Thus, several percent of all the O-B supergiants are in binary X-ray systems. These are *close* binary systems and it is not likely that more than a small fraction of all supergiants are in close binary systems. Thus, we come up with the possibility that *virtually all O-B supergiants which are formed with close binary companions are X-ray sources.* Compare this with the earlier observation that on the average only one star in 10^9 is an X-ray source.

What this means, simply stated, is that once such a system is formed, it will eventually evolve a configuration having a compact star that becomes a strong X-ray emitter. Van den Heuvel and Heise (1972; see also Gursky and

Figure 5-19. Evolutionary scheme for X-ray sources containing a bright supergiant such as Centaurus X-3 or Cygnus X-1. The dashed figures are the Roche lobes and the shaded areas are the regions occupied by the star. The illustrated stellar system is "born" at $t = 0$ with a binary period P of 3 days. The two stars begin with masses of 16 and 3 $M_\odot$ respectively, as shown. At stage B, the 16 $M_\odot$ star has swelled to fill its Roche lobe and begins to transfer mass to its companion. This first mass transfer is completed at stage C, leaving behind a 4 $M_\odot$ helium star, which explodes at stage D. The supernova remnant may be a neutron star or black hole which becomes an X-ray source when the second stage of mass transfer begins at stage E. Adapted from Gursky and Van den Heuvel (1975).

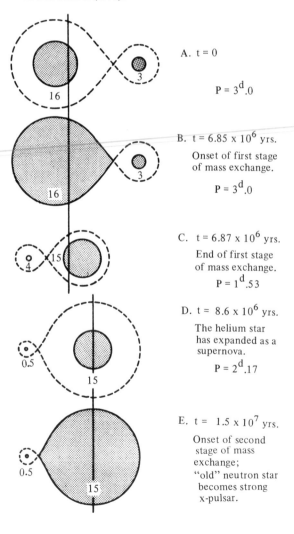

A. $t = 0$

$P = 3^d.0$

B. $t = 6.85 \times 10^6$ yrs.

Onset of first stage of mass exchange.

$P = 3^d.0$

C. $t = 6.87 \times 10^6$ yrs.

End of first stage of mass exchange.

$P = 1^d.53$

D. $t = 8.6 \times 10^6$ yrs.

The helium star has expanded as a supernova.

$P = 2^d.17$

E. $t = 1.5 \times 10^7$ yrs.

Onset of second stage of mass exchange; "old" neutron star becomes strong x-pulsar.

Van den Heuvel, 1975) were the first to propose a scheme, illustrated in Figure 5-19, to account for this fact. Beginning with a system comprising a massive star and a less massive companion, the massive star evolves (burns its nuclear material) more rapidly. At some point, it enters a giant phase and begins to expand. As noted, however, the expansion in a binary system is limited by the Roche surface, and material, instead, is transferred through the inner Lagrange point, to the companion. Eventually, all that is left of the original star is a hot massive core composed of helium and heavier elements. These are the remains of the nuclear burning. This core will eventually undergo a supernova explosion, leaving behind a neutron star or a black hole. Remembering that these are very young systems, we can probably relate the supernovae of type II to these events. The details of the process are not entirely clear. The mass of the core probably depends on the mass and exact chemical composition of the original star and the remains of the supernova explosion will depend on the nature of that core. Nevertheless, what will be present after the explosion is a binary system composed of a compact star (but not a white dwarf) and a star that now suddenly finds itself very massive. This massive star now begins to evolve very rapidly and, if its mass is in the range 20 $M_\odot$ or more, will be seen to be an O-B type supergiant. It is just this combination that we want in order to make a binary X-ray system like Cyg X-1.

How does the matter actually come to be accreted on the compact star? The process is not necessarily Roche-lobe overflow; rather, it may simply be capture by the compact star of some small fraction of the stellar wind which normally flows out from the surface of supergiants (Davidson and Ostriker, 1973). The stellar wind flow amounts to $\sim 10^{-5}$ $M_\odot$/year; thus capture efficiencies of 10^{-3} and less are sufficient to power the most luminous X-ray sources.

This scheme, then, can account in a natural way for the class of X-ray sources accompanied by an O-B supergiant. However, there is at least a second kind of X-ray source present that does not have a massive companion and cannot be accounted for by this scheme. As shown in Table 5-4, for the sources Her X-1, Cyg X-2, Cyg X-3, and Sco X-1, the stellar companion cannot exceed a few solar masses. In the case of Her X-1, we can measure the mass and radius of this star more or less directly through a measurement of the binary elements. In the case of Cyg X-3 and Sco X-1, the binary period is too short to accommodate a massive star, and in the case of Cyg X-2, we see the companion star and it is not an O-B supergiant. Furthermore, the numbers in Table 5-4 do not give a proper indication of the relative number of binary systems with massive companions that are easy to find by virtue of the brightness of the star and the high probability of there being an X-ray eclipse. The distribution of X-ray sources in Figure 5-12, as we discussed, indicates that the average distance of these objects from the plane of the Galaxy is in the range 500–1000 pc. This large a distance is characteristic of "old-disk" stars, like the sun, which are formed after the plane of the Galaxy was formed but are still as old as 10^{10} years. The young massive stars typically lie at distances less than 100 pc from the plane. Thus it is likely that binary X-ray stars with massive companions are a small fraction of all the X-ray sources.

What is the origin, then, of this second class of binary X-ray stars? Because of the existence of Her X-1, we know that the model comprising accretion onto a neutron star or black hole companion must still be considered a favored hypothesis. It is possible that these systems form by a variety of schemes. Sometime during the life of a binary system, one of the stars explodes as a supernova, leaving behind the neutron star. At some later time the companion star begins to evolve off the main sequence, fills its Roche lobe, and dumps matter onto its compact companion, thereby starting the production of X-rays.

The difficulty with this kind of "random" process is that theoreticians can only make massive stars (> 4–$5 M_\odot$) blow up as supernovae. The implications of this is simply that if we start with a binary system comprising stars of $1 M_\odot$ and $5 M_\odot$, the $5 M_\odot$ star, when it explodes, will probably disrupt the binary system. (There is a general theory that if more than half the mass of a binary system is lost during a supernova explosion, the system will become unbound.)

The key to understanding these low-mass binary X-ray stars may be the existence of the type I supernova, which in spite of the theoretical difficulties appears to occur in old, low-mass stars. One way out of the theoretical difficulties, as first proposed by Schatzman (1963), is to start with a white dwarf and then slowly add matter to it until it exceeds its mass limit and collapses. As we have been discussing, the binary systems provide a natural means for adding matter to stars. Binary systems in which matter is being transferred to a white dwarf are well known. They form a class of highly variable stars (including the novae, SS Cygni stars, U Geminorum stars, and others) known collectively as cataclysmic variables.

It is possible that the continued evolution of such a white-dwarf binary system results in a type I supernova, leaving behind a neutron star, which then later will become an X-ray source (Gursky, 1976). This scheme would account for many of the observed characteristics of both X-ray sources and type I supernovae.

Summary
The binary X-ray stars must be considered a prominent constituent of the Galaxy, despite their small numbers. It is likely that the X-ray emission arises from a mass-accreting neutron star or black hole. Being in binary systems, these highly compact stars can yield certain kinds of information, such as mass, which is not available from single stars. For example, we may be able to determine empirically the mass limit for neutron stars, which is an important parameter in determining the equation of state of bulk neutron matter. Also, by studying details of the X-ray emission, we may be able to determine something about the transition region in a black hole, just before processes become totally invisible.

Also, because of the binary nature of these systems, we can get information on their origin, and we appear to have established one well-defined scheme for making X-ray sources.

Conclusions

These three fields—supernovae, pulsars, and binary X-ray stars—are as distinct observationally as one can imagine. Supernovae are studied with conventional optical techniques and ground-based telescopes; most pulsar research comprises low-frequency radio observations; and in order to find the binary X-ray stars, instruments must be carried into space on rockets and satellites.

The three fields are now intimately tied together through the phenomena they involve, in ways that no one could have imagined until only recently. We appear to have established the existence of neutron stars with some certainty, and the existence of black holes with a high degree of probability, and the fact that both originate in supernova explosions. Thus we can, at least as a working hypothesis, complete the story of the evolution of stars. It had been known that many stars end their lives as white dwarfs. We know that certain stars end as neutron stars, visible as pulsars and X-ray sources, and as black holes, visible as X-ray sources. There is also a strong presumption that there are many more neutron stars and black holes in existence, but as yet undiscovered. Also, we have some information about the structure of neutron stars.

This field of astronomy is still very new and further progress requires the best of our theoretical and technical abilities. However, the potential for new knowledge is impressive. Neutron stars and black holes lie in the domain of general relativity and quantum mechanics, testing these theories in a unique way. In the purely astronomical domain we are seeing the completion of a chapter in our understanding of stars, a story that began a hundred years ago when the physical basis for stellar investigations was established.

I have mentioned a number of the primary and secondary references throughout this chapter. I call attention here to recently published books which contain more thorough discussions of this material: *Black Holes, Gravitational Waves and Cosmology* (Rees, Ruffini, and Wheeler, 1974); *Neutron Stars, Black Holes and Binary X-ray Stars* (Gursky and Ruffini, 1975); and *Astrophysics and Gravitation,* (Pines, 1975).

References

Allen, C. W. 1973. *Astrophysical Quantities*, 3rd edition, Athlone Press, London.

Baade, W. 1942. The Crab Nebula, *Astrophys. J. 96*, 188–198.

Baade, W., and Zwicky, F. 1934. Supernovae and cosmic rays, *Phys. Rev. 45*, 138.

Blauuw, A., 1965. Concept of stellar population, in A. Blauuw and M. Schmidt, eds., *Stars and Stellar Systems*, vol. 5, *Galactic Structure*, pp. 435–453, University of Chicago Press, Chicago.

Bok, B. J., and Bok, P. F. 1974. *The Milky Way*, 4th edition, Harvard University Press, Cambridge, Mass.

Bolton, C. T. 1972. Identification of Cygnus X-1 with HDE 226868, *Nature 235*, 271–273.

Bowyer, C. S., Byram, E. T., Chubb, T. A., and Friedman, H. 1965. Cosmic X-ray sources, *Science 147*, 394–398.

Boynton, P. E., Groth, E. J., Partridge, R. B., and Wilkinson, D. T. 1969. Precision measurement of the frequency decay of the Crab Nebula Pulsar, NP 0532, *Astrophys. J. (Letters) 157*, L197–201.

Boynton, P. E., Groth, E. J., Hutchinson, D. P., Nanos, G. P., Partridge, R. B., and Wilkinson, D. T. 1972. Optical timing of the Crab pulsar, NP 0532, *Astrophys. J. 175*, 217–241.

Bregman, J., Butler, D., Kemper, E., Kaski, A., Kraft, R. P., and Stone, R. P. S. 1973. On the distance to Cygnus X-1 (HDE 226868), *Astrophys. J. (Letters) 185*, L117–120.

Burbidge, E. M., Burbidge, G., Fowler, W. A., and Hoyle, F. 1957. Synthesis of the elements in stars, *Rev. Mod. Phys. 29*, 547–650.

Burbidge, G. 1972. Binary stars as X-ray sources, *Comments Astrophys. Space Phys. 4*, 105–111.

Cameron, A. G. W. 1958. Element evolution of the stars: Introductory report, *Mem. Soc. Roy. Sci. Liege, 5th Ser. 3*, 163–171.

Chandrasekhar, S. 1931. The highly collapsed configurations of a stellar mass, *Monthly Not. Roy. Astron. Soc. 91*, 456–466.

Chandrasekhar, S. 1932. Some remarks on the state of matter in the interior of stars, *Z. Astrophys. 5*, 321–327.

Chandrasekhar, S. 1935. The highly collapsed configurations of a stellar mass (II), *Monthly Not. Roy. Astron. Soc. 95*, 207–225.

Chandrasekhar, S. 1969. The Richtmyer Memorial Lecture: Some historical notes, *Amer. J. Phys. 37*, 577–584.

Chandrasekhar, S. 1972. The increasing role of general relativity in astronomy, *Observatory 92*, 160–174.

Comella, J. W., Croft, H. D., Lovelace, R. V. E., Sutton, J. M., and Tyler, G. L. 1969. Crab Nebula Pulsar NP 0532, *Nature 221*, 453–544.

Davidson, K., and Ostriker, J. 1973. Neutron star accretion in a stellar wind: Model for a pulsed X-ray source, *Astrophys. J. 179*, 585–598.

Fowler, R. H. 1926. On dense matter, *Monthly Not. Roy. Astron. Soc. 87*, 114–122.

Gamov, G. 1937. *Structure of Atomic Nuclei and Nuclear Transformations,* The Clarendon Press, Oxford.

Giacconi, R., Gursky, H., Paolini, F. R., and Rossi, B. R. 1962. Evidence for X-rays from sources outside the solar system, *Phys. Rev. Letters 9*, 439–443.

Giacconi, R., Gorenstein, P., Gursky, H., Usher, P. D., Waters, J. R., Sandage, A., Osmer, P., and Peach, J. 1967. On the optical search for the X-ray sources Cyg X-1 and Cyg X-2, *Astrophys. J. (Letters) 148*, L119–127.

Giacconi, R., Murray, S., Gursky, H., Kellogg, E., Schreier, E., Matilsky, T., Koch, D., and Tananbaum, H. 1974. The third Uhuru catalog of X-ray sources, *Astrophys. J. Suppl. 27*, 37–64.

Gold, T. 1968. Rotating neutron stars as the origin of the pulsating radio sources, *Nature 218*, 731–732.

Gold, T. 1969. Rotating neutron stars and the nature of pulsars, *Nature 221*, 25–27.

Grindlay, J. 1972. Detection of pulsed gamma-rays of $\sim 10^{12}$ eV from the pulsar in the Crab Nebula, *Astrophys. J. (Letters) 174*, L9–17.

Groth, E. 1975. Observational properties of pulsars, in H. Gursky and R. Ruffini, eds., *Neutron Stars, Black Holes and Binary X-Ray Sources,* pp. 119–173, Dordrecht and Boston.

Gursky, H. 1976. Binary X-ray stars and supernovae of Type I, in S. Mitton and J. Whelan, eds., *Close Binary Systems,* IAU Symposium No. 73, to be published.

Gursky, H., and Ruffini, R., eds. 1975. *Neutron Stars, Black Holes and Binary X-ray Stars,* Reidel, Dordrecht and Boston.

Gursky, H., and Schreier, E. 1975. Galactic X-ray sources, in H. Gursky and R. Ruffini, eds., *Neutron Stars, Black Holes and Binary X-ray Sources,* pp. 175–220, Reidel, Dordrecht and Boston.

Gursky, H., and Van den Heuvel, E. 1975. X-ray emitting double stars, *Sci. Amer. 232*, 24–35.

Gursky, H., Paolini, F. R., Giacconi, R., and Rossi, B. R. 1963. Further evidence for the existence of galactic X-rays, *Phys. Rev. Letters 11*, 530–535.

Guseinov, O., and Zel'dovich, Ya. B. 1966. Collapsed stars as components of binaries, *Sov. Astron. 10*, 251–253.

Hawking, S., and Ellis, G. F. R. 1973. *The Large-Scale Structure of Space Time,* p. 303, Cambridge University Press.

Hewish, A. 1970. Pulsars, *Annu. Rev. Astron. Astrophys. 8*, 265–296.

Hewish, A., Bell, S. J., Pilkington, J. P. H., Scott, P. F., and Collins, R. A. 1967. Observation of a rapidly pulsating radio source, *Nature 217*, 709–713.

Hulse, R. A., and Taylor, J. H. 1975. Discovery of a pulsar in a binary system, *Astrophys. J. (Letters) 195*, L51–53.

Kirshner, R. P., Oke, J. B., Penston, M. V., and Searle, L. 1973. The spectra of supernovae, *Astrophys. J. 185*, 303–322.

Kristian, J.   1970. On the optical identification of the Vela Pulsar: Photoelectric measurements, *Astrophys. J. (Letters) 162,* L103–104.

Landau, L. D.   1932. On the theory of stars, *Phys. Zeit. 1,* 285–308.

Landau, L. D., and Lifshitz, E. M.   1958. *Quantum Mechanics,* Addison-Wesley, Reading, Mass.

Margon, B., Bowyer, S., and Stone, P. S.   1973. On the distance to Cygnus X-1, *Astrophys. J. (Letters) 185,* L113–116.

Mayall, N. U., and Oort, J. H.   1942. Further data bearing on the identification of the Crab Nebula with the supernova of 1054 A.D., *Publ. Astron. Soc. Pacific 54,* 95–104.

Minkowski, R.   1941. Spectra of supernovae, *Publ. Astron. Soc. Pacific 53,* 224–225.

Minkowski, R.   1942. The Crab Nebula, *Astrophys. J. 96,* 199–213.

Minkowski, R.   1964. Supernovae and supernova remnants. *Annu. Rev. Astron. Astrophys. 2,* 247–266.

Moore, E. P.   1974. Predetonation lifetime and mass of supernovae from density-wave theory of galaxies, *Publ. Astron. Soc. Pacific 85,* 564–567.

Oda, M., Gorenstein, P., Gursky, H., Kellogg, E., Schreier, E., Tananbaum, H., and Giacconi, R.   1971. X-ray pulsations from Cygnus X-1 observed from Uhuru, *Astrophys. J. (Letters) 166,* L1–7.

Oke, J. B., and Shipman, H. L.   1971. Effective temperature of white dwarfs, in W. J. Luyten, ed., *White Dwarfs,* IAU Symp. No. 42, pp. 67–76, Reidel, Dordrecht-Holland.

Oppenheimer, R. J., and Snyder, R.   1939. On continued gravitational contraction, *Phys. Rev. 56,* 455–459.

Oppenheimer, R. J., and Volkoff, G. M.   1938. On massive neutron cores, *Phys. Rev. 55,* 374–381.

Pacini, F.   1968. Rotating neutron stars, pulsars and supernova remnants, *Nature 219,* 145–146.

Penrose, R.   1969. Gravitational collapse: The role of general relativity, *Riv. Nuovo Cim. 1,* No. 1.

Pines, D.   1975. Observing neutron stars: Information on stellar structure from pulsars and X-rays, in *Astrophysics and Gravitation,* Proceedings of the 16th Solvay Conference, pp. 147–173, Editions de l'Universite de Bruxelles, Brussels.

Pringle, J., and Rees, M.   1972. Accretion disc models for compact X-ray sources, *Astron. Astrophys. 21,* 1–9.

Rees, M., Ruffini, R., and Wheeler, J. A.   1974. *Black Holes, Gravitational Waves and Cosmology,* Gordon and Breach, New York.

Richards, D. W., and Comella, J. W.   1969. The period of Pulsar NP 0532, *Nature 222,* 551–552.

Ruffini, R.   1975. The physics of gravitationally collapsed stars, in H. Gursky and R. Ruffini, eds., *Neutron Stars, Black Holes and Binary X-ray Sources,* pp. 59–118, Reidel, Dordrecht and Boston.

Russell, H. N.   1914. Relations between the spectra and other characteristics of the stars, *Pop. Astron. 22,* 275–294.

Salpeter, E.   1964. Accretion of interstellar matter by massive objects, *Astrophys. J. 140,* 796–799.

Sandage, A. R., Osmer, P., Giacconi, R., Gorenstein, P., Gursky, H., Waters, J. R., Bradt, H., Garmire, G., Sreekantan, B. V., Oda, M., Osawa, K., and Jukago, J.   1966. On the optical identification of Sco X-1, *Astrophys. J. 146,* 316–322.

Schatzman, E.   1963. White dwarfs and type I supernovae, in L. Gratton, ed., *Star Evolution,* Proc. Internat. School Phys. "Enrico Fermi," pp. 389–393, Academic Press, New York.

Schreier, E., Levinson, R., Gursky, H., Kellogg, E., Tananbaum, H., and Giacconi, R.   1972. Evidence for the binary nature of Centaurus X-3 from Uhuru X-ray observations, *Astrophys. J. (Letters) 172,* L79–89.

Shklovskii, I. S.   1968. The nature of the X-ray source Sco X-1, *Sov. Astron. 11,* 749–753.

Staelin, D. H., and Reifenstein, E. C.   1968. Pulsating radio sources near the Crab Nebula, *Science 162,* 1481–1483.

Tamman, G. A.   1974. Statistics of supernovae, in C. B. Cosmovici, ed., *Supernovae and Supernova Remnants,* pp. 155–185, Reidel, Dordrecht and Boston.

Van den Heuvel, E., and Heise, J.   1972. Centaurus X-3, possible reactivation of an old neutron star by mass exchange in a close binary, *Nature Phys. Sci. 239,* 67–69.

Wade, N.   1975. Discovery of pulsars: A graduate student's story, *Science 189,* 358–364.

Webster, B., and Murdin, P.   1972. Cygnus X-1: A spectroscopic binary with a heavy companion, *Nature 235,* 37–38.

Zeldovich, Ya. B.   1964. The fate of a star and the evolution of gravitational energy upon accretion, *Sov. Phys. Dokl. 9,* 246–249.

Zwicky, F.   1974. Review of research on supernovae, in C. B. Cosmovici, ed., *Supernovae and Supernova Remnants,* pp. 1–16, Reidel, Dordrecht and Boston.

6

Infrared Astronomy

G. G. Fazio

The infrared region of the electromagnetic spectrum covers the extensive wavelength range from the red region of the optical spectrum (1 μm or 10^{-4} cm) to the very shortest wavelengths in the microwave radio region (1000 μm or 1 mm). We shall divide the spectrum into two regions, the "near" infrared which extends from 1 to 25 μm and the "far" infrared region from 25 to 1000 μm. This division is based on the use of different observational techniques in the two regions.

We can trace the origin of infrared astronomy back to the early nineteenth century when William Herschel first used a simple thermometer and prism to observe the infrared spectrum of the sun. Even until recent times, the only infrared objects observed were the sun, the moon, and the bright planets. The primary reason for the late development of this field was the lack of sensitive detectors. A secondary reason was that predicted infrared fluxes, based on extrapolation of radio observations at the shortest wavelengths, seemed below detectable limits. In the late 1950s and early 1960s the advent of cooled photoconductive and bolometric detectors, accompanied by the development of adequate filters and cryogenic technology, greatly increased the ability of astronomers to detect new sources. Atmospheric absorption has limited most observations to portions of the near-infrared region that are observable from the ground. However, the recent growth in airborne, high-altitude balloon, and satellite technology has now permitted the entire infrared spectrum to be open for exploration.

Within the infrared spectral region lies a wealth of data which, until now, have barely been explored. For example:

1. Astronomical objects with temperatures less than 4000 K radiate most

of their energy in the infrared. They include the moon, the planets, and several kinds of very cool galactic objects, particularly protostars.

2. The infrared region of the spectrum offers a unique opportunity to see far into highly obscured dust regions, such as the center of our Galaxy, H II regions, and dense dark clouds.

3. Several extragalactic objects have been observed emitting most if not all of their energy at far-infrared wavelengths.

4. Many objects, such as unusual types of galaxies and quasi-stellar sources, should emit strong, nonthermal radiation in the infrared.

5. The vibrational and rotational line spectra of many astrophysically important molecules, such as H_2, fall in this wavelength region.

Infrared observations are therefore essential in solving several very important astronomical problems:

1. The processes by which stars form from the collapse of dense clouds of dust and gas and their subsequent evolution.

2. The source of energy for the extraordinarily high luminosity of the nuclei of some galaxies.

3. The origin of the high luminosity of the nucleus of our Galaxy.

4. The over-all distribution of gas and dust in our Galaxy and its relation to the stellar content and galactic structure.

5. The abundances, distribution, and chemical kinetics of the constituents of the interstellar medium.

Although still a relatively new field of research, infrared astronomy has rapidly become one of the most important areas of observational astronomy. This chapter briefly reviews the observational methods used and then discusses recent infrared observations and their interpretation.

Observational Methods in Infrared Astronomy

Radiation Laws

Black-body radiation. The intensity of radiation in erg sec^{-1} cm^{-2} ster^{-1} Hz^{-1} as a function of frequency within an enclosure in thermal equilibrium at temperature T is given by the Planck function

$$B_\nu(T) = \frac{2h\nu^3/c^2}{e^{h\nu/kT} - 1},$$

where h is Planck's constant, k is Boltzmann's constant, c is the velocity of light, and ν is frequency in Hz ($= \sec^{-1}$). The corresponding intensity in erg $\sec^{-1}$ cm^{-2} ster^{-1} cm^{-1} as a function of wavelength λ, in cm, is

$$B_\lambda(T) = \frac{2hc^2/\lambda^5}{e^{hc/\lambda kT} - 1}.$$

In Figure 6-1, $B_\lambda(T)$ is plotted against λ for various temperatures. A solid, black surface also radiates according to these equations. A star or gas cloud radiates as a black body when T characterizes both the gas kinetic temperature and the microscopic properties of the radiating atoms and molecules.

The maximum of $B_\nu(T)$ occurs at the wavelength $\lambda(\text{cm}) = 0.51/T(\text{K})$ while the maximum $B_\lambda(T)$ occurs at $\lambda(\text{cm}) = 0.29/T(\text{K})$. Thus at room temperature (~ 300 K), B_ν peaks at a frequency corresponding to 17 μm and B_λ near 10 μm.

For the two extreme ranges of $h\nu/kT$ the following approximations can be made to the Planck function. When $h\nu/kT \gg 1$ we have the Wien approximation

$$B_\nu(T) = \frac{2h\nu^3}{c^2} e^{-h\nu/kT},$$

or

$$B_\lambda(T) = \frac{2hc^2}{\lambda^5} e^{-hc/\lambda kT}.$$

The Rayleigh-Jeans approximation for $h\nu/kT \ll 1$ is

$$B_\nu(T) = 2 \left(\frac{\nu}{c}\right)^2 kT$$

or

$$B_\lambda(T) = 2ckT/\lambda^4.$$

Integration of B_ν over all ν gives

$$B = (2\pi^4 k^4/15c^2 h^3)T^4 = (\sigma/\pi)T^4,$$

where σ is the Stefan-Boltzmann constant. Integration of B_λ over all λ gives the same result. This equation shows, for example, that a spherical black body having the surface area 1 m^2 and temperature 300 K emits in all directions 4.6×10^9 erg/sec or 460 watts of power.

Often a source of radiation having a temperature T does not emit radiation according to $B_\nu(T)$. For example, the surface layers of the source may freely transmit or scatter radiation originating in deeper layers having a higher temperature. The emissivity $\epsilon(\nu,T)$ of a source is often defined as the ratio $B'_\nu/B_\nu(T)$, where B'_ν is the emitted intensity. In the same way, $\epsilon(T) = B'/(\sigma/\pi)T^4$ is the average emissivity.

Figure 6-1. The brightness $B_\lambda(T)$ of a black-body radiation as a function of λ and T (Neugebauer and Becklin, 1973).

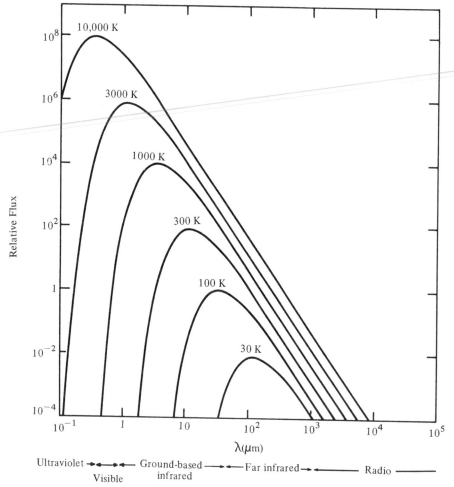

The integral of the intensity of an observed source over the solid angle subtended by the source is the flux

$$F_\nu = \int_{\text{source}} \epsilon(\nu,T)B_\nu(T)\, d\Omega.$$

A common flux unit is the Jansky: $1\ \text{Jy} = 10^{-26}\ \text{watt m}^{-2}\ \text{Hz}^{-1} = 10^{-29}\ \text{erg}$ $\text{sec}^{-1}\ \text{cm}^{-2}\ \text{Hz}^{-1}$.

Emission and absorption of radiation. For a beam of radiation passing through an absorbing medium, the change in intensity dI_ν is proportional to the intensity I_ν and the distance dr, i.e.,

$$dI_\nu = -a_\nu I_\nu\, dr,$$

where the constant of proportionality a_ν is the absorption coefficient per unit length. If we define $d\tau_\nu = a_\nu\, dr$, this equation can be written

$$dI_\nu + I_\nu\, d\tau_\nu = 0,$$

which is equivalent to

$$d(I_\nu e^{\tau_\nu}) = 0,$$

so that

$$I_\nu e^{\tau_\nu} = \text{constant.}$$

Then

$$I_\nu = I_{\nu 0}e^{-\tau},$$

where $I_{\nu 0} = I_\nu(\tau_\nu = 0)$. In the case of a cloud of thickness L, τ_ν is the optical depth

$$\tau_\nu = \int_0^L a_\nu\, dr,$$

$I_{\nu 0}$ is the incident intensity, and I_ν is the emergent intensity.

If the absorbing medium also emits radiation, then

$$dI_\nu = -a_\nu I_\nu\, dr + j_\nu\, dr,$$

where j_ν is the energy emitted per unit frequency, volume, and solid angle. We let $S_\nu = j_\nu/a_\nu$ (S_ν is called the source function) so that

$$dI_\nu + I_\nu\, d\tau_\nu = S_\nu\, d\tau_\nu.$$

Following the same procedure as before, we obtain the solution

$$I_\nu = I_{\nu 0}e^{-\tau_\nu} + S_\nu(1 - e^{-\tau_\nu}).$$

The first term is the incident intensity attenuated by the cloud, and the second term is the intensity generated within the cloud. There are two limiting cases: an optically thick cloud ($\tau_\nu \gg 1$) for which $I_\nu = S_\nu$, and an optically thin cloud ($\tau_\nu \ll 1$) for which $I_\nu = I_{\nu 0} + (S_\nu - I_{\nu 0})\tau_\nu$.

Source temperatures. Several different kinds of source temperatures are defined, depending on the type of observational data. The brightness temperature T_b is defined as that temperature of a black body which would give the same energy output per unit frequency interval as the source at a given frequency. For example, we can use the Rayleigh-Jeans law, and write

$$T_b(\nu) = \frac{c^2 I_\nu}{2\nu^2 k},$$

where I_ν is the intensity emitted by the source. The brightness temperature is a convenient parameter used to denote the intensity of radiation at a given wavelength. It has the advantage of involving only one unit (that of temperature) instead of five (energy, time, area, band width, and solid angle).

The color temperature T_c of a source is a parameter defined by the ratio of the observed intensities I'_{λ_1} and I'_{λ_2} at two neighboring wavelengths λ_1 and λ_2:

$$\frac{I_{\lambda_1}}{I_{\lambda_2}} = \left(\frac{\lambda_2}{\lambda_1}\right)^5 \cdot \frac{e^{hc/\lambda_2 k T_c} - 1}{e^{hc/\lambda_1 k T_c} - 1}.$$

Thus T_c is the temperature of the Planck function having the same wavelength dependence as the observed distribution in the interval (λ_1, λ_2).

The effective temperature T_{eff} of a source is the temperature of a spherical black body having the same radius as the source and the same total energy output L, i.e.

$$L = 4\pi R^2 \sigma T_{\text{eff}}^4.$$

Atmospheric Effects

Because of atmospheric absorption, ground-based infrared observations are limited almost entirely to several "windows" in the near-infrared region of the spectrum. The atmospheric attenuation of the signal is due primarily to molecular absorption bands in water vapor (H_2O) and carbon dioxide (CO_2), with additional minor contributions due to a number of other constituents, such as O_3, CO, N_2O, and CH_4. Fortunately, the two major constituents of the atmosphere, O_2 and N_2, are homonuclear molecules, possessing neither a permanent nor an induced dipole moment, and hence exhibit no molecular absorption bands. The windows available for observation are designated as J(1.24 μm), H(1.6 μm), K(2.2 μm), L(3.6 μm), M(5.0 μm), N(10.6 μm) and Q(21 μm). These bands will be discussed further in the section on photometry. In the region above 25 μm, except for partial windows at 34 μm, and 350 μm, all observations must be made from aircraft, high-altitude balloons, rockets, or spacecraft. Figure 6-2 shows the atmospheric transmission at several altitudes corresponding to a mountain top (4.2 km), aircraft altitudes (14 km) and two balloon altitudes (28 km and 41 km). The concentration of atmospheric absorbers varies significantly from site to site and in time. This is particularly true of water vapor. Extinction caused by water vapor at a site is usually quoted in terms of precipitable millimeters of water in the path (pr. mm H_2O), which is the thickness of the H_2O column that would result if all the water were condensed at one end of the optical path.

Absorption is not the only atmospheric effect influencing infrared observations. Deflection and scattering by particles in the air path can also occur which blur the image ("seeing") and limit the minimum usable field of view. In addition, gases and particles themselves can be sources of radiation. The earth's atmosphere itself radiates, but not as a black body; it has an emissivity, $\epsilon(\lambda, T)$, as shown in Figure 6-3. This radiation determines the ultimate lower limit to the background noise for an observation. An additional source of noise is rapid fluctuations in atmospheric density and temperature, e.g., turbulence; this source is termed "sky noise" and is also highly variable.

Detectors

There are two broad classes of detectors: coherent and incoherent. Coherent detectors are, in general, those that preserve frequency and phase information during the receiving process. Incoherent detectors are those in which a current or voltage is generated in direct proportion to the amount of received power, i.e., they have a square-law response. Here we shall discuss only the incoherent detectors, which can be further divided into two main groups: quantum detectors and thermal detectors (listed in Table 6-1).

Thermal detectors. In a thermal detector, radiation is absorbed on a nearly black surface and converted efficiently into heat. The ideal thermal detector has uniform sensitivity at all wavelengths. The most commonly used de-

Figure 6-2. Transmission of the atmosphere at infrared wavelengths at four altitudes (Traub and Stier, 1976).

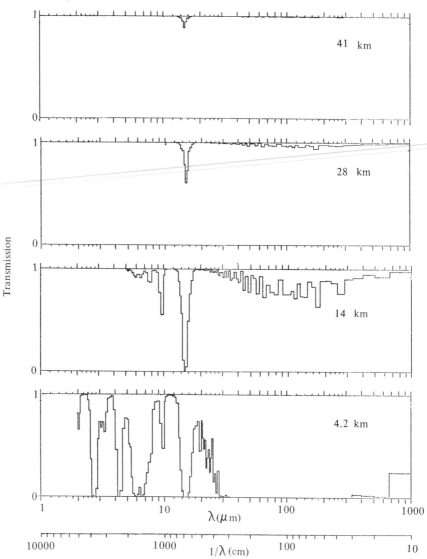

Figure 6-3. Emissivity of the atmosphere at infrared wavelengths at four altitudes (Traub and Stier, 1976).

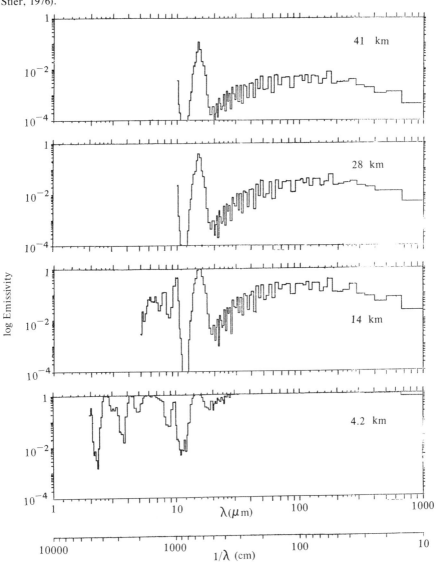

Table 6-1   Types of incoherent detectors.

Quantum Detectors
Photon detectors
Photoconductive
Photovoltaic
Josephson effect
Thermal Detectors
Room temperature
Pneumatic (Golay)
Thermistor bolometer
Thermocouple
Pyroelectric
Cyrogenic
Bolometers (carbon, germanium, silicon)
InSb free-carrier-absorption bolometer
Superconducting bolometer
Diode detectors (room temperature)
Semiconductor
Metal

From Arams (1973).

tector is the cryogenic gallium-doped germanium bolometer, developed by F. J. Low. In this detector the temperature change of a crystal of germanium with added gallium impurity is measured by a change in resistance:

$$R(T) = R_0(T_0/T)^A$$

where $A \simeq 4$. Considerable advantages can be gained by operating the crystal at very low temperatures (~ 2 K). The main advantage comes about from reductions in the fluctuations of the temperature of the crystal and the voltage across it. These fluctuations generate voltage variations in the output signal (called "noise") and reduce the ultimate sensitivity of the detector. The crystal is operated in contact with a pumped liquid helium bath; reducing the pressure on liquid helium lowers its temperature. To measure the change in resistance a bias current is applied to the crystal and a large load resistor in series, and the voltage change is measured across the crystal.

Quantum detectors. Another type of detector that has played an important role in infrared astronomy, particularly at wavelengths less than 10 μm, is the photon detector. In the photoconductive device (e.g., lead sulfide, PbS), photons are absorbed in the crystal and excite holes and electrons to the conduction band, changing the electrical conductivity of the crystal. In the photovoltaic device (e.g., indium antimonide, InSb) a large-area p-n junction exists. When the junction is illuminated with radiation, holes and electrons are generated and subsequently separated by the junction. A photovoltaic potential is then produced which is capable of sustaining a current.

In photodetectors the wavelength response is not uniform at all wavelengths, but a threshold energy exists, determined by the semiconductor energy gap between bound and free carriers. Therefore, only photons with energy greater than $h\nu_{gap}$ are detected. Figure 6-4 shows the relative wavelength response of photodetectors and bolometers.

The PbS photoconductive device, cooled to liquid nitrogen temperatures (77 K), is used in the region 1 to 3 μm and the InSb photovoltaic device, also cooled by liquid nitrogen, is used in the region 1 to 5 μm. The latter system is the primary detector now used in near-infrared astronomical research because it has faster time response and greater sensitivity than the PbS device.

For longer wavelengths a number of extrinsic photoconductive devices exist. These are crystals to which an impurity has been added in controlled amounts. The most common are Ge crystals activated by appropriate shallow-level impurities, which result in a much smaller threshold energy, and hence long wavelength response. Thus with Ge as the host lattice, impurities such as Au yield a response to 10 μm; Cu to 30 μm; Ga to 110 μm; and B to 130μm. Doped silicon detectors are also being used more fre-

Figure 6-4.   Relative wavelength response of an ideal photodetector and bolometer.

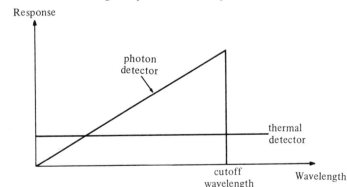

quently, particularly in integrated infrared sensor arrays. These extrinsic photoconductive devices are usually operated at 4 K.

Narrow-energy-gap intrinsic photoconductors in the region 8 to 14 μm are also rapidly being developed. In these detectors no impurities are added. These include the ternary alloys $Hg_{1-x}Cd_xTe$, and $Pb_{1-x}Sn_xTe$. The band gaps can be adjusted over a considerable range by varying the molar fraction x.

Detector parameters. To categorize a detector, the primary parameters usually quoted are the spectral response, the time constant, and the noise equivalent power or detectivity. In actual practice other parameters are also desirable, such as noise bandwidth, magnitude of the signal obtained, temperature characteristics, and also their interrelation.

The spectral response has been discussed previously. It is an inherent property of the material from which the detector is constructed, and can be varied significantly only by variation in temperature of the sensing element.

The time constant is a measure of the speed of response of the detector. In thermal detectors this response depends on the heat capacity C of the detector, and the thermal conductance G between the detector and its heat sink. In photon detectors, the response is determined by the rate at which charge carriers become immobilized. In general, photon detectors have a response time many orders of magnitude faster than thermal detectors.

The minimum intensity of radiant power incident on the detector that will give rise to a signal voltage equal to the noise voltage is called the "noise equivalent power" (*NEP*). The value of *NEP* is normalized to a 1-Hz electrical bandwidth, and since the *NEP* is proportional to the square root of the bandwidth, it is expressed in units of watt/Hz$^{1/2}$. It is a measure of the smallest signal that can be detected and for astronomical purposes it is the most relevant parameter for comparison among detectors.

In most cases the *NEP* is also proportional to the square root of the detector area. Therefore a figure of merit independent of detector area and electrical bandwidth is the detectivity (D^*):

$$D^*(T_0, f_c, \Delta f) = A^{1/2}/NEP \qquad (cm\ Hz^{1/2}/watt).$$

Here D^* is quoted for a given value of the black-body temperature (T_0) of the incident radiation, the sinusoidal frequency (f_c) of the incident radiation, and the post-detection electrical bandwidth, Δf, The larger the value of D^*, the more sensitive the detector is. Caution should be exercised in the use of D^* to make sure that the above proportionalities remain true.

Detector noise. As the input signal level on a detector is decreased, ultimately a point is reached where the random fluctuations in the output signal (called "noise") dominate, so that the output signal cannot give an accurate measure of the input signal. The amount of this fluctuation is a function of fundamental physical quantities and, for a given detector element, represents a level of sensitivity that cannot be exceeded. Many detectors now used in infrared astronomy operate near this ultimate limit.

The total noise generated at the point of an infrared detector which is coupled to a telescope is a combination of the "photon noise," which is due to the statistical fluctuations in the rate of arrival of photons in the background radiation, "sky noise," and "detector noise." In the section on infrared telescopes we will discuss photon noise in greater detail. Detector noise, which is noise associated with the sensing element itself, can be of several kinds. The most important types are:

1. Johnson noise—the thermal noise of a resistor in thermal equilibrium at a given temperature; even when no current is flowing through a resistor, a small fluctuating voltage exists across its output terminals because of random motion to the electrical charge in the resistive element.

2. Photon noise—noise caused by statistical fluctuations in the emission of radiation by the detector element.

3. Phonon or temperature noise—noise due to statistical interchange of thermal energy between detector and its surroundings.

4. Current or shot noise—random fluctuations due to discreteness of electrical charges in the flow of DC current.

5. Generation-recombination noise—noise in a semiconductor detector generated by fluctuations in the number of thermally generated carriers.

6. Modulation noise—noise in a semiconductor due to resistance fluctuations.

7. Contact noise—noise due to resistance fluctuations at a contact point.

8. Flicker noise—low-frequency noise that is in excess of shot noise.

The magnitude of each type of noise listed above can be computed from elementary statistical principles. In each case we are dealing with the statistical deviation ΔN of a large random number N of discrete quantities (photons, electrons, or phonons) from their mean value.

As an example, in a Ge:Ga bolometer cooled to 2 K, Johnson noise and phonon noise dominate, and the *NEP* of the detector is given by the expression

$$NEP \simeq 4T_0(kG)^{1/2},$$

where T_0 is the bath temperature, G is the thermal conductance between the bath and the bolometer element, i.e., the conductance of the leads, and k is Boltzmann's constant. Note that the NEP is independent of the dimension of the detector element, and therefore the value of D^* cannot be used.

In general detector noise can be reduced by cooling the detector in its enclosure; increasing the detector responsivity, i.e., volts output per watt of incident power; reducing the field of view with a cooled aperture stop; reducing the spectral bandwidth with a cooled filter; and reducing the detector's thermal conductance to the heat sink (this applies specifically to bolometers).

Telescopes

To make useful astronomical observations the detector must be coupled to a telescope. At wavelengths less than 5 μm, conventional ground-based optical telescopes can be used, subject to the limited wavelength bands where atmospheric transmission occurs. However, above 5 μm one must also contend with the thermal emission from both the sky and the telescope. Both have a temperature in the range of 260 to 300 K and hence the peak emission is at 10 μm, in a primary atmospheric window. At 10 μm the emissivity of the sky is about 0.1, whereas for a conventional telescope it is 0.5. Therefore, thermal emission from the telescope is the primary background and not thermal emission from the sky; in the 21 μm band the sky emissivity is higher and closer to the emissivity of the conventional telescope. Therefore observing at 10 μm with a ground-based telescope is equivalent to optical observations made with a telescope and dome well illuminated by flickering light. The background radiation on the detector is of the order of 10^{-7} watt, while the source signal is of the order of 10^{-14} watt.

The noise level observed at the output of the detector caused by this background radiation is the "photon noise" caused by the statistical fluctuations in the thermal radiation from the sky and telescope. Since the fundamental problem in infrared astronomical observations is detection of a weak source signal against this noisy background, reduction of photon noise is the dominant factor in the design of a telescope.

Fortunately, significant reductions in the background radiation can be made by proper telescope design. The main source of this radition is high emissivity (i.e., black) surfaces in the beam path, such as the sky baffles, and the central primary hole. Low and Rieke (1974) have shown that an infrared telescope with the following characteristics would be a high-quality instrument: f/45 Cassegrain design with no sky baffle, thin secondary supports, small primary hole, undersized secondary mirror to prevent viewing the edges of the primary, and low-emissivity mirror surfaces.

To determine what minimum source intensity can be detected with such a telescope we must compute the noise equivalent flux density (*NEFD*), which is the source flux at the top of the atmosphere which would result in a signal-to-noise ratio of one in a 1-Hz bandwidth. It is given by:

$$\frac{\text{signal intensity}}{\text{noise intensity}} = 1 = \frac{(\text{NEFD}) \cdot T_0 \cdot A_0}{(NEP)}$$

$$NEFD = \frac{NEP}{A_0 \cdot T_0 \cdot \Delta f} \frac{\text{watt}}{\text{cm}^2 \, \mu\text{m}},$$

where NEP = noise equivalent power (watt $\text{Hz}^{-1/2}$), A_0 = effective collecting area (cm^2), T_0 = optical efficiency, and Δf = spectral wavelength bandwidth (μm). In this case the value of NEP is determined by the photon noise due to the sky and telescope background radiation.

The background radiation incident on the detector is also proportional to the solid angle-area factor, $A\Omega$, of the telescope. Thus the smallest possible field of view must be used to reduce the noise and obtain maximum sensitivity. However, there is a limit to the field of view that can be used, which is ultimately determined by the diffraction limit of the telescope:

$$\alpha(\text{arcsec}) = 50.3 \left(\frac{\lambda}{D}\right),$$

where D is the telescope diameter (cm) and λ is the wavelength (μm). Here the value of α is measured by the diameter of the first null in the diffraction pattern. However, other factors affecting the image, such as atmospheric seeing, wind, stiffness of the telescope and the telescope drives, may limit the field of view to an even larger angle. Decreasing the field of view below this angle decreases the signal-to-noise ratio. For some observations it is necessary to reduce the sky background radiation even more. This can be done by placing the telescope at higher altitudes.

The emissivity of the atmosphere decreases rapidly with altitude; e.g., for every 10 km above a mountain top altitude (4.2 km) the emissivity decreases by a factor of 10, and in space the background radiation is even much lower with an effective temperature of only a few degrees Kelvin. Thus photon noise can be reduced significantly by observations from an aircraft, high-altitude balloon, or spacecraft.

A significant additional advantage to the telescope in a spacecraft is that the mirror can be cooled to the temperature of liquid helium, dramatically reducing the background emission. This is difficult to do in the atmosphere because of problems with water condensation.

We have thus far discussed the sensitivity of infrared telescopes considering the only noise to be photon noise due to thermal emission by the atmosphere and the telescope. However, under certain observing conditions it is possible that the limiting sensitivity could be determined by sky noise and/or detector noise. For example at 20 μm, when a large field of view (~ 1 arcmin) is used to view an extended source, the limit of sensitivity is determined by sky noise.

Because of the extremely high ratio of background signal to source signal it is necessary, during an observation of a weak infrared source, to accurately subtract the background signal from the sum of the source-plus-background signal. One technique would involve moving the entire telescope alternately from sky plus source to sky and taking the difference. The problem with this technique is that low-frequency noise will not be eliminated and changes in telescope and sky background noise during the sample

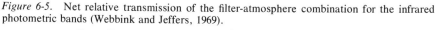

Figure 6-5. Net relative transmission of the filter-atmosphere combination for the infrared photometric bands (Webbink and Jeffers, 1969).

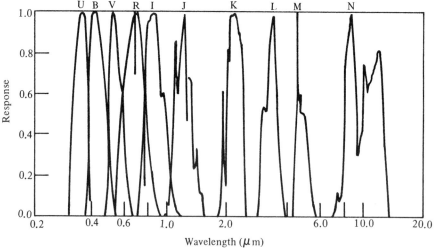

period will not be cancelled. An alternative scheme, which is most often used today on a Cassegrain telescope and which is equivalent to beam-switching by moving the telescope, is wobbling the secondary mirror in a square wave pattern at a frequency of 10 to 20 Hz, displacing the beam by a distance of a few beam widths. Such a mirror is often referred to as a "chopping" secondary mirror. This modulation technique also permits AC amplication techniques to be used for the detector signal.

Infrared Photometry

In order to compare infrared observations, particularly the spectral content of the observed object, a system of photometric bands has been defined in the infrared region similar to the U, B, V system in optical astronomy. The system is now defined out to a wavelength of 20 μm and the bands are designated R, I, J, K, L, M, N, and Q. The wavelength intervals in these bands are defined to take advantage of the atmospheric transmission windows. Figure 6-5 shows the net relative transmission of the combination of spectral filter plus atmosphere.

The effective wavelength, λ_0, for these bands is defined as

$$\lambda_0 = \frac{\int \lambda \phi(\lambda) d\lambda}{\int \phi(\lambda) d\lambda},$$

where $\phi(\lambda)$ is the product of filter response function and atmospheric transmission. To a first-order approximation the observed flux density is equivalent to a monochromatic flux density at the wavelength λ_0. Table 6-2 gives the characteristics of the infrared photometric bands.

Table 6-2 Characteristics of the infrared photometric bands.

Photometric band	λ_1 Cut-on wavelength (μm)	λ_2 Cut-off wavelength (μm)	λ_0 Effective wavelength (μm)
H	1.45	1.8	1.63
K	1.9	2.5	2.22
L	—[a]	—[a]	3.5
L'	3.05	4.1	3.6
M	4.5	5.5	5.0
N	7.9	13.2	10.6
Q	17	28	21

From Low and Rieke (1974).
[a] Defined by Johnson (1965).

Absolute measurements in infrared astronomy are difficult because of the high background levels and poorly defined instrument efficiency. As a result, photometric measurements are done with respect to a system of standard stars. The ratio of fluxes at two different wavelengths can also be compared for two different sources. These ratios are called "color indices" or "colors" of the object and are expressed as the difference of magnitudes in two photometric bands. Thus the $K - L$ color index is the difference in magnitudes measured in the K and L bands. It is equivalent to the ratio of the K and L fluxes times a factor which is related to the absolute calibration magnitude.

Johnson (1965) originally set the zero point of the magnitude scale such that for an average A0 V star, all magnitudes are the same, i.e., all color indices are zero. Low and Rieke (1974) give the flux levels for a 0.0-mag star in each of the infrared bands. They are based on a 10,000 K black-body curve fitted to a zero point at 3.6 μm. These values differ only slightly from those of Johnson. They are given in Table 6-3; it also gives the limiting magnitudes for a 61-inch telescope for an integration time of 1 hour and a field of view 6 arcsec in diameter. It is a best estimate of the present state of the art with a pointed telescope.

For wavelengths greater than 20 μm the fluxes from any of the standard stars are too small to be useful for calibration purposes. Instead, the bright planets, Venus, Mars, Jupiter, and Saturn, are used as standard black-body sources against which to compare flux measurements.

Table 6-3 Flux levels for 0.0 mag in the infrared bands and limiting magnitudes.

Photometric band	Effective wavelength (μm)	Flux for 0.0 mag (watt cm^{-2} μm^{-1})[a]	Typical limiting magnitude (61-inch telescope)[b]
K	2.22	4.14×10^{-14}	10.0
L	3.6	6.38×10^{-15}	9.0
M	5.0	1.82×10^{-15}	7.5
N	10.6	9.7×10^{-17}	6.0
Q	21	6.5×10^{-18}	3.0

[a] From Low and Rieke (1974).
[b] For an integration time of 1 hour and a 6-arcsec field of view.

Infrared Observations and Interpretation

Stars with Infrared Excess

The surface temperature of the sun is about 6000 K, and hence its continuous spectrum peaks at about 0.5 μm. As a star becomes cooler, its peak radiation approaches infrared wavelengths and for objects with color temperatures less than 4000 K, more than 50% of their energy is emitted beyond 1 μm. At 2 μm the brightest star in the sky is Betelgeuse, which is only the 12th-brightest star at optical wavelengths. Betelgeuse is a luminous supergiant star with a surface temperature of about 3000 K; it is a star that, on completion of its hydrogen burning, has cooled and expanded. Most of the brightest objects at 2 μm are these cool, ordinary stars (e.g., supergiants, giants, or long-period variables) in which the radiation mechanisms are relatively well understood.

Of particular interest are those stars that show an excess of infrared radiation. This excess appears in a given wavelength region and is above the radiation intensity expected from a black body with a color temperature determined from the star's spectral class. Many such objects have been detected, exhibiting a wide variety of types and a wide range of values of excess radiation. The present evidence is that excess intensity originates from a dust shell surrounding a central star. The dust cloud absorbs ultraviolet and optical radiation from the central star, is heated, and then re-emits the absorbed energy as infrared radiation. The origin of the dust cloud is still open to considerable speculation. It could be that the central star has reached a highly evolved state, i.e., is a late-type star, and is losing mass. The dust shell is the condensate of the mass ejected by the star. An alternative interpretation is that the central star is an early-type star, i.e., a protostar, which is evolving toward the main sequence by accretion of mass from a surrounding dense dust cloud, leaving a remnant shell.

In some cases, spectral classification of the central star can determine whether it is of the late or early type, or whether it is ejecting mass. But in many cases, owing to obscuration by the dust, the properties of the central star are uncertain or impossible to determine. Thus far, infrared observations conclude only that a dust cloud exists.

Let us now consider some examples of both types of central stars.

The spectra of two of the brightest late-type stars with infrared excess are given in Figure 6-6. The stars are primarily late-type M supergiants, M- and

S-type Mira variables, and carbon stars. Their energy distribution can be interpreted in terms of black-body radiation of the photosphere (~ 2000 K) plus excess radiation at 8–14 μm, and in some cases at 20 μm, which is characteristic of silicate materials (e.g., Mg_2SiO_4 and Al_2SiO_5). Since spectroscopic evidence indicates that most stars of this class are losing mass, and that silicates can condense in the region surrounding oxygen-rich stars, the most likely interpretation of the excess infrared radiation is that it is emitted by an optically thin circumstellar shell of silicate grains. It should also be noted that in all extreme cases of infrared excess, radio astronomers have also detected hydroxyl (OH) emission from these stars.

As a particular, although extreme, example of a late-type star, let us consider in detail IRC +10216. The designation indicates that it was observed in the California Institute of Technology infrared survey (Neugebauer and Leighton, 1969) as the 216th source observed in the declination band centered on +10°. IRC +10216 is believed to be a carbon star ($T \sim 2000$ K) surrounded by an extensive dust envelope that scatters some of the starlight,

Figure 6-6. Infrared spectral distribution of the energy of two late-type stars with infrared excess. The spectral distribution of the energy from the sun is shown for comparison; the flux scale is not to be compared (Neugebauer and Becklin, 1973).

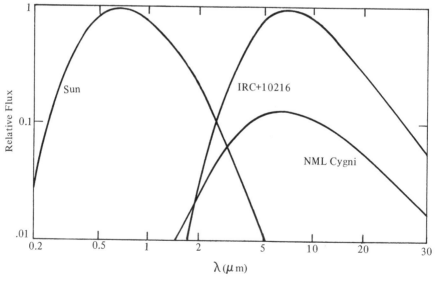

but absorbs and thermally re-emits most of it. The infrared spectrum indicates no evidence for the silicate peaks, and thus the dust grains in the shell are most likely composed of graphite. In optical photographs it appears as an extended object fainter than 18th magnitude. The radiation emitted by this object peaks at about 5 μm and is similar to that of a 600 K black body, with the exception that at wavelengths greater than 10 μm the intensity exceeds the black body values. At 2 μm the source is variable with a period of about 600 days. From lunar occultation experiments at infrared wavelengths, it has been determined that the dust shell can be represented by two components, a bright, optically thick, central component with a diameter of about 0.4 arcsec and at 600 K, and a less luminous, optically thin component with a diameter of 2 arcsec and a temperature of 375 K. Assuming the distance to IRC + 10216 is about 1000 light years, the radius of the central shell is about 60 AU, or about the diameter of Neptune's orbit. Radio observations at millimeter wavelengths indicate that a relatively large molecular cloud also surrounds the star. CO has been detected from an extended halo $\sim$ 2.3 arcmin in diameter, or 345 times larger than the central dust shell. In addition, CS, CN, HCN, C_2H, SiS, SiO, and HC_3N have also been observed. The HCN appears to come from a much smaller region than the CO. The molecular line observations also indicate that the shell is expanding at about 12 km/sec. It appears that pumping of the rotational levels of the molecules by near infrared radiation may dominate the excitation of the molecules.

Infrared excess in intermediate and early-type stars differs from that in late-type stars in that the excess radiation appears as an exception rather than the rule. Early-type stars such as emission-line B stars, high-luminosity stars such as F, G, and K supergiants, T Tauri stars, and novae have exhibited evidence for this excess infrared radiation.

T Tauri objects are the most likely candidates to have observable remnants of a protostellar dust cloud. R Monocerotis, an object related to T Tauri stars, is an extreme case. Yet, even in this case the situation is not clear. The spectral types of central stars cannot be well determined; thus it is difficult to test whether the optical and ultraviolet energy absorbed by the dust is sufficient to supply the observed infrared flux. In studying the evolution of pre-main-sequence stars, it has been shown that for several stars of T Tauri type, it is possible to explain the observed energy distribution with the protostar dust-cloud model. But other data such as optical spectra indicate that there is an outward mass flow from the central star on a scale sufficient to provide a dust shell.

Another outstanding candidate for protostar formation in a dust cloud is the BN (Becklin-Neugebauer) source (or Becklin's star) in the Orion Nebula. It is an unresolved source with a color temperature of about 600 K that lies in a cluster of infrared sources near the center of a large molecular cloud and close to several sources of OH and H_2O emission. The energy distribution of the star exhibits strong absorption lines at 3.1 μm due to H_2O ice and at 10 μm because of silicates. Another interpretation of this source is that it is a highly luminous F supergiant star obscured by cold dust, producing some 80 mag of visual extinction. A third possibility is that the source is an evolved star with a hot dust shell. But because the BN star is located in the Orion Nebula, a region rich in gas, dust, and young stars, the protostar idea seems most attractive at this time.

For the emission-line B stars with infrared excess at 3.5 μm, the excess radiation appears to be consistent with free-free emission from ionized hydrogen, and probably originates in a circumstellar cloud of ionized hydrogen.

Galactic novae (e.g., Nova Serpentis 1970, and Nova Aquilae 1970) are another example of the ejection of matter by a stellar outburst and the later condensation into a circumstellar shell of dust. Significant infrared excess from the shell does not become apparent, however, until some 50 days after the outburst, when the ejected matter, moving outward at 1000 km/sec, has had time to condense. The infrared radiation reaches a peak intensity at ~ 90 days, with the dust at ~ 900 K, then cools rapidly and decays with a half-life of ~ 100 days. At 100 days the temperature approaches 600 K. At the peak intensity, the infrared luminosity is approximately 90% of the total stellar luminosity.

The infrared emission observed from novae is similar in many respects to the infrared excess observed from stars with a constant rate of mass loss. Grain formation in the two cases may be very similar.

Another outstanding example of a nova-type object is η Carinae. It is a variable star that flared dramatically, in 1843, becoming one of the brightest visual stars in the sky. It faded slowly, and today it appears as a faint nebulosity almost invisible to the naked eye. Yet at 20 μm it is one of the brightest sources outside the solar system. Again the evidence indicates that the infrared excess comes from heated dust ejected at the original outburst.

Galactic Sources

H II regions. Gaseous nebulae, which are extended, optically bright objects, can be divided into two main classes: diffuse nebulae or H II regions, and planetary nebulae. The observed emission originates from ultraviolet radiation from one or more hot stars ($T > 3 \times 10^4$ K) in or near the nebula. Hydrogen is the predominant element in the nebula, with helium the next most abundant. Ultraviolet photons with wavelengths <912 Å ionize the hydrogen gas and are absorbed, with the excess energy going into the electron kinetic energy. Helium is ionized at wavelengths <504 Å. Electron-electron and electron-photon collisions redistribute the energy such that the electron temperature is in the range 5000 K to 20,000 K. In addition, neutral atoms and molecules are formed in excited states that radiate in the optical and radio region, and emission lines are also observed from electron recombination by ions.

In particular, H II regions are nebulae that are excited by an O- or early B-type star or cluster of stars. These are hot, luminous stars that have recently been formed from matter in the nebula. The effective temperature of the stars is about 3 to 5×10^4 K and throughout the nebula, hydrogen is completely ionized and helium singly ionized. Particle densities in the region are of the order of 10 to 10^4 cm^{-3}. Internal motions are of the order of 10 km/sec. The hot ionized gas tends to expand into the cooler neutral gas surrounding it.

In the late 1960s it was observed that many H II regions radiate most of their energy at wavelengths of the order of 100 μm, and the infrared radiation was in excess of the electron free-free emission observed over the wavelength interval from 5 μm to 100 μm (Figure 6-7). The current interpretation of the infrared excess is that it is energy reradiated from cool (~ 100 K) dust in and around the ionized gas. It is known that large amounts of dust are associated with H II regions from comparison of radio and optical maps, from stellar reddening, from the presence of molecules, and the observed deep 3.1-μm absorption feature of H_2O ice and the broad 10-μm absorption feature due to silicate grains.

Infrared sources associated with H II regions can be grouped in one of three classes: (1) those associated with high-emission-measure "compact" H II regions, (2) those imbedded within massive molecular clouds which are associated with H II regions, and (3) those associated with OH/H_2O maser sources.

Let us now investigate some of the theoretical details of a model H II
region and determine what can be learned from infrared observations.

An O-type star with a surface temperature of 5×10^4 K emits $\sim 2/3$ of its
energy at wavelengths less than 921 Å (Lyman continuum radiation). Each of
these photons that is absorbed in the hydrogen gas around the star produces
a Lyman α (Ly α) photon (1216 Å), a Balmer photon, plus several lower-

Figure 6-7. Spectrum of the excess infrared emission from an H II region. The dotted line is a
black-body curve at 70 K and the solid line the theoretical curve for emission from electron
transitions in a plasma (Wynn-Williams and Becklin, 1974).

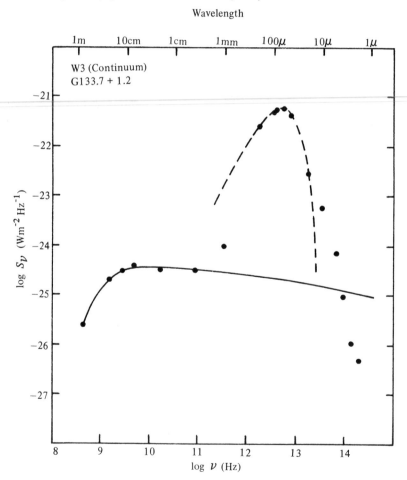

energy photons. About one-third of the energy from the O star is converted into Ly α photons, which remain trapped in the ionized region and are eventually absorbed by dust grains in the ionized volume. The amount of dust in the H II region can vary considerably. If the optical depth of the dust is small (the depleted case), the ultraviolet stellar photons can escape the ionized region; if the optical depth is large (the undepleted case), the dust absorbs the Ly α photons as well as the direct ultraviolet stellar photons. Infrared observations at 10 and 20 μm show brightness distributions very similar to the maps of free-free electron emission at radio wavelengths, which demonstrate the coexistence of dust ($T \sim 100\text{--}200$ K) with this ionized gas.

Of the total infrared emission, about two-thirds is observed at wavelengths greater than 25 μm. This emission cannot be explained by "hot" dust within the H II region. An outer shell of cold (50--70 K) gas and dust must also exist which will absorb the remaining stellar photons as well as the photons which emerge from the bounded ionized region. Because of the dust in and around the H II region, all of the energy from the star is converted into infrared radiation without hindering the ionization process.

To understand the details of this model and to derive some of the more important properties of the region, let us consider the observational data. One set consists of the infrared flux density and the intensity.

The infrared flux density of a source is given by

$$F_{IR}(\nu) = \int_{\text{source}} I_{IR}(\nu) \, d\Omega,$$

where $I_{IR}(\nu)$ is the infrared intensity. The total luminosity is

$$L_{IR} = 4\pi D^2 \int_0^\infty F_{IR}(\nu) \, d\nu \leq L_*,$$

where D is the distance to the source and L_* is the total stellar luminosity. L_{IR} is usually a lower limit to the value of L_*; the equality is true if all of the photons emitted by the star are absorbed by the dust, either in the H II region or the surrounding shell.

From the intensity measurements at two different wavelengths, we can obtain the color temperature T_d of the dust. If we take this temperature to be the actual temperature of the dust, the source function (cf. section on Emission and Absorption of Radiation) is given by the Planck function $B_\nu(T_d)$.

Therefore

$$I_{IR}(\nu) = (1 - e^{-\tau_\nu})B_\nu(T_d),$$

and for the optically thin cloud, $\tau_\nu \ll 1$,

$$\tau_\nu = \frac{I_{IR}(\nu)}{B_\nu(T_d)},$$

from which we can determine the optical depth of the dust at infrared wavelengths.

The second set of data consists of the radio flux density and intensity. The radio flux density, $F_R(\nu)$, at frequencies where the optical depth in the H II region is small, yields the total number of stellar Lyman continuum (Ly$_c$) photons, N_c', which are absorbed by the gas, i.e.,

$$D^2 F_R(\nu) \propto N_c' \le N_c,$$

where N_c' is the lower limit to the total number of stellar Lyman continuum photons emitted, N_c. The equality is true if the H II region is ionization-bounded and no Lyman continuum photons are absorbed by the dust grains.

The intensity of radio emission in an optically thin H II region does not depend strongly on frequency, but is proportional to its optical emission, i.e.,

$$\frac{I_R(\nu)}{I_0(H\beta)} \approx \text{constant.}$$

This relation is very useful in determining the total absorption by dust particles, which weakens Hβ emissions, but leaves the radio emission completely unaffected.

At lower frequencies the radio emission from free-free transitions is no longer constant but decreases. From the frequency where the break in the curve occurs, we can determine the emission measure, $\int n_e^2\, ds$, in the H II region. n_e is the electron density (i.e., the number of electrons per unit volume).

The correlation between the infrared luminosity and the radio continuum luminosity is then given by (Petrosian, Silk, and Field, 1972)

$$L_{IR} = L_\alpha + (1 - f)\langle h\nu \rangle_{Ly_c} \cdot N_c + L_{\nu < Ly_c}(1 - e^{-\tau_0'}).$$

where $\langle h\nu \rangle_{Ly_c}$ is the average energy of the stellar Lyman continuum photons, $L_{\nu < Ly_c}$ is the stellar luminosity below the Lyman continuum, and τ_0' is the effective absorption optical depth of the dust in the H II region and in the surrounding shell for photons >912 Å; N_c is the number of photons <912 Å emitted per sec by the star, and f is the fraction of photons <912 Å absorbed by the gas in the ionized region. The value of f can be approximated by $e^{-\tau}$ where τ is the optical depth of the dust for Ly_c photons in the ionized region.

The Ly α luminosity of the nebula is given by

$$L_{Ly\alpha} = (\tfrac{2}{3})h\nu_{Ly\alpha} \cdot N_c' = (\tfrac{2}{3})h\nu_{Ly\alpha} \cdot N_c \cdot f,$$

assuming the nebula is optically thick to Ly α photons. The factor $\tfrac{2}{3}$ appears because only two-thirds of the atomic recombinations result in Ly α quanta. The value of Ly α can be inferred either from the Hβ observations or from the radio continuum flux.

The measured value of L_{IR} is a good estimate of the total luminosity, L_*, of the exciting star. From L_* and the continuum radio flux, a value of f can be derived, which in turn is related to the optical depth of the dust for Ly_c photons in the ionized region.

In summary:

$$\left.\begin{matrix} L_{IR} \longrightarrow L_* \longrightarrow N_c \\ L_R \longrightarrow \\ H\beta \longrightarrow \end{matrix}\right\} Ly\,\alpha \longrightarrow N_c' \left.\begin{matrix} N_c' \\ \overline{N_c} \end{matrix}\right\} \longrightarrow f \longrightarrow \tau.$$

The optical depth of the dust can then be compared with optical depth at Hβ. If the stellar spectrum is known, the albedo of the dust can be obtained, yielding some insight into the physical properties of the dust.

Figure 6-8 is a plot of total infrared luminosity versus the flux of Ly_c photons. The infrared flux was based on results of the 40-cm balloon-borne telescope of the University College, London. The theoretical boundary curve for zero-age main-sequence stars (ZAMS) is taken from Panagia (1973). Since the infrared luminosity is a lower limit to the stellar luminosity and the Ly_c flux can only be less than the flux from the star without any dust absorption, the data points on the graph are to the left and below their position on the ZAMS curve. The horizontal displacement of a point is, in fact, the fraction f, or τ. For luminosities greater than those of an O4 star, a cluster of OB stars is assumed.

To produce a more detailed model of the H II region, several input parameters concerning the dust have to be assumed:

1. The absorption coefficient for H- and He-ionizing photons and for photons with $\lambda > 912$ Å.

2. The dust-to-gas mass ratio M_d/M_g.

3. The absorption coefficient in the infrared, Q_{IR}.

4. Perhaps the most important input parameter, the gas density as a function of position inside and outside the ionized region.

The absorption coefficient for dust at a wavelength λ is given by the expression;

$$Q_\lambda = \frac{\sigma_\lambda{}^{ab}}{\pi a^2} = \frac{A_n}{\lambda^n},$$

Figure 6-8. Total luminosity plotted against the flux of Lyman continuum photons as deduced by radio observations. The data points represent the total infrared luminosity of HII regions as measured by the University College, London, balloon-borne telescope (adapted from Jennings, 1975).

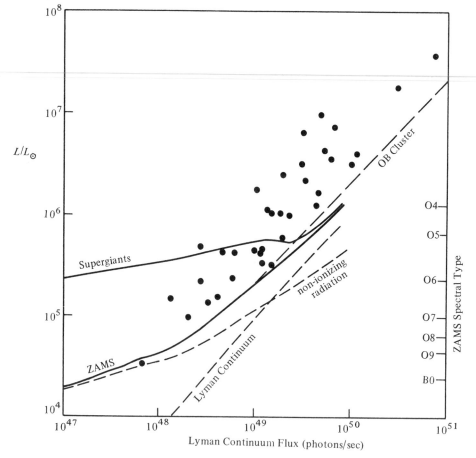

where a is the grain radius, $\sigma_\lambda{}^{ab}$ is the absorption cross section at wavelength λ, A_n is a constant, and n is a constant with values of 0, 1, or 2. The optical depth τ_λ is related to Q_λ by

$$\tau_\lambda = \int_R n_d Q_\lambda \pi a^2 dr.$$

For an optically thin isothermal dust cloud the relationship between the total luminosity and the dust temperature T_d can then be written as

$$L_{IR} \propto \int_0^\infty \tau_\lambda B_\lambda(T_d) dv \propto \int_0^\infty \frac{1}{\lambda^n} B_\lambda(T_d) dv,$$

where $B_\lambda(T_d)$ is the Planck function. Therefore

$$L_{IR} \propto T^{4+n}$$

and we see that the amount of thermal energy emitted by the grains varies very rapidly with the grain temperature, with the actual exponent depending on the function of Q_λ assumed. The wavelength dependence of Q_λ also affects the spectrum of the observed infrared radiation since

$$I_\lambda d\lambda \propto \lambda^{-n} B_\lambda(T_d) d\lambda$$

and

$$I_v dv \propto v^n B_v(T_d) dv.$$

In the particular spectral region where the Rayleigh-Jeans approximation is true, usually far-infrared and submillimeter wavelengths, we have

$$I_\lambda \propto \lambda^{-(4+n)},$$
$$I_v \propto v^{2+n}.$$

Thus we see that the spectral region longward of the peak in the Planck curve can yield information about the dust absorbtivity.

Because the value of τ_{IR} can be determined from observations, Q_{IR} can be determined if the thickness of the radiating layer is known. In a simple model this thickness is determined by $\tau_{UV} \approx 1$, i.e., most of the ultraviolet radiation that heats the dust is absorbed at one optical depth. Since $Q_{UV} \approx 1$, the thickness of the heated layer can then be determined, and hence Q_{IR}. Typically, Q_{IR} has values ranging from 10^{-2} to 10^{-3}. These values of Q_{IR} can then be compared with theoretical values derived assuming various dust materials, e.g., silicates, graphite, or silicate or graphite with ice mantles.

Another important quantity is the gas-to-dust mass ratio in the ionized region of nebula. The mass of dust in a spherical nebula is given by the equation

$$M_d = (\tfrac{4}{3})\pi R^3 n_d (\tfrac{4}{3})\pi a^3 \rho,$$

where ρ is the grain density. Since

$$\tau_{UV} = n_d \pi a^2 Q_{UV} R,$$

then

$$M_d = \frac{16}{9} \pi R^2 \rho a \frac{\tau_{UV}}{Q_{UV}}.$$

The total mass of ionized gas is given by the expression

$$M_g = (\tfrac{4}{3})\pi R^3 n_e m_H (X + 4Y)$$

where n_e is the electron density derived from the radio measurements of the emission measure, m_H is the mass of hydrogen atom, X and Y are fractional abundances by number of hydrogen and helium. Thus,

$$\frac{M_d}{M_g} = \frac{4\rho a \tau_{UV}}{3 Q_{UV} n_e R m_H (X + 4Y)}.$$

For $\rho = 2.3$ gm/cm^3, $a = 0.05$ μm,

$$Q_{UV} = 1, X = 0.9 \text{ and } Y = 0.1,$$
$$M_d/M_g \approx 10^{-2}.$$

We must remember that the above discussion assumed uniformly mixed dust and ionized gas filling a spherical volume. This is not always the case. In the dust-depleted model of Wright (1973) a low dust density exists in the ionized region, which is now surrounded with a dense neutral shell of gas and dust. In this model, little of the Ly$_c$ radiation is absorbed by the interior dust.

A good example of an H II region is the optical nebula IC 1795. It is a powerful galactic radio source (W3) that is approximately 3 kpc from the sun in the Perseus spiral arm. It is heavily obscured by dust and hence our knowledge of it is based primarily on radio and infrared observations. The radio map of the continuum radiation is shown in Figure 6-9. The total spectrum is given in Figure 6-10. High-resolution maps of the central region show that it consists of a number of sources; some are present at both radio and infrared wavelengths, others are not (Figure 6-11).

The radio source W3(A) has a size similar that seen at 20 μm, ~40 arcsec. IRS 2 and a recently observed nearby source IRS 2a are the highly reddened

Figure 6-9.   Radio continuum map of the H II region W3 (Schraml and Mezger, 1969).

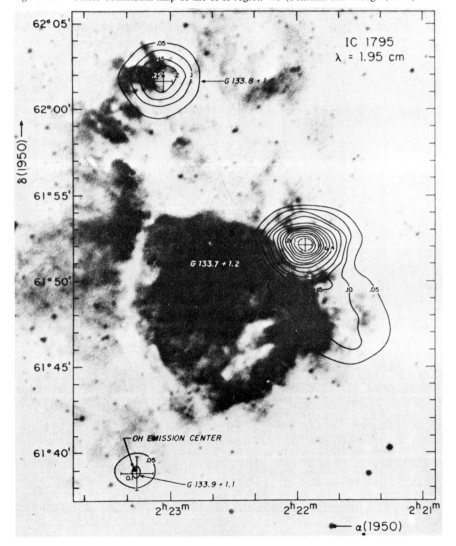

exciting stars, with 10 to 15 mag of extinction. IRS 1/W3A appears to be a shell source around these exciting stars. At far-infrared wavelengths (69 μm) the source is somewhat larger, ~ 1.9 arcmin (Figure 6-12).

Mezger (1975) has proposed that the main component of W3 is an O-star association, in an early stage of evolution, where star formation has occured in subgroups each with $\sim 1000\ M_\odot$ of stars. The O-stars form last in the group and reach the main sequence imbedded in a remnant dust shell. At first, all stellar radiation is absorbed by dust and the shell is observed as an infrared source, peaking at $\sim 20\ \mu$m, with no related radio source. For gas densities less than 10^5 cm^{-3}, a compact H II region can occur, still surrounded by the dust shell. We then observe an ionization-bounded H II

Figure 6-10.   Total spectrum of the H II regions W3(A) and W3(OH) (Mezger, 1975).

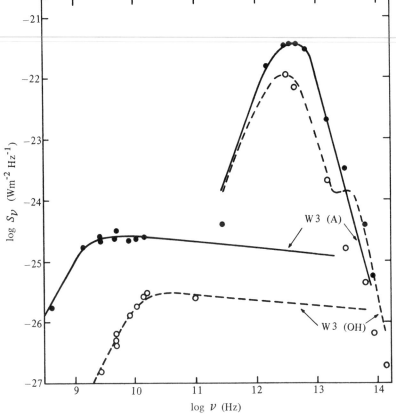

region with a defined Strömgren sphere and infrared source peaking at 100 μm (present state of W3A).

Figure 6-13 gives the density (n_p) profile for W3(A) which best fits the data. Mezger proposes that the H II region consists of a shell of ionized gas ($n_e \sim 8 \times 10^3$ cm^{-3}) which is imbedded in a shell of neutral gas of the same

Figure 6-11. High-resolution maps of the main component of W3 (Wynn-Williams and Becklin, 1974).

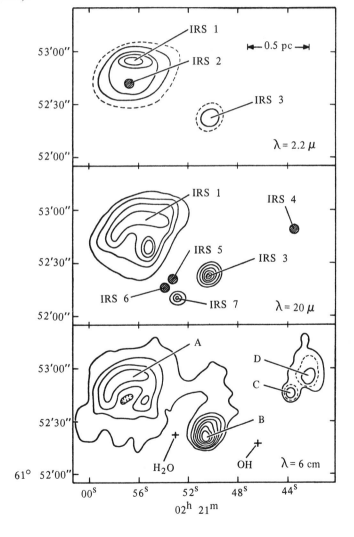

density. The width of the neutral shell should be large enough to absorb all photons >912 Å. The shell structure is also indicated by the observed brightness distributions of W3(A) at radio and infrared wavelengths. If all the infrared radiation from W3(A) were from dust grains uniformly distributed over 40 arcsec, the spectrum would be considerably broader since the dust grains would have widely different temperatures.

The H II region continues to expand and eventually the neutral shell is fully ionized. The ionization front progresses quickly into the surrounding,

Figure 6-12.   Far-infrared map of W3 region (Fazio et al., 1975).

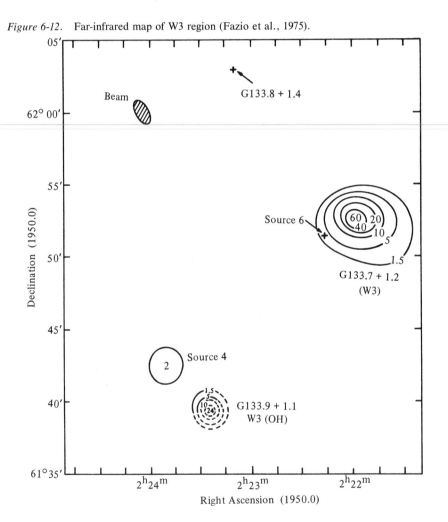

low-density, interstellar matter, and the dust shell becomes transparent to stellar radiation >912 Å. At this stage one finds a compact H II region imbedded in an extended low density H II region, with most of the radio emission coming from the extended source. The extended source is no longer associated with far infrared emission. The southern extension of W3 might be at this stage.

W3(OH) is a very compact H II region located 17 arcmin southeast of W3(A). It is in association with several H_2O/OH maser sources. The diameter of the source as measured at radio wavelengths and at 20 μm is 1.7 arcsec. At far-infrared wavelengths the size is <1 arcmin. The spectrum is given in Figure 6-10.

Mezger has proposed that W3(OH) is a compact H II region, very depleted of dust in the ionized region; he quotes a dust-to-gas mass ratio equal to 10^{-2} of the value in the interstellar medium. The low value of $f = N_c'/N_c$ measured for this source implies that the ionizing O-star has not yet reached the main sequence, and therefore the Ly_c flux is lower than that of a ZAMS star of the

Figure 6-13. Density profile assumed for W3(A) (Mezger, 1975).

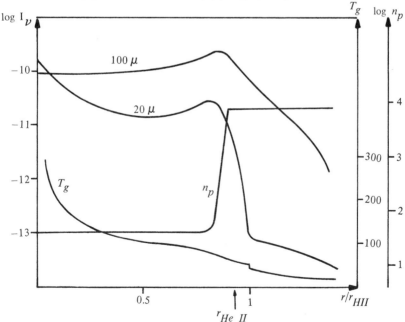

same luminosity. To produce the far-infrared flux the dust-depleted compact H II region must be surrounded by a dense shell of gas and dust. W3(OH) probably represents an earlier stage of evolution than W3(A).

Another example of a complex H II region, and perhaps one of the best studied, is the Orion Nebula (Figure 6-14). It is located only ~500 pc from the sun.

The radio continuum map is shown in Figure 6-15. Near the center of this source there are two main regions of interest (Figure 6-16). The first is centered on the Trapezium stars, the main exciting stars of the optical nebula. These stars also coincide with the peak of the ionized hydrogen region as determined from the radio maps. The second region is 1 arcmin northwest of the Trapezium stars. It is a cluster of infrared sources that are optically invisible. This second region also corresponds with the center of a large molecular cloud.

In the first region there exists a 20-μm source, 30 arcsec in diameter, centered on the Trapezium star farthest to the east, called the Ney-Allen source. It also exhibits a 10-μm emission feature due to heated silicate dust within the ionized region. The 69-μm map (Figure 6-17) shows an extended region of emission that is similar to the 11-cm radio distribution throughout the nebula. The dust temperature appears to be about 100 K. For this H II region there is no apparent boundary corresponding to a Strömgren sphere. Instead there is a continuous flow of ionized gas resulting in an extended, low-density, H II region. Several ionization fronts appear in the radio and infrared maps, the most prominent being the bar which is 2 arcmin southwest of the Trapezium. It appears to be an H I-H II transition zone, with $T \sim 80$ K, and gas density $\sim 10^4$ cm^{-3}.

The main group of infrared sources is located in the second region, at the center of the molecular cloud. The region is otherwise undistinguished at optical and radio wavelengths, but it does contain a number of OH and H_2O maser sources.

In the 3- to 10-μm range the brightest source is the BN object, which has a spectrum corresponding to a 530-K black body, with absorption at 3.1 μm due to ice and at 10 μm due to silicate dust. Its diameter is less than 2 arcsec. At far-infrared wavelengths the brightest source is the Kleinmann-Low nebula, which is located 12 arcsec south of the BN object. Its temperature is ~70 K. About 30% of the total infrared flux of the nebula comes from this cluster of sources. The cluster appears to be located in the molecular

Figure 6-14.   Optical photograph of the Orion Nebula (Lick Observatory photograph).

cloud, and the molecular cloud appears to be just behind the H II region. The Trapezium stars are ~ 0.1 pc from the cloud.

In summary, let us consider a possible evolution of an H II region where dust plays an important role. The following description is quoted from Mezger (1975), and is based on the work of Davidson and Harwit (1967) and Krügel (1974):

> When a massive proto-star approaches the main sequence, its surface temperature is low and it cannot ionize the ambient gas, but it has already acquired its full luminosity. The radiation pressure acts on the dust and drives it at high speed (~ 10 km s^{-1}) outwards, without dragging along the neutral gas, because friction between the gas and dust by elastic collisions is weak. The dust piles up in a front until its density there is so high that coupling between the gas and dust becomes effective. The dust front is optically thick at visible wavelengths and hides the star from the optical observer. The dust front takes up the whole momentum of the stellar radiation. Its drift velocity through the gas is small (about 1 km s^{-1}) compared to the speed at which it is driven away

Figure 6-15. Radio continuum map of the Orion Nebula (Schraml and Mezger, 1969).

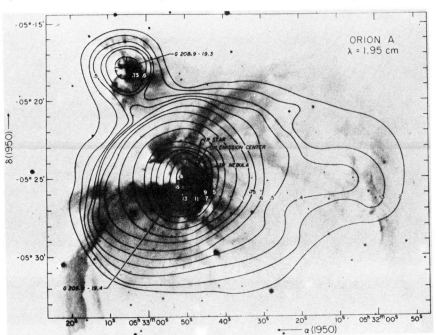

from the star (about 10 km s^{-1}). Therefore, the dust front acts like a supersonic piston on the surrounding gas leaving in its wake a highly rarified region.

When the star becomes hot enough, it ionizes (part of) its surrounding gas from which the dust has been expelled, thus forming a dust-depleted H II region, which is surrounded by a low-density region (which has been cleared by the dust front), and further out by the dust front itself. This may be the stage at which we observe W3(OH).

In the further evolution, the ionization front overruns the compact H II region surrounding the star and proceeds rapidly through the rarified gas until it reaches the dust front. The grains there are exposed to ener-

Figure 6-16. Near-infrared map of the Orion Nebula (Wynn-Williams and Becklin, 1974).

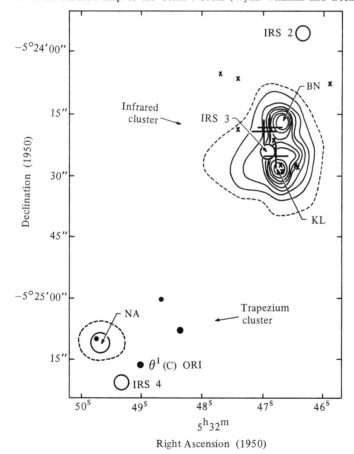

Right Ascension (1950)

getic Lyman continuum radiation and are (probably) charged by the photo-electric effect. The ambient gas is ionized, and the grains are effectively frozen into the gas by Coulomb forces. When the compressed material in the dust front is ionized, its pressure rises and it expands. Now the situation is very similar to W3(A): a low density inner region (the central condensation may have dispersed) surrounded by a dense shell with enhanced dust-to-gas ratio.

At this stage, the shell-like H II region is still surrounded by dense neutral gas and therefore, has a well defined outer boundary. In its further evolution, more and more of this neutral shell gets ionized while the density of the H II region decreases. Eventually, all of the neutral gas is ionized and at that stage the H II region appears as a condensation of medium electron density embedded in an extended low-density H II region. Since the optical absorption depths in the ionized gas are small, H II regions at this evolutionary stage are no longer observable as

Figure 6-17.   Far-infrared map of the Orion Nebula (Fazio et al., 1974).

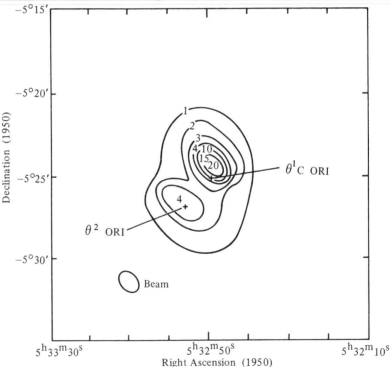

strong *IR* sources. The compact H II regions in W3, southern extension, appear to represent this evolutionary stage.

If the formation of a dust front is a typical stage in the pre-MS evolution of O-stars, it provides a simple explanation of the similarity of most far *IR* spectra of galactic H II regions; providing the thermostat which keeps the grain temperature at about 70–80 K. The dust surrounding O-stars would always form a shell of typical radius of the order 10^{18} cm; this radius may depend somewhat on the stellar luminosity. This is of the right order to yield the observed colour temperatures of dust grains in and around H II regions.

Dark dust clouds. The absence of stars, or an obvious decrease in the number of stars in a given region of the sky, indicates the presence of nonluminous, absorbing dust clouds in interstellar space. Recent renewed interest in star formation from dense dust clouds and early stages of stellar evolution has caused a re-examination of the properties of dark interstellar clouds. In fact, the processes of star formation in these clouds should bear a strong similarity to the processes occurring in dense molecular clouds (e.g., the Orion Nebula).

Because optical observations are impossible, speculation on the stellar content in dark clouds long received little attention. However, at infrared wavelengths the region is less opaque, and cool protostars are expected to emit most of their energy in the infrared. The discovery of a cluster of stars, observed at 2 μm in the prominent dark-cloud complex south of the star Rho Ophiuchi, has stimulated new observations at infrared and radio wavelengths. Vrba and associates (1975) found 67 point sources, which were heavily obscured. At visual wavelengths the extinction was estimated to be between 10 and 40 mag. If these sources are on the main sequence they may be between stellar types B3 and F5. These authors suggest that the infrared sources represent the upper main sequence of a young cluster and strongly suggest that dark clouds contain star clusters even several million years after their formation. It appears that star formation does not occur in a burst or in the sudden collapse of a single molecular cloud, but takes place over times as long as 10^7 years. Dark clouds are found to contain extremely young objects (10^4–10^5 years old) in the dense regions and much older stars over a more distributed volume.

Radio observations have detected the presence of molecular clouds of various species, e.g., CO, SO, H_2CO, and CS. The CO measurements indicate that at the core of the dust cloud densities may exceed 10^5 cm^{-3}. The gas temperature is of the order of 30 K. Under these conditions the most abundant molecule is molecular hydrogen, although it has not been directly detected.

Previous assumptions that ultraviolet radiation in the dark cloud is negligible seem no longer true. Radio continuum radiation (free-free emission by electrons) has been detected as well as C^+ recombination lines from sources in the cloud. This radiation probably originates from an ionized region, generated by ultraviolet emission from a early-type star embedded in the cloud. These stars also provide the heating for the cloud. A far-infrared source, with an extent of 2.5 arcmin, has been found in the region of the two brightest infrared stars and near the peak of the CO emission. This radiation is probably due to heated dust grains.

Future measurements must lead to better determination of the temperature, density, and gas velocities in the cloud complexes. Understanding the mechanism of cloud collapse and fragmentation into stars still remains a problem.

Planetary nebulae. Planetary nebulae are a type of isolated gaseous nebulae possessing a fair degree of bilateral symmetry. They consist of shells of gas and dust that have been ejected in the recent past by the central star. The central stars in planetary nebulae are old, much hotter ($T \sim 5 \times 10^4$ K to 3×10^5 K), and often less luminous than galactic O-stars. There is also a higher degree of ionization than in H II regions and large amounts of He^{++} are present. The central star appears to be rapidly evolving toward the white-dwarf stage. The shell is rapidly expanding outward at a velocity of ~ 25 km/sec and the mean lifetime is estimated to be about 10^4 years. Densities in the shell can range from 10^2 to 10^4 cm^{-3}, with total masses varying from 0.1 to 1.0 $M_\odot$.

Infrared radiation in the 8- to 14-μm region has been observed from planetary nebulae and found to be in excess of the expected free-free emission by a factor ranging from 25 to 160. The spectrum rises steeply with increasing wavelength, similar to that observed in H II regions. In the region 1.65 to 3.4 μm the excess is not always present but appears to be correlated with high-density, compact nebulae; this observation however could be due to a selection effect.

Excess infrared continuum radiation is a common phenomenon among planetary nebulae but only a few are strong infrared sources, the brightest being NGC 7027. The most probable origin of this radiation is dust grains in the shell heated to about 200 K by the intense central source. In the few cases for which the infrared distribution across the nebula has been measured, it is identical to the radio continuum maps, indicating that the dust is well mixed with the gas. The origin of the dust remains a mystery. It could be formed by condensation at the time of ejection of the nebula or it could be a remnant of the atmosphere of the preplanetary star.

Line emission has been proposed as a major source of infrared radiation and weak line emission has been observed at 12.8 μm (Ne II), 10.5 μm (S IV), and 9.0 μm (Ar III), but the present evidence is that line emission does not contribute significantly to the total flux.

X-ray sources. A few X-ray sources have been observed at near-infrared wavelengths. These observations have been made primarily at 1.6 and 2.2 μm with the 200-inch Hale telescope. Although the fluxes are weak, in those cases where infrared observations have been possible, they combine with radio and X-ray observations to place strong limitations on any model of the source.

Scorpius X-1 is the brightest X-ray source in the sky in the 1–10 Å wavelength region. No periodic pulsations have been observed from this source, but there have been intensity variations of the order of 50% on a time scale of hours and days. Optical observations indicate that Sco X-1 is a 12th- or 13th-magnitude star that may be one component of a binary system at a distance of about $\gtrsim 500$ pc. The X-ray flux is consistent with an optically thin gas at 4 to 10×10^7 K. Infrared measurements have been made only at 1.6 and 2.2 μm; the data are consistent with a gas at $\sim 5 \times 10^7$ K that becomes optically thick at 1 μm. The optical and infrared spectrum is consistent with a flat spectrum extrapolated from the X-ray region. At wavelengths greater than 1 μm the spectrum decreases, following a ν^2 dependence into the infrared. Models of the infrared source, based on the above data, give a source diameter of about 10^4–10^5 km, which is about the size of a white-dwarf star. The optical and infrared fluxes may be produced in a region of this size, but the X-ray flux may be produced in a much smaller inner region.

Cygnus X-3 is a binary X-ray source located at a distance of ~ 10 kpc. The X-ray flux varies smoothly with a 4.8-hour period. The radio flux is extremely variable, exhibiting erratic outbursts. At infrared wavelengths, Cygnus X-3 has been observed at 1.6 and 2.2 μm, exhibiting a flat spectrum. The radio and infrared sources appear to be coincident to within 2 arcsec. No optical counterpart has been observed. Simultaneous X-ray and infrared observations have exhibited the 4.8-hour period at both wavelengths. This object is the first to show binary or rotation phenomena in the infrared. No simple radio or infrared correlation exists, although both regimes are active at the same time.

In the infrared region there is evidence for several types of behavior: simple periodic variations, but no long-term periodic structure, although this may be masked by the outbursts; short-term outbursts (~ 2 minutes); long-term outbursts (~ 1 to 2 hours); long-term variability. The infrared and X-ray periodic structure strongly suggest that Cygnus X-3 is an eclipsing binary system. The high infrared surface brightness and similarity of the X-ray and infrared eclipse curves suggest that the X-rays and infrared radiation come from the same hot object.

The Crab Nebula (Figure 6-18) is a special case of an X-ray source; it is the remnant of a supernova explosion in A.D. 1054. Over the past 25 years it has been one of the principal celestial objects for astrophysical research. It was the first radio source to be identified with an optical object, the first source to be identified with synchrotron radiation, and the first X-ray source to be identified with a supernova remnant. At its center is a pulsar with the fastest period known, and the only pulsar that is observed at radio, infrared, optical, X-ray, and gamma-ray wavelengths.

There are very few observations of the Crab Nebula at infrared wavelengths. The Nebula has been observed from 1 to 10 μm with upper limits only for the far-infrared flux. The infrared spectrum can, as in the radio region, be expressed as a power law of the form $\nu^{-\alpha}$. From optical and infrared data, corrected for extinction, the value of α ranges from 0.65 to 1.5, depending on the value of the extinction, A_v, used (0.9 to 1.5). In any case, these values of α show that the radio spectrum, with $\alpha = 0.3$, does not continue with the same slope into the infrared and optical regions, with the change in slope occurring at $\sim 10^{14}$ Hz (3 μm) or less. As with the radio radiation, the infrared spectrum is explained by synchrotron emission from high-energy electrons. If these electrons are continually injected into the nebula, a break in the spectrum is predicted at a frequency that is dependent only on the characteristic lifetime of the electrons and the mean magnetic field. Taking the characteristic lifetime to be the age of the nebula (~ 900 years) and the frequency break at 10^{14} Hz, we derive a magnetic field value $\geq 3.5 \times 10^{-4}$ gauss and a predicted change in the spectral index α, from 0.3 to 0.8.

Figure 6-18. Optical photograph of the Crab Nebula in red light (Lick Observatory photograph).

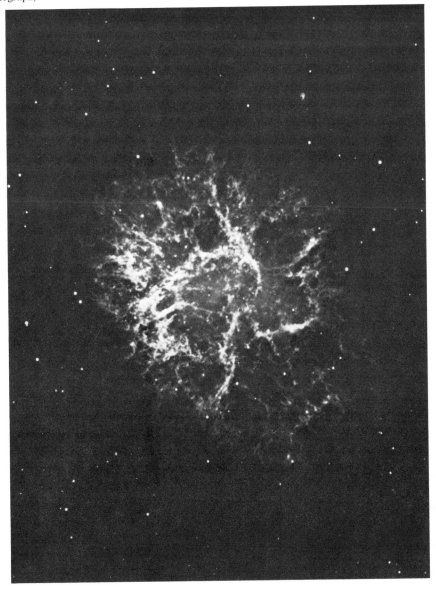

The flux measurement at 10 μm lies above the spectrum extrapolated from shorter wavelengths. If this value is correct, the excess might be a result of dust in the nebula that is heated by synchrotron radiation. A definite flux measurement at ~100 μm would be most important in determining whether this dust exists.

At the center of the Crab Nebula is a pulsar (NP 0532) with a 33-msec period. With the 200-inch Hale telescope, the pulsar has been observed at 1.6, 2.2, and 3.5 μm. At 2.2 μm the pulse shape is identical with the optical pulse. The spectrum of the pulsar peaks in the optical region and decreases in the infrared. Since the optical signal is highly polarized, the most likely origin is synchrotron radiation. The decrease of the flux in the infrared has been interpreted as due to synchrotron self-absorption as well as to a transition from synchrotron to gyroradiation. It is important to continue observations of the pulsar in the infrared to better determine its shape, intensity, polarization, and the nature of the infrared cutoff.

The Center of our Galaxy

The galactic center is a rich and complex source of infrared radiation in which interesting phenomena occur in great variety. The region can be described as having two parts: a large extended emission region, which exists over several degrees along the galactic plane, and a complex set of sources and intense emission in the central 1-arcmin area at the approximate position of the radio source Sgr A.

As observed in the far-infrared, the galactic center appears similar to an intense H II region, with a spectrum peaking at about 100 μm. Heavy visual obscuration by dust ($A_v \geq 27$ mag) is present and silicate absorption appears in the spectrum at 10 and 20 μm. Early maps made by Hoffmann, Federick, and Emery (1971) in the far-infrared (75–125 μm) with a small balloon-borne telescope had poor resolution (15 arcmin), but showed an extended emission over a region $2° \times 4°$, with peak emission over a region 15×38 arcmin. The total flux was of the order of 10^6 Jy. Later, higher-resolution (5.6 arcmin) maps by Alvarez and colleagues (1974) showed that many features observed in radio maps of the same region are reproduced in the far-infrared (40–350 μm) and the main radio sources are also observed: Sgr A, Sgr B$_2$, (an H II region), Sgr C, and Sgr D (Figure 6-19). These sources all appear to be extended. On this scale, the far-infrared emission and CO emission also appear to be correlated. The 2.2-μm emission is also extended, and elongated along the plane, over 1°, with the central region near Sgr A and the peak infrared emission showing high surface brightness. This radiation is probably the result of a high spatial density of stars.

The most interesting phenomena, however, are observed in the highest-resolution maps of the central 1 arcmin. At a distance of 10 kpc this corresponds to a region ~3 pc in diameter. Becklin and Neugebauer (1975) have recently published very-high-resolution (2.5 arcsec) maps of this region at 2.2 and 10 μm.

Figure 6-19.   Far-infrared map of galactic center region (Furniss et al., 1974).

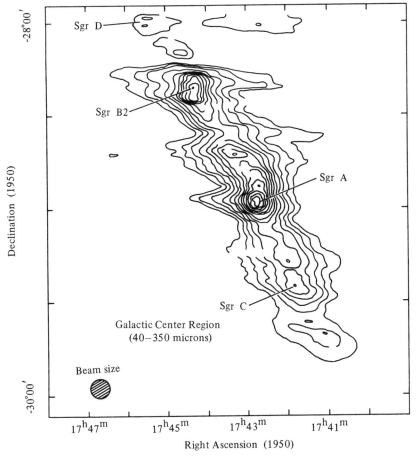

The 2.2-μm radiation from the central 30-arcsec region comes primarily from discrete sources, but has three components. About one-third of the radiation comes from very bright point sources, which are also seen at 10 μm and are probably very luminous stars. Another one-third comes from fainter discrete sources, of which only two appear to be extended. The final one-third comes from the general diffuse background.

At 10 μm an almost entirely different picture emerges. Nine discrete sources are observed in addition to a curved ridge of extended emission. About one-fourth of the flux is from discrete sources; about one-half from the emission ridge, and about one-fourth from low-surface-brightness radiation. Radio continuum maps also show emission along the ridge, indicating the presence of ionized gas. The infrared radiation from the ridge probably arises from heated dust uniformly mixed with the ionized gas. Five of the 10-μm discrete sources are located along the ridge (Figure 6-20).

An intense, subarcsec radio source falls on the northern edge of the 10-μm ridge, coincident with an extended 2.2-μm source. This point could be the position of highest stellar density. However, in general, at this resolution scale, there is no simple correspondence between radio continuum and infrared sources.

To better understand this complex region and the origin of this intense radiation, whether it is thermal or nonthermal, maps of even higher resolution are needed in the near-infrared and particularly at far-infrared wavelengths. Aircraft observations with a 90-cm telescope with 1-arcmin resolution, using a multicolor far-infrared photometer (Harvey, Campbell, and Hoffmann, 1976) have recently been made; a more extensive map, with similar resolution, has also been made, using observations from the 1-m balloon-borne telescope with a broad-band far-infrared photometer (Fazio, Wright, and Low, 1975).

Extragalactic Objects

One of the most important astronomical problems being pursued at the present time is the nature of infrared emission from galaxies. If major advances in infrared astronomy are to occur, they will probably be in this area. It is interesting to note that X-ray astronomy is in a similar position.

At a wavelength of 2 μm, infrared observations of galaxies provide no surprises. In most cases they can be explained as due to stellar emission, in which the flux decreases at longer wavelengths. However, measurements at

Figure 6-20. High-resolution map of the galactic center at 10-μm wavelength (Becklin and Neugebauer, 1975).

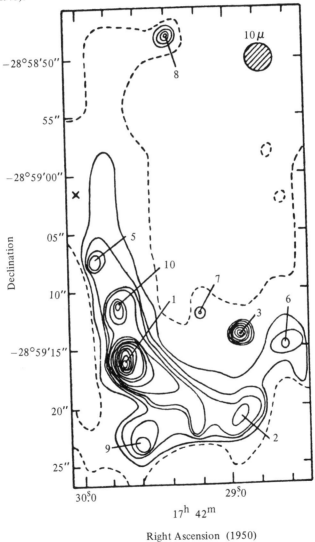

wavelengths greater than 10 μm indicate there is a group of galaxies that are extremely luminous in the infrared and measurements at 20 μm indicate that, for some, their spectra must peak at far-infrared wavelengths. Rieke and Low (1972, 1975) have now detected nearly a hundred extragalactic sources at 10 μm, but only seven have been observed at wavelengths greater than 20 μm: M 82, NGC 253, NGC 1068, NGC 4151, MKN 231, NGC 1275, and NGC 5253, and only three (M 82, NGC 253, and NGC 1068) at wavelengths greater than 40 μm.

The extragalactic objects observed at 10 μm can be divided into four classes: quasi-stellar objects (e.g., 3C 273), ultrahigh-luminosity infrared galaxies (MKN 231, I Zw 1, Cyg A, NGC 1614, II Zw 136), Seyfert and related galaxies (e.g., NGC 253 and NGC 5253), and others.

The most numerous group is the Seyfert and related galaxies, where the data indicate that the infrared flux is proportional to the radio flux from the nuclei of these galaxies.

The ultrahigh-luminosity galaxies have luminosities well above those of the Seyfert-type galaxies, and approach values of $10^{13} L_\odot$. This is to be compared to our galactic center emitting $\sim 10^9 L_\odot$ and the brightest H II regions in our Galaxy with 10^5–$10^6 L_\odot$. The three brightest galaxies of this group also show permitted Fe lines in emission. The most luminous, Markarian 231, is also one of the most luminous galaxies known in the visual region, with a visual magnitude of -23, yet its infrared energy output is greater by an order of magnitude, and lies in the range of quasi-stellar objects.

Of the quasi-stellar objects only 3C 273 is easily observed at 10 μm. Its luminosity is $\sim 10^{13} L_\odot$. A considerable number of quasars have been observed at 2 μm.

In about one-third of the normal galaxies observed, strong 10-μm emission was also observed, indicating that many normal galaxies are more luminous than our own by a factor of 10 or more.

At the present time there is no satisfactory explanation for the origin of infrared radiation from galaxies. The mechanism could be thermal (heating of dust) or synchrotron radiation (nonthermal). The thermal models are most popular, consisting of a dust cloud in the nucleus of the galaxy heated radiatively by an internal source or sources. The central source, however, could be nonthermal or a hot plasma. The synchrotron radiation models have the problem of explaining the sharp cutoff at low frequencies.

Spectroscopy in the 10 μm region on several galaxies, e.g., NGC 1068,

NGC 253, and M 82, has shown an absorption feature at 10 μm caused by silicate dust. Structure on the scale of 2 to 4 arcsec has also been seen in M 82 and NGC 253, and several galaxies have been found to be extended on a scale of 5 to 15 arcsec.

Variability of 10 μm extragalactic fluxes on a time scale of months would require a nonthermal origin. The variable radio source OJ 287 has been observed to be variable at 10 μm also, with closely correlated variations. Evidence for the variation in NGC 1068 at 10 μm is contradictory at the present time.

Future Prospects

Our present over-all knowledge of celestial infrared sources has been severely restricted by the relatively few all-sky surveys that have been performed. Our knowledge of the general properties of these sources is, therefore, very biased, particularly at far-infrared wavelengths, by these selected observations. An unbiased survey is important because it will increase the population of known sources, from which interesting objects can be selected for more detailed study; permit better correlation with optical, radio, and X-ray properties; and uncover many new types of sources. We shall first review what surveys have been performed and then discuss future possibilities.

The first ground-based survey, made at 2.2 μm by the California Institute of Technology (Neugebauer and Leighton, 1969), covered about three-fourths of the sky to a limiting flux densiy of about 40 Jy. About 5600 sources were found, most of which could be identified as stars of spectral type later than K5, i.e., the majority are late-type giant stars. However, ~50 sources have 0.8- 2.2-μm color temperatures of ~1000 K and most of these have not been identified with optical objects.

The Air Force Cambridge Research Laboratory (AFCRL) performed a survey at 4, 11, and 20 μm during 1971 and 1972 (Walker and Price, 1975), using a small, cyrogenically cooled, rocket-borne telescope. About 79% of the sky was surveyed to a limiting flux of ~100 Jy at 11 μm. About 3200 sources were observed: 2507 at 4.2 μm, 1441 at 11 μm and 873 at 19.8 μm.

At infrared wavelengths greater than 20 μm only a small percent of the sky has been surveyed. This was done primarily by the NASA Goddard Institute for Space Studies with a small balloon-borne telescope (Hoffmann,

Frederick, and Emery, 1971). Only ~100 sources have been observed to a limiting magnitude of ~10^4 Jy (Hoffmann and Aannestad, 1974). These sources lie predominantly along the galactic plane. The wavelength gap from 20 μm is relatively large; hence at 100 μm there is not much of a clue as to what to expect.

Furniss, Jennings, and Moorwood (1974) at University College, London, using a 15-inch balloon-borne telescope, have detected more than 50 H-II regions in the 40–250 μm region.

Friedlander, Goebel, and the Joseph (1974) have detected, during a partial sky survey, 12 extremely strong sources ($\geq 3 \times 10^{-12}$ W/cm^2) in the spectral band (50–500 μm). However, at the present time these sources have not been verified.

Future balloon-borne far-infrared surveys include an 8-inch low-emissivity telescope, by F. J. Low, and a 16-inch cryogenically cooled telescope, by the Cornell/Arizona group.

High-resolution mapping and sensitive photometry of selected regions in the far-infrared will continue to be performed on the 90-cm airborne telescope and the 102-cm balloon-borne telescope.

An infrared astronomical satellite has been proposed for the late 1970s by a joint Netherlands–United States group, to survey the sky in the region 30–200 μm. This telescope would be 60 cm in diameter and have a resolution of about 2 arcmin. At 11 μm its sensitivity should be ~10^3 times better than the AFCRL survey; but at 100 μm its sensitivity will be detector-limited and it will have approximately the same sensitivity as the 102-cm balloon-borne telescope.

A large (2.4-m) space telescope, the LST, is being planned for 1982. Although primarily an optical and ultraviolet telescope, a dual-beam infrared photometer, in the region 2–1000 μm, is one of seven instruments being proposed. With this telescope it should be possible to determine the number and location of all sources similar to the Orion Nebula, W3(A), and M 17 in each of the galaxies in the local group. It should be possible to monitor any time variations in the quasi-stellar source 3C 273 and possible to make a far-infrared survey of all 1149 galaxies in the Shapley-Ames Catalogue (Kleinmann, 1975).

Another area where significant advances are going to be made in the near future is high-resolution ($\Delta\lambda/\lambda \sim 10^{-3}$ to 10^{-4}) spectroscopy. The vibration and rotational line spectra of many cosmologically important molecules fall in this region of the spectrum, as well as the atomic emission lines of abundant elements. To date, very little high-resolution spectroscopy has been performed.

High-resolution spectral studies of the planets could yield important data on the composition and structure of the atmosphere, particularly the minor constituents and composition of the surface. For example, the Jovian spectrum between 20 and 200 μm will yield information on the He/H$_2$ ratio, the temperature structure of the upper atmosphere, and the abundance of H$_2$O. On Venus the chemical interactions among sunlight, cloud particles, and atmospheric gases will be better understood when we know the abundance of the chemically active minor constituents, many of which are detectable in the far-infrared, e.g., H$_2$O, HCl, HF, CO, and H$_2$O$_2$.

Far-infrared spectra of galactic sources, such as H II regions, are needed to measure atomic and molecular abundances and to better define the electron and particle densities in regions that are inaccessible at shorter wavelengths. Several atomic emission lines in the far-infrared provide the principal heat-loss mechanism for gas in interstellar clouds, so that a detection of these lines would be of particular interest.

Far-infrared spectroscopy with high resolution ($\Delta\lambda/\lambda \sim 10^{-3}$) will soon be possible from both aircraft and balloon-borne telescopes. Infrared spectroscopy will be of particular interest in determining the structure and composition of extragalactic objects. A limited amount of spectroscopy is available now in the 8- 13-μm region on only the few brightest objects. Extension of this ground-based work as well as far-infrared spectroscopy will be extremely valuable.

Infrared astronomy is still in its infancy, but many of the technological difficulties that limited its advancement are rapidly being overcome, and the future prospects for this area of astronomy are particularly bright.

References

References cited in text

Alvarez, J. A., Furniss, I., Jennings, R. E., King, K. J., and Moorwood, A. F. M. 1974. Far infrared observations of W51 and the galactic center, in A. F. M. Moorwood, ed., *H II Regions and the Galactic Center*, Proc. 8th ESLAB Symposium (ESRO SP-105), pp. 69–77, Frascatti, Italy.

Arams, F. R. 1973. *Infrared-to-millimeter wavelength detectors* Artec House, Inc., Dedham, Mass.

Becklin, E. E., and Neugebauer, G. 1975. High-resolution maps of the galactic center at 2.2 and 10 microns, *Astrophys. J. (Letters) 200*, L71–74.

Davidson, K., and Harwit, M. 1967. Infrared and radio appearance of cocoon stars, *Astrophys. J. 148*, 443–448.

Fazio, G. G., Kleinman, D. E., Noyes, R. W., Wright, E. L., Zeilik, M. II, and Low, F. J. 1974. A high resolution map of the Orion Nebula region at far infrared wavelengths, *Astrophys. J. (Letters) 192*, L23–25.

Fazio, G. G., Kleinman, D. E., Noyes, R. W., Wright, E. L., Zeilik, M. II, and Low, F. J. 1975. A high resolution map of the W3 region at far infrared wavelengths, *Astrophys. J. (Letters) 199*, L177–179.

Fazio, G. G., Wright, E. L., and Low, F. J. 1975. Flight performance of the 102-cm balloon-borne far infrared telescope, in Proc. Conf. on Far-Infrared Astronomy, Windsor Great Park, England, to be published.

Friedlander, M. W., Goebel, J. H., and Joseph, R. D. 1974. Detection of new celestial objects at far-infrared wavelengths, *Astrophys. J. (Letters) 194*, L5–8.

Furniss, I., Jennings, R. E., and Moorwood, A. F. M. 1974. Far infrared photometry of H II regions, in A. F. M. Moorwood, ed., *H II Regions and the Galactic Center*, Proc. 8th ESLAB Symposium (ESRO SP-105), pp. 61–68, Frascatti, Italy.

Harvey, P. M., Campbell, M. F., and Hoffmann, W. F. 1976. High resolution far-infrared observations of the galactic center region, *Astrophys. J. (Letters)*, in press.

Hoffmann, W., and Aannestad, P. A. 1974. 100 micron surveys in the northern and southern hemispheres, in C. Swift, F. C. Wittebom, and A. Shipley, eds., *Telescope Systems for Balloon-Borne Research*, 7–15, NASA TM X-62, 397.

Hoffmann, W. F., Frederick, C. L., and Emery, R. J. 1971. 100-micron survey of the galactic plane, *Astrophys. J. (Letters), 170*, L89–97.

Jennings, R. E., ed., 1975. *H II Regions and Related Topics*, Proc. of the European Physical Society, to be published.

Johnson, H. L. 1965. Interstellar extinction in the Galaxy, *Astrophys. J. 141*, 923–942.

Kleinman, D. E. 1975. The use of a large telescope in the infrared, in Proc. Conf. on Far-Infrared Astronomy, Windsor Great Park, England, to be published.

Krügel, E. 1974. Ph. D. thesis, University of Göttingen, unpublished.

Low, F. J., and Rieke, G. H. 1974. The instrumentation and techniques of infrared photometry, in N. Carleton, ed., *Methods of Experimental Physics,* vol. 12, part A, pp. 415–462, Academic Press, New York and London.

Mezger, P. G. 1975. The distribution of ionized gas and dust in W3(A) and W3(OH), Proc. Conf. on Far-Infrared Astronomy, Windsor Great Park, England, to be published.

Neugebauer, G., and Becklin, E. E. 1973. The brightest infrared sources, *Sci. American 228,* 28–40.

Neugebauer, G., and Leighton, R. B. 1969. *Two-Micron Sky Survey, A Preliminary Catalog,* NASA SP-3047.

Panagia, N. 1973. Some physical parameters of early-type stars, *Astron. J. 78,* 929–934.

Petrosian, V., Silk, J., and Field, G. B. 1972. A simple analytic approximation for dusty Strömgren spheres, *Astrophys. J. (Letters) 177,* L69–73.

Rieke, G. H., and Low, F. J. 1972. Infrared photometry of extragalactic sources, *Astrophys. J. (Letters) 176,* L95–100.

Rieke, G. H., and Low, F. J. 1975. Measurements of galactic nuclei at 34 microns, *Astrophys. J. (Letters) 200,* L67–69.

Schraml, J., and Mezger, P. G. 1969. Galactic H II regions. IV. 1.95-cm observations with high angular resolution and high positional accuracy, *Astrophys. J. 156,* 269–301.

Traub, W. A., and Stier, M. T. 1976. Theoretical atmospheric transmission in the mid- and far-infrared at four altitudes, *Applied Optics,* in press.

Vrba, F. J., Strom, K. M., Strom, S. E., and Grasdalen, G. L. 1975. Further study of the stellar cluster embedded in the Ophiuchus dark cloud complex, *Astrophys. J. 197,* 77–84.

Walker, R. G., and Price, S. D. 1975. *AFCRL Infrared Sky Survey,* Air Force Cambridge Research Laboratories Technical Report AFCRL-TR-75-0373.

Webbink R. F., and Jeffers, W. Q. 1969. Infrared astronomy, *Space Sci. Rev. 10,* 191–216.

Wright, E. L. 1973. Infrared brightness distribution of dusty H II regions, *Astrophys. J. 185,* 569–572.

Wynn-Williams, C. G., and Becklin, E. E. 1974. Infrared emission from H II regions, *Proc. Astron. Soc. Pacific 86,* 5–25.

Additional references of general interest

Allen, D. A. 1971. Blue stars in the infrared, *Astrophys. Space Sci. 1,* 232–235.

Hudson, R. D. Jr., and Hudson, J. W., eds. 1975. *Infrared Detectors,* Halsted Press, New York.

Hyland, A. R. 1971. Galactic infrared astronomy, *Proc. Astron. Soc. Australia 2,* 14–20.

Kruse, P. W., McGlauchlin, L. D., and McQuistan, R. B. 1962. *Elements of Infrared Astronomy,* John Wiley and Sons, New York.

Low. F. J.   1969. Infrared astrophysics, *Science 164,* 501–505.

Neugebauer, G., and Leighton, R. B.   1968. The infrared sky, *Sci. American 219,* 50–65.

Neugebauer, G., Becklin, E., and Hyland, A. R.   1971. Infrared sources of radiation, *Annu. Rev. Astron. Astrophys. 9,* 67–102.

Smith, R. A., Jones, F. E., and Chasman, R. P.   1968. *The Detection and Measurement of Infrared Radiation,* 2nd edition, Oxford University Press, Oxford.

7

Gaseous Nebulae and Their Interstellar Environment

Eric J. Chaisson

Conceived from the collapsing fragments of an interstellar cloud, a young hot star radiates sufficient energy to alter significantly the future course of its parent region. Nearby portions of an otherwise neutral atomic and molecular cloud transform into gaseous plasma. Temperatures rise by orders of magnitude. The gas becomes turbulent. Shock fronts are established. And, an initially collapsing interstellar cloud reverses its gravitational infall and expands outward into the relatively quiescent interstellar environment, thereby altering the subsequent history of the region. Such tumultuous, often visually spectacular, interstellar regions surrounding youthful stars are known as gaseous nebulae.

Energy is transported within gaseous nebulae by ultraviolet photons emitted from a hot O- or B-type star having an effective surface temperature in excess of 30,000 K. These ultraviolet photons impart energy to the gas by photoionizing hydrogen, the excess energy appearing as kinetic energy of newly created photoelectrons. The behavior of these unbound electrons is critical. They perform three principal tasks: electrons collide with other electrons, they recombine, and they excite other ions.

Inelastic collisions among photoelectrons rapidly distribute their initially gained energy throughout the region of ionized gas, establishing a Maxwellian velocity distribution characterized by a kinetic temperature of approximately 10^4 K. The nebular thermal structure is then generally maintained by an equilibrium between heating by photoionization and cooling by recombination and by other processes that produce low-energy radiation capable of escaping the nebula.

Occasionally, free electrons are recaptured into excited atomic states, decaying to lower and lower energy levels by radiative transitions, eventually reaching the ground level. In this process, photons are emitted at specific frequencies characteristic of the atom, giving rise to the well-known optical lines of the H and He Balmer series. Radio-frequency photons can also be detected as electron transitions between two atomic levels of high principal quantum number ($n > 30$), providing a powerful tool for the study of nebulae in the absence of interstellar scattering that often plagues optical observations.

And finally, unbound electrons can collisionally excite fine-structure energy levels of certain ions which, although metastable, have sufficient time to relax radiatively in the low-density ($< 10^5$ cm^{-3}) interstellar environment. Consequently, gaseous nebulae also emit visually observed forbidden lines such as those of [N II] in the red, [O II] in the violet, and [O III] in the green.

Thus knowledge of gaseous nebulae and their interstellar environment accumulates solely from the detection of electromagnetic radiation emitted and absorbed by electrons, atoms, and molecules. The operational astrophysicist categorizes these electromagnetic data according to four fundamental parameters, namely direction, frequency, intensity, and polarization, in order to characterize the physical and chemical state of the gas in terms of temperature, density, composition, and velocity.

This chapter attempts to present a broad explanation of the current state of nebular affairs in terms of the fundamental physical principles known to third- and fourth-year students concentrating in physics and astronomy. It is not a review article or a complete guide to all the recent literature. Instead, it comprises a pedagogical package of select theoretical and observational scenarios that exemplify the fundamental principles of nebular radiation as we now understand them. Emphasized are topics from optical and radio astronomy for which a substantial base of observational data exists. Infrared studies of gaseous nebulae are discussed in Chapter 6.

The discussion here will be divided into four major areas: fundamental concepts, continuous nebular radiation, the spectroscopy of gaseous nebulae, and the spectroscopy of nebular environments.

Fundamental Concepts

Atomic Physics

The element hydrogen comprises 90% of the matter in gaseous nebulae. Consequently an enormous amount of nebular information can be gleaned from an analysis of the physics of this atom.

The first of the two Bohr postulates that form the foundation of modern atomic theory, namely that electrons of mass m_e, charge e, and velocity v can reside only in discrete orbits at distance r from the nucleus of charge Ze and mass M, enables us to write two conservation equations, one for the electron's energy,

$$E = \frac{-Ze^2}{r} + \frac{m_e v^2}{2} \tag{7-1}$$

and another for the electron's momentum,

$$m_e v r = n \hbar. \tag{7-2}$$

Here, n is known as the principal quantum number. When coupled with an expression equating the electron's Coulomb and centripetal forces in the stationary orbit,

$$\frac{Ze^2}{r^2} = \frac{m_e v^2}{r}, \tag{7-3}$$

these expressions lead to Bohr's second postulate: photons of discrete energy are emitted or absorbed when electrons jump between different quantum states having energy

$$E = \frac{-RchZ^2}{n^2}. \tag{7-4}$$

Here c is the velocity of light and $R = 2\pi^2\mu e^4 c^{-1} h^{-3}$ is the Rydberg constant, a factor that depends on the atomic system's reduced mass $\mu = m_e M/(m_e + M)$.

Figure 7-1 shows the simplest energy-level diagram for the H atom and a schematic representation of the H spectrum for the three lowest atomic series. Only the Balmer series, transitions terminating in level $n = 2$, has spectral features coincident with the conventional visual spectrum ($\lambda\lambda$ 3700–7300 Å). Prominent features among the Balmer series are the optical recombination lines Hα ($n = 3 \to 2$) at λ 6563, Hβ ($n = 4 \to 2$) at λ 4861, Hγ ($n = 5 \to 2$) at λ 4340, Hδ ($n = 6 \to 2$) at λ 4102, and the series limit ($n = \infty \to 2$) near λ 3646. Shortward of 3646 Å, the H atom emits a continuous spectrum through free-bound transitions. The entire Lyman series, beginning at λ 1216 ($n = 2 \to 1$) and terminating at λ 912 ($n = \infty \to 1$), lies deep in the ultraviolet, absorbed by the earth's atmosphere, while the Paschen, Brackett, and other series terminating in levels of higher $n < 30$ are located in the infrared. Extremely low-energy transitions between adjacent levels involving $n > 30$ occur in the radio-frequency domain, as illustrated in the larger view of the electromagnetic spectrum shown in Figure 7-2.

Relativistic motion of the electron produces fine structure among the energy levels,

$$E = \frac{-RchZ^2}{n^2}\left[1 + \frac{\alpha^2 Z^2}{n}\left(\frac{1}{l} - \frac{3}{4n}\right)\right]. \tag{7-5}$$

Figure 7-1. Top, Energy levels for the H atom; some of the strongest transitions are indicated. Bottom, Detailed H spectrum about the visible domain as a function of wavelength λ, frequency ν, and principal quantum number n.

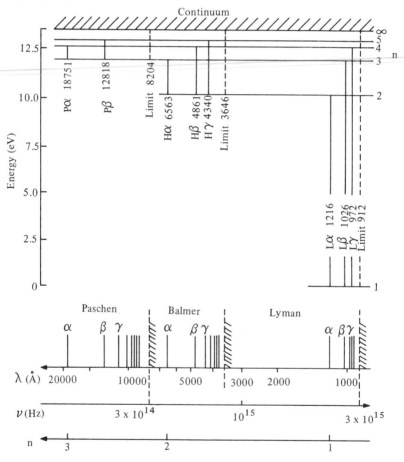

Here l is the azimuthal quantum number, characteristic of the atom's angular momentum in units of $\hbar$, and $\alpha \ (= 1/137)$ is the fine-structure constant. Thus each principal level with a given n actually consists of n sublevels spaced very close to one another. The H energy-level diagram can be redrawn like that in Figure 7-3 to demonstrate this atomic fine structure. The terms S, P, D, . . . represent the net electronic angular momentum of the entire atomic system $\sum_i l_i = 0, 1, 2, \cdots$. The superscript represents the level multiplicity, $2 \sum_i s_i + 1$, which for H is always 2, a doublet, since there is only one

Figure 7-2. Energy difference between adjacent n in H, giving rise to α lines throughout the electromagnetic spectrum.

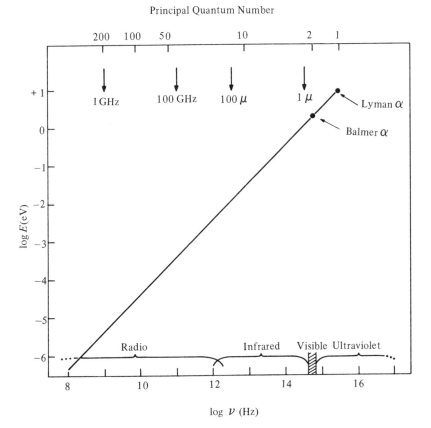

electron of spin $s = 1/2$. The subscript denotes the total atomic angular momentum $J = \sum_i l_i + \sum_i s_i$.

Only certain transitions are allowed among the fine-structure levels; the Bohr correspondence principle for large quantum numbers demands that for the atomic system $\Delta \sum_i l_i = 0, \pm 1$, and for the individual electron undergo-

Figure 7-3.   Schematic diagram of H atom fine structure. Allowed transitions are shown as solid lines. The oval insert shows the ground-state hyperfine splitting that gives rise to the λ21-cm line.

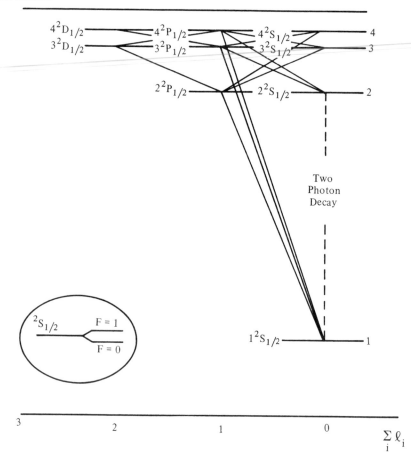

ing the transition, $\Delta l = \pm 1$ only. Actually, these selection rules are not iron-clad. Quantum theory dictates that they be obeyed only by electric dipole radiation. Weaker electric quadrupole and magnetic dipole radiations are allowed, especially between those levels forbidden by electric dipole rules. Electric quadrupole and magnetic dipole radiation is weaker simply because of the small probability of jumping between two atomic levels n and m. That is, if the transition probability is denoted by the conventional Einstein coefficient A_{mn} (sec^{-1}), then the emitted intensity of a nebular spectral line, given by $N_m A_{mn} h \nu_{mn}$, is small for electric quadrupole and magnetic dipole radiation; N_m here is the number of atoms in an upper state m. Numerically, the time normally required for an excited atomic system to emit electric dipole radiation spontaneously is about $A_{mn}^{-1} \simeq 10^{-8}$ sec. Transitions between levels forbidden by electric dipole rules, but allowed by electric quadrupole or magnetic dipole rules, take much longer, on the order of 1 and $> 10^4$ sec respectively. These long-duration states are called metastable states, and lines emitted from them are called forbidden lines.

For example, the Lyman-α transition, $2^2P_{1/2} \rightarrow 1^2S_{1/2}$, emits electric dipole radiation rapidly, $A^{-1} \simeq 10^{-8}$ sec, whereas the other candidate Lyman-α transition, $2^2S_{1/2} \rightarrow 1^2S_{1/2}$, violates the electric dipole selection rule $\Delta l = \pm 1$, and therefore can emit only electric quadrupole radiation, $A^{-1} \simeq 0.1$ sec. The most famous forbidden line is the λ 21-cm line of neutral hydrogen. Shown within the insert in Figure 7-3, the $1^2S_{1/2}$ state is actually split by an incredibly small 6 μ eV because of the electron-proton magnetic-moment interaction. Although the electric dipole selection rule $\Delta l = \pm 1$ is violated, making $A^{-1} \simeq 10^{14.5}$ sec, the universally large abundance of H ensures that magnetic dipole radiation at 1420 MHz (λ 21 cm) is emitted in substantial amounts throughout the Milky Way.

The long time required for metastable excited levels to decay spontaneously makes it nearly impossible to observe forbidden lines in the laboratory. Even in the best laboratory vacuum, collisional de-excitation ensures depopulation of excited metastable levels before they have a chance to radiate electric quadrupole or magnetic dipole features. However, in most gaseous nebulae and many other regions of the interstellar medium, densities are so low that the time scale for collisional de-excitation is comparable to or even longer than that for radiative processes, permitting cosmic forbidden lines to be observed. In fact, some of the most important lines observed by optical astronomers result from forbidden transitions within low-lying energy levels of abundant nebular ions. As we discuss later, these ions are critically important as a cooling mechanism because their forbidden radiation can readily escape the nebula, carrying with it energy input by hot exciting stars.

Consider a notable example of forbidden-line radiation prominent in gaseous nebulae. Figure 7-4 shows the lowest energy-level configuration for doubly ionized oxygen, O^{++}, normally denoted by its chemical symbol O III. The excited levels have the same electronic configuration as the ground level, namely, six bound electrons, $1s^2 2s^2 2p^2$, so that electric dipole radiation is forbidden. In the case of forbidden lines, the chemical symbol is usually bracketed, [O III]. Weaker electric quadrupole radiation is permitted between $^1S_0$ and $^1D_2$, producing a λ 4363 spectral line; magnetic dipole radiation is also permitted for the low-lying pairs of lines, $^1D_2 \rightarrow {}^3P_1$ at λ 4959 and $^1D_2 \rightarrow {}^3P_2$ at λ 5007. These forbidden optical lines are indeed observed from numerous gaseous nebulae. Other forbidden lines arising from transitions within the ^3P state occur in the infrared, whereas the $^1S_0 \rightarrow {}^3P$ lines occur in the ultraviolet. None of the invisible lines have yet been detected.

Figure 7-4 shows the low energy-level configurations for several other reasonably abundant ions that also act as principal nebular coolants. In later sections, we shall discuss these species in some detail, for it is these low-energy ($h\nu < 4$ eV) lines capable of escaping gaseous nebulae that are most readily detectable at earth. Forbidden spectral lines arising from abun-

Figure 7-4. Energy-level diagram for the lowest terms of several abundant ions. Splitting of the doublet and triplet terms has been exaggerated for clarity. The optical emission lines emitted by gaseous nebulae are denoted by dashed lines. Note how [O II] and [S II] as well as [O III] and [N II] are isoelectronic and have similar fine structure levels.

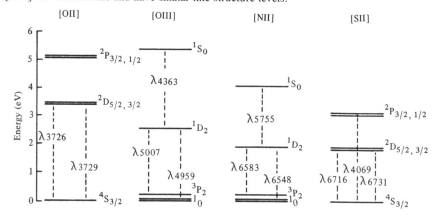

dant ions, together with permitted spectral lines of hydrogen and helium, provide an enormous wealth of information concerning the physics of gaseous nebulae.

Radiative Transfer

The nebular radiation field incident on the earth is characterized by the passage of energy as a function of position x, direction k, time t, and frequency ν. The specific intensity I_ν is defined such that $I_\nu d\nu d\omega dA dt$ is the total energy within a time dt passing through a frequency interval $d\nu$ and an area dA located at position x, perpendicular to the photon direction k about which there subtends a solid angle $d\omega$ (cf. Figure 7-5). The transfer equation specifies the positive or negative alteration of I_ν when photons interact with matter along some path dx,

$$\frac{dI_\nu}{dx} = -\kappa_\nu I_\nu + j_\nu, \tag{7-6}$$

where j_ν, the emissivity, and κ_ν, the absorptivity, are defined so that $j_\nu d\nu d\omega dV dt$ and $\kappa_\nu I d\nu d\omega dV dt$ are the respective energies emitted and absorbed by an element of volume dV ($= dAdx$), of time dt, of solid angle $d\omega$ and of frequency $d\nu$. With the conventional definition of differential optical depth,

$$d\tau = \kappa_\nu \, dx, \tag{7-7}$$

Figure 7-5. Transfer of radiation of intensity I through a gaseous element of differential area dA, depth dx, and solid angle $d\omega$ along a path k toward earth.

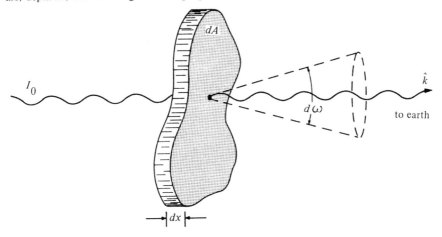

we have, upon integration of equation (7-6),

$$I_\nu = I_0 \, e^{-\tau} + \int_0^\tau \frac{j_\nu}{\kappa_\nu} \, e^{-\tau'} \, d\tau', \tag{7-8}$$

where I_0 denotes the radiation intensity incident on the more distant side of the gaseous element (cf. Figure 7-5).

If thermodynamic equilibrium prevails, then the absorption and emission of radiation of intensity I_ν, given by the universal temperature T function known as Planck's law,

$$B_\nu(T) = \frac{2h\nu^3}{c^2} \left(\exp \frac{h\nu}{kT} - 1 \right)^{-1}, \tag{7-9}$$

is balanced separately at each frequency according to Kirchhoff's law,

$$B_\nu(T) = j_\nu/\kappa_\nu. \tag{7-10}$$

Thus equation (7-8) becomes, for an isothermal case,

$$I_\nu = I_0 \, e^{-\tau} + B_\nu \, (1 - e^{-\tau}). \tag{7-11}$$

We shall return to this equation many times in this chapter.

Ionization Equilibrium

Gaseous nebulae are produced by the photoionization of an interstellar cloud by ultraviolet photons from a young, hot star or cluster of stars. By ultraviolet, we generally mean photon energies that exceed $h\nu_1 = 13.6$ eV, the binding energy of the H atom in the $1^2S_{1/2}$ ground level. In a steady state, the number of photoionizations per unit volume per unit time equals the corresponding number of recombinations. Restricted for the moment to hydrogen, by far the most abundant element, the equation of ionization equilibrium becomes

$$N(\text{H}) \, \varphi = N_e N(\text{H}^+)\alpha(\text{H},T), \tag{7-12}$$

where $N(\text{H})$, N_e, and $N(\text{H}^+)$ are the respective volume number densities of H atoms, electrons, and protons. We discuss separately the physical significance of φ and $\alpha \, (\text{H},T)$, the ionization- and recombination-rate coefficients. First, the ionization-rate coefficient (sec^{-1}) equals the rate of energy absorbed when an atom jumps from one ionization stage to another, and is given by the expression

$$\varphi = \int_{\nu_1}^\infty \frac{4\pi J_\nu}{h\nu} \, \sigma(\text{H},\nu)d\nu. \tag{7-13}$$

Here, $4\pi J_\nu/h\nu$ represents the total number of stellar photons of mean intensity $J_\nu \equiv \dfrac{1}{4\pi} \displaystyle\int I_\nu d\Omega$ incident on a unit volume in unit time in unit frequency interval, while $\sigma(H,\nu)$ is the photoionization cross section for H by photons with energy above the threshold $h\nu_1$. The discussion of $\sigma(H,\nu)$ is restricted here to energies larger than 13.6 eV because virtually all interstellar neutral hydrogen is in the $1^2S_{1/2}$ ground level. Recombinations from the continuum to any level $n > 1$ are followed by rapid radiative transitions downward at a rate given by $A \simeq 10^4 - 10^8$ sec^{-1}. Thus, the radiative lifetime A^{-1} of excited H levels is exceedingly short compared with the mean photoionization lifetime ($\simeq 10^8$ sec) of the H atom. Consequently, the photoionization cross section $\sigma(H,\nu)$ need be considered only for the $1^2S_{1/2}$ level; its precise calculation is very complicated, but Figure 7-6 shows that its frequency dependence for photons of $\nu \geq \nu_1$ varies approximately as $(\nu_1/\nu)^3$ for ν not far above thresh-

Figure 7-6. Frequency distribution of $\sigma(H, \nu)$ about the threshold $\nu_1 = 3.3 \times 10^{15}$ Hz, $\lambda_1 = 912$ Å.

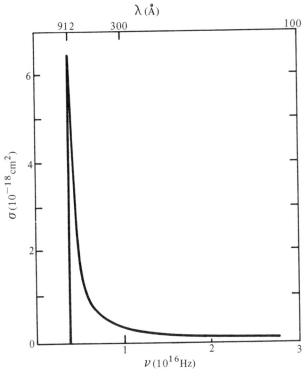

old. Note that this cross section and, for that matter, the extent of ionization depend solely on the atomic properties of hydrogen and the number of sufficiently energetic photons, while they are independent of the thermodynamic properties of the gas.

On the other hand, the degree of radiative recombination of electrons is a sensitive function of the nebular gas temperature T, in addition to the number of electrons available for recombination. The total recombination coefficient (cm³/sec) to all atomic levels n except $1^2S_{1/2}$ is given by the equation

$$\alpha(H,T) = \sum_{n=2}^{\infty} \int_0^{\infty} \sigma_n v f(v) dv, \tag{7-14}$$

where σ_n is the H recombination cross section for electrons having velocity v prior to capture, and $f(v)$ is the Maxwell-Boltzmann distribution of electron velocities,

$$f(v) = 4\pi(m_e/2\pi kT)^{3/2}v^2 \exp[-(m_e v^2/2kT)]. \tag{7-15}$$

(Recombination to the $1^2S_{1/2}$ ground level has no net effect on the overall ionization balance since each recombination to this level immediately produces a photon capable of ionizing another H atom.) For a typical range $T = (5-20) \times 10^3$ K, σ_n approaches 10^{-20} cm², considerably smaller than the geometrical cross section of an H atom. Since the cross sections vary inversely as the electron energy (or therefore as v^{-2}) and since $f(v)$ goes like $v^2 T^{-3/2}$, then $\alpha(H,T)$ varies approximately as $T^{-1/2}$. Figure 7-7 displays the variation of $\alpha(H,T)$ as a function of T for a range of nebular temperatures. Large deviations from ionization equilibrium are unlikely since a typical recombination time, $[\alpha(H,T)N_e]^{-1} \simeq 10^5 N_e^{-1}$ years, is short compared with typical nebular lifetimes ($\simeq 10^6-10^7$ years) even for low N_e ($\simeq 10$ cm⁻³).

What does the ionization structure of a nebula look like? Consider a region very close to an exciting star or group of stars. The density of Lyman continuum (Ly$_c$) photons ($h\nu > 13.6$ eV), the only quanta possessing sufficient energy to ionize H from the $1^2S_{1/2}$ level, is reasonably high. Neutral atoms formed by recombination are promptly reionized by another Ly$_c$ photon. However, this radiative flux becomes diluted at some distance from the star, not only because of the inverse-square law but also because some Ly$_c$ photons are used to ionize new H atoms as the ionized zone, quickly at first and then slowly, expands into the surrounding neutral region. On the order of 10^4 years or less, the ionization structure reaches a virtual steady state, becoming completely ionized within a certain sphere of radius R and remaining almost entirely neutral beyond. This picture can be made more quantitative. If f denotes the fractional ionization, $N(H^+)/N(H) + N(H^+)$, then equation (7-12) yields

$$f(1 - f)^{-1} = \frac{N(H^+)}{N(H)} = \frac{\varphi}{\alpha(H,T)N_e} \qquad (7\text{-}16)$$

or, upon evaluation, $f(1 - f)^{-1} \simeq 10^{4.5}N_e^{-1}$ which, for reasonable N_e, demonstrates the total ionization of the gas within a certain radius R.

But how large is this radius of ionized material? Letting $N(Ly_c)$ be the total number of Ly_c photons flowing per second through a spherical shell of radius R, we can set the differential passage of $N(Ly_c)$ through the shell thickness dR equal to the number of recombinations per second in that shell:

$$\frac{dN(Ly_c)}{dR} = 4\pi R^2 N(H^+)N_e\alpha(H,T). \qquad (7\text{-}17)$$

Integration then establishes the steady-state condition that the rate of Ly_c

Figure 7-7. Temperature dependence of the recombination coefficients $\alpha(H, T)$ and $\beta(H, T)$.

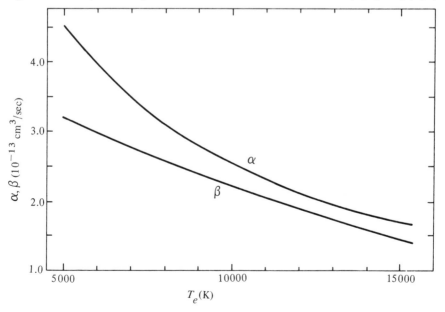

photons emitted into a uniformly dense sphere of radius R equals the rate of recombination occurring in that sphere,

$$N(Ly_c) = \tfrac{4}{3}\pi R^3 N_e^2 \alpha(H,T). \qquad (7\text{-}18)$$

Equation (7-18) can be solved for R, often known as the critical or Strömgren radius. Columns 4 and 5 of Table 7-1 list, for $T = 10^4$ K, a general value of $RN_e^{2/3}$ and a specific value of R for a representative density $N_e \simeq N(H) + N(H^+) = 300$ cm^{-3} for stars of different spectral type. Associated with each spectral designation are values of T_{eff}, defined as that black-body (surface) temperature whose emitted flux shortward of the Lyman limit equals that from the star, and $N(Ly_c)$, the total emission rate of Ly_c photons, computed from the Planck formula. Agreement between the theoretical notions of Table 7-1 and the observational picture discussed later may not be perfect for several reasons: the representative nebular density of 300 cm^{-3} may not apply to each region, the density may not be uniform throughout the Strömgren sphere, and the value of $N(Ly_c)$ is enhanced slightly at large R by a "diffuse radiation field" produced by numerous recombinations onto elements other than hydrogen.

It is nevertheless important to realize that, although interstellar regions near hot stars are fully ionized, any group of exciting stars can produce only a finite number of ultraviolet photons and thus can ionize only a finite volume. Of course, if there is a lack of neutral material capable of becoming ionized near a hot star, then the size of this "density-bounded" nebula becomes limited by the amount of neutral material present, and its generally irregular

Table 7-1 Characteristics of stellar ultraviolet radiation.

Stellar spectral type	T_{eff}(K)	log $N(Ly_c)$[a] (sec^{-1})	$R\,N_e^{2/3}$ (pc/cm^2)	R (pc, for $N_e = 300$ cm^{-3})
O4	52,000	50.0	148	3.3
O5	50,200	49.8	125	2.8
O6	48,000	49.4	92	2.0
O7	45,200	49.0	68	1.5
O8	41,600	48.7	54	1.2
O9	37,200	48.4	43	0.9
B0	32,200	47.6	23	0.5
B1	22,600	45.2	4	0.2

[a] Taken from Churchwell and Walmsley (1973).

shape takes on the geometrical distribution of interstellar material near the star. In recent years, however, we have come to realize that most nebulae are embedded in much larger and more massive neutral atomic and molecular clouds, which provide a cocoon for "radiation-bounded" nebulae. Such nebulae often have some symmetry about their exciting sources. The thickness of the transition region between the almost completely ionized H II gas within the nebula and the predominantly neutral H I gas outside is exceedingly sharp compared with the overall dimensions of the nebula. Since photoionization is an absorption phenomenon, this thickness is approximately one mean free path of a Ly_c photon. Evaluation of equation (7-7) for the case $\tau(Ly_c) \simeq 1$ yields a transition-zone thickness

$$\delta x \simeq \kappa^{-1} \simeq \{[N(\text{H}) + N(\text{H}^+)]\sigma(\text{H},\nu)\}^{-1}, \tag{7-19}$$

which for a total density of about 300 cm^{-3} yields $\delta x \simeq 10^{-3}$ pc.

A gaseous nebula, then, is a region of nearly complete interior ionization separated by an exceedingly thin transition region from an outer, predominantly neutral atomic and molecular interstellar cloud. Figure 7-8 illustrates a family of curves for various coupling conditions between the effective stellar temperature and the total nebular density.

Figure 7-8. Ionization structure for differing conditions in uniformly dense, pure H, model nebulae. Curve *a* corresponds to either an O9-type star and $N_e = 300$ cm^{-3} or O6 and 3000 cm^{-3}; curve *b* is valid for O7 stars and 300 cm^{-3} or O5 and 3000 cm^{-3}; curve *c* represents O6 and 300 cm^{-3} or O4 and 3000 cm^{-3}.

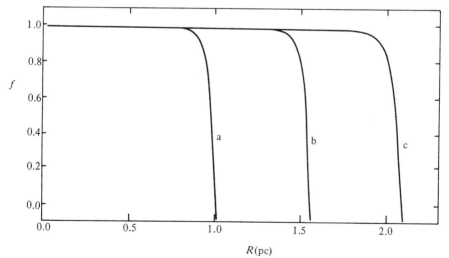

This discussion of nebular photoionization equilibria has been restricted to ionization and recombination of hydrogen. But other elements, especially the next most abundant element, helium, do have some effect. In particular, two different types of ionization structure are possible, depending on the spectrum of ionizing radiation and the abundance of He. Because the first ionization potential of helium, 24.6 eV, is considerably larger than the 13.6 eV required for hydrogen, an O9-type star with a spectrum concentrated within the range $13.6 < h\nu < 24.6$ eV creates a small central He II zone surrounded by a larger H II and eventually H I region. But if the stellar input spectrum contains a large fraction of photons with $h\nu > 24.6$ eV, like that from an O6 or hotter-type star, then the H II and He II regions can be approximately coincident. Numerical calculations for nebulae that include He are complicated primarily because recombination to the ground level of He can ionize either H or He; photons capable of ionizing H are emitted even via cascade from excited levels to the ground level of He. The ionization-equilibrium equations for H and He are therefore coupled by radiation fields having appreciable numbers of photons with $h\nu > 24.6$ eV. Figure 7-9 shows the relative ionization structure for model nebulae containing these two most abundant elements. Note that for exciting stars of spectral type O6 or hotter, the H II and He II zones are virtually coincident.

Figure 7-9. Ionization structure for uniformly dense nebulae, including H and He with respective Strömgren radii R(H) and R(He). (Numerical values taken from Rubin, 1969.)

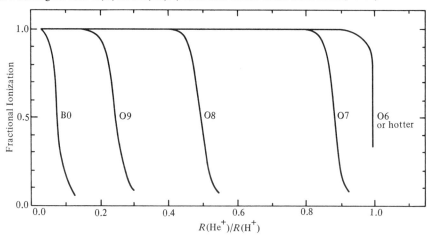

Even the hottest O stars emit hardly any photons with $h\nu > 54.5$ eV, the second ionization potential of He. Indeed, there is no observational evidence for doubly ionized He in gaseous nebulae (although in the more highly excited planetary nebulae, optical and radio recombination spectra have been detected from He III zones).

Heavier but less abundant elements such as oxygen, nitrogen, neon, carbon, and sulfur tend to be mostly singly ionized near the periphery of nebulae, though regions near the central star(s) can have substantial amounts of doubly ionized matter. These heavy elements have negligible effects on the overall ionization structure of nebulae. However, this passive ionization role of the heavy elements contrasts sharply with the critically important influence they have on the nebular thermal balance that we now discuss.

Thermal Equilibrium

Precise thermodynamic equilibrium may never be achieved in a microscopic realm within gaseous nebulae. In what follows, we consider the macroscopic thermal properties of the gas throughout the interstellar region ionized by an exciting star or group of stars. The discussion here is limited to the nonadiabatic interaction between radiation (the exciting photons) and the matter (the nebular gas), and excludes dynamic discontinuities or shock fronts that can also provide additional sources of heat.

We begin with the first law of thermodynamics. Consider an amount of heat ∂Q given to one gram-molecule of gas having a volume differential ∂V and thermal differential ∂T. The first law is then

$$\partial Q = p \partial V + C \partial T, \tag{7-20}$$

where p is the gas pressure, given by the ideal-gas equation of state,

$$pV = RT/\mu, \tag{7-21}$$

and C is the specific heat, expressed as

$$C = 3R/2\mu; \tag{7-22}$$

R is the gas constant and μ the mean molecular weight. Now, examine the variation of this heat with time. Let Γ be the amount of heat absorbed (gained) by the gas per unit volume and per unit time, and let Λ be the amount of heat radiated (lost) by the gas under the same conditions; we then have from equations (7-20), (7-21), and (7-22) an expression for the net heat transfer,

$$\partial Q = \Gamma - \Lambda = p \left(V^{-1} \frac{\partial V}{\partial t} + \frac{3}{2} T^{-1} \frac{\partial T}{\partial t} \right). \tag{7-23}$$

To solve equation (7-23) for the nebular thermodynamic structure, the exchange rates of energy per unit volume, Γ and Λ, must be specified. The gas is heated by photoionization, liberating a photoelectron having an initial energy $\frac{1}{2}m_e v_1^2 = h(\nu - \nu_1)$ where ν_1 is the Lyman-limit frequency and m_e and v_1 are the electron's mass and initial velocity. These free electrons immediately establish a Maxwell-Boltzmann energy distribution since the cross section for inelastic scattering collisions among electrons (of order $4\pi(e^2/mv^2)^2 \simeq 10^{-13}$ cm²) exceeds by a large factor all other nebular cross sections, including recombination. Thus the electron distribution function can be regarded as Maxwellian—hence the origin of equation (7-15)—characterized by some nebular temperature, often called the electron temperature T_e.

Now, recall that in ionization equilibrium, the number of photoionizations is balanced by an equal number of recombinations. Consequently, a thermal electron having energy $\frac{1}{2}m_e v_2^2$ is lost through recombination. The electron's velocity v_2 before recombination may be considerably less than its initial value v_1 after photoionization, since some energy of order $\frac{1}{4}m_e v_1^2$ is generally lost through inelastic collisions within the typical time scale for recombination, $\simeq 10^5 N_e^{-1}$ years. The net difference between the mean energies of the ensemble of newly created electrons and of the recombining electrons represents the net energy gained by the electron gas by the ionization-recombination process. However, the electron gas also loses energy by other processes, which include inelastic collisions between electrons and protons involving no bound states (thermal bremsstrahlung), and electron collisional excitation of low-lying fine-structure levels of abundant ions. Each of these processes emits photons of low energy ($h\nu \lesssim 4$ eV) capable of escaping and therefore cooling the nebula. We now consider in some detail each of these gain and loss mechanisms in order to evaluate equation (7-23) numerically.

In a pure hydrogen nebula, the energy input by photoionization per unit volume per unit time is, from equation (7-13)

$$\Gamma(H) = N(H)\varphi h(\nu - \nu_1). \tag{7-24}$$

In ionization equilibrium, this may be re-expressed, from equation (7-12), as

$$\Gamma(H) = N_e N(H^+)\alpha(H,T)h(\nu - \nu_1),\qquad(7\text{-}25)$$

where $h(\nu - \nu_1) = \frac{3}{2}kT$ represents the initial kinetic energy of the newly created photoelectrons. Equation (7-25) emphasizes the fact that nebular heating is dependent on the *form* (i.e., the frequency dependence) of the ionizing radiation field, and not on the absolute strength of the radiation.

As part of this same process, some of the electron's kinetic energy is lost per unit time via recombination,

$$\Lambda_R(H) = N_e N(H^+)\beta(H,T)kT_e,\qquad(7\text{-}26)$$

where, in analogy with equation (7-14), the effective recombination coefficient is averaged over the kinetic energy,

$$\beta(H,T) = \sum_{n=2}^{\infty} \frac{1}{kT_e} \int_0^{\infty} \sigma_n \nu f(\nu)\tfrac{1}{2}m_e \nu^2 d\nu.\qquad(7\text{-}27)$$

Recombination directly to the ground level ($n = 1$) has no cooling effect since the emitted photon immediately ionizes another hydrogen atom. Figure 7-7 displays the variation of $\beta(H,T)$ for a range of nebular temperatures.

Photoionization heating and recombination cooling rates for He and heavier elements can, of course, be included in a straightforward manner for more realistic nebular models. However, since the above equations show that these rates are proportional to the densities of the ions involved, their effect on the thermal balance in nebulae is of relatively minor consequence. The solid line of Figure 7-10 shows the net input rate ($\Gamma - \Lambda_R$) for the photoionization-recombination process for a typical case of an O5-type star having an effective surface temperature of about 50,000 K.

While waiting to recombine, unbound electrons also lose some of their kinetic energy through inelastic collisions with other charged particles. Figure 7-11 schematically illustrates an electron-ion encounter, known as a free-free transition. The electron of mass m_e and charge e enters the Coulomb field of the heavier ion of charge Ze (usually a proton with $Z = 1$) with an impact parameter b and a nonrelativistic approach velocity ν_3, and is scattered or decelerated to a lower configuration, $\frac{1}{2}m_e \nu_4^2$, without recombining. The emitted energy is called "bremsstrahlung," i.e., "braking radiation." This small amount of energy, $\frac{1}{2}m_e(\nu_3^2 - \nu_4^2) \ll 1$ eV generally, is emitted as a photon capable of escaping the nebula. Precise calculation for the energy

Figure 7-10. Relative net thermal gain (solid line) and loss (dashed line) for a nebula surrounding an O5-type star. The point of intersection is the equilibrium temperature. The dash-dot curves show the relative contribution of some of the fine-structure coolant transitions (adapted from Osterbrock, 1974).

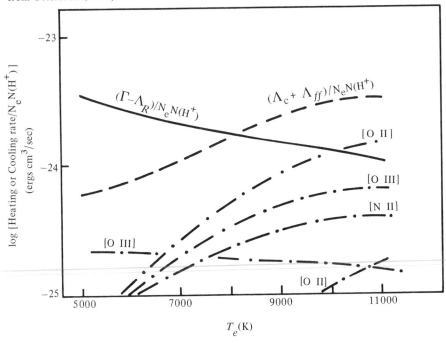

Figure 7-11. The geometry of an electron-ion encounter.

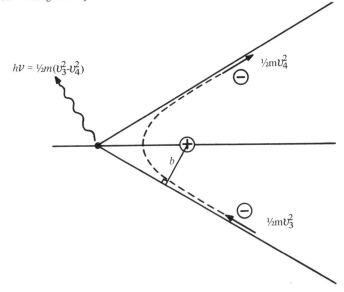

emitted in a single electron-ion encounter is well established, but the summation over the ensemble of encounters is not straightforward and requires some approximations.

Specifically, we seek the emissivity j_{ff} of free-free encounters, for this will specify the isotropic energy loss per unit volume per unit time,

$$\Lambda_{ff}(Z) = 4\pi \int_0^\infty j_{ff} \, d\nu. \tag{7-28}$$

This extraordinarily complex calculation, whose precise solution has still not been unanimously agreed on by researchers, can only be sketched here. In a classical two-body approximation, the microscopic rate of spontaneous emission of a single encounter is obtained by computing the deceleration of the electron in its hyperbolic orbit about the ion. For total bremsstrahlung emission from the plasma, integration over two parameters is necessary, the first a straightforward integration over the distribution of all electron velocities, the second a more difficult one over the distribution of impact parameters. The difficulty with the second integration occurs because of uncertainties in the upper integration limit of the impact parameter b_{max}; at large values of b, the two-body approximation breaks down, owing to interaction with neighboring charged particles. For reasonably high-density ($N_e \simeq 10^{19}$ cm^{-3}) plasma appropriate to laboratory conditions, b_{max} is given by the Debye length, but even the most dense ($N_e \simeq 10^5$ cm^{-3}) parts of gaseous nebulae are considerably more tenuous than the best laboratory vacuum. Fortunately, the more frequent, distant encounters correspond to hyperbolic orbits that are almost straight lines, for which reasonable estimates of b_{max} can be phenomenologically made. If we assume a Maxwellian velocity distribution for the electrons, the emissivity finally becomes

$$j_{ff} = \frac{32Z^2e^6}{12\pi c^3 m_e^2} \, N_e N(\mathrm{H^+ + He^+}) \left(\frac{2\pi m_e}{kT_e}\right)^{0.5} \ln \left[\frac{(2kT_e)^{1.5}}{\pi e^2 \gamma^{2.5} m_e^{0.5} Z\nu}\right], \tag{7-29}$$

where γ is the Euler constant ($= 1.78$) and $N(\mathrm{H^+ + He^+})$ denotes the volume number density of protons and singly ionized helium.

Averaged over a variety of impact parameters and electron approach velocities, a continuous spectrum of energy is emitted, predominantly in the radio-frequency domain. When j_{ff} is integrated over all frequencies and the atomic constants numerically evaluated, the cooling rate becomes

$$\Lambda_{ff}(Z) = 1.4 \times 10^{-27}Z^2 T_e^{-0.5} N_e N(\text{H}^+ + \text{He}^+). \tag{7-30}$$

As we shall discuss later, the thermal bremsstrahlung mechanism is important, not so much because of its rather minor contribution to the cooling rate, but rather because this low-energy radiation escapes the nebula, to be detected primarily at radio frequencies where the effects of interstellar extinction are negligible. It provides for us a powerful means of studying the thermodynamic properties of gaseous nebulae.

Ions of reasonably abundant elements such as oxygen, nitrogen, and sulfur provide the dominant cooling in gaseous nebulae. As noted earlier, many common ions have low-lying energy levels with excitation potentials of order kT_e capable of being collisionally excited. Figure 7-12 shows the Maxwell-Boltzmann distribution for a 10^4 K electron gas with a mean energy $kT_e \simeq 1$ eV. The particles in the extended tail of the distribution are the ones having sufficient energy ($\simeq 2$–4 eV) to excite collisionally the first few levels of the predominantly observed ions O^+, O^{++}, N^+, and S^+ (cf. Figure 7-4).

Figure 7-12. Maxwell-Boltzmann distribution of an ensemble of electrons having a given energy corresponding to a 10^4 K electron gas.

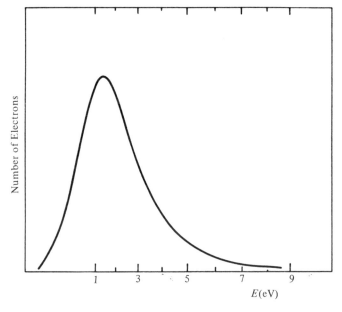

But there are hardly any particles with enough energy ($\approx$ 10–20 eV) to excite the more abundant hydrogen and helium species to their first excited states. Consequently, it is the trace elements having abundances more than a thousand times less than hydrogen or helium that serve to control the nebula thermostatically.

Consider the simplest case of an ion having a single excited level m in addition to its ground level n. The cooling rate per unit volume equals the amount of energy per unit volume per unit time escaping the nebula by photon emission as the atom relaxes back to the ground configuration; to wit,

$$\Lambda_c = N_m A_{mn} h \nu_{mn}. \qquad (7\text{-}31)$$

Note the similarity with the recombination cooling rate Λ_R given by equation (7-26). Recall from our earlier discussion that these particular transition probabilities A_{mn} are small (< 1 sec^{-1}), since radiative transitions between excited levels and the ground level are forbidden by electric-dipole selection rules. For this very reason, then, collisional de-excitations from the excited level compete with radiative processes. If we define $N_e N_n q_{nm}$ as the collisional excitation rate per unit volume from the ground level to the excited level, then the statistical equilibrium equation for excitation to and de-excitation from the excited level becomes

$$N_e N_n q_{nm} = N_e N_m q_{mn} + N_m A_{mn}. \qquad (7\text{-}32)$$

The first term on the right denotes the depopulation rate of the excited level via collisions, while the second term expresses the de-excitation rate of that level via radiative relaxation. Solving equation (7-32) for N_m and substituting into equation (7-31), we can distinguish two limiting cases for the cooling rate per unit volume:

$$\Lambda_c = \begin{cases} N_e N_n q_{nm} h \nu_{mn} & (\text{for } N_e \to 0), \\ N_n q_{nm} q_{mn}^{-1} A_{mn} h \nu_{mn} & (\text{for } N_e \to \infty). \end{cases} \qquad (7\text{-}33)$$

Here, q_{mn} is found from evaluation of the total collisional excitation rate per unit volume,

$$N_e N_m q_{nm} = N_e N_m \int_0^\infty v \sigma_{nm}(v) f(v) dv, \qquad (7\text{-}34)$$

where σ_{nm}, the cross section for an excitation from the ground level having statistical weight $g_n (= 2J + 1)$ by electrons with velocity v, is given by a bimodal distribution separated by an energy threshold at $h\nu_{mn}$:

$$\sigma_{nm}(v) = \begin{cases} 0 & \text{(for } \tfrac{1}{2}m_e v^2 < h\nu_{mn}), \\ \pi\hbar^2(m_e^2 v^2 g_n)^{-1}\Omega_{nm} & \text{(for } \tfrac{1}{2}m_e v^2 \geq h\nu_{mn}). \end{cases} \qquad (7\text{-}35)$$

Note that σ_{nm} is inversely proportional to the electron's kinetic energy once the threshold has been achieved; in other words, as might be expected, slower electrons above threshold collide with the ion most readily. Substitution of this result into equation (7-34), followed by an integration over the Maxwell-Boltzmann distribution $f(v)$, yields

$$q_{nm} = 8.6 \times 10^{-6}\, g_n T_e^{-0.5}\Omega_{nm}, \qquad (7\text{-}36)$$

where Ω_{nm} is known as the collision strength, taken here to be independent of the electron velocity. The principle of detailed balancing demands a completely symmetrical relation between Ω_{nm} and q_{mn}. These collision strengths must be calculated quantum mechanically for each individual transition. Of course, for more realistic cases like those shown in Figure 7-4, more than two levels must be considered. Laborious numerical computations performed on computers often show that an average Ω_{nm} is sufficient for a particular transition, despite rapidly varying energy resonances superposed on a smoothly varying energy component. The penultimate column of Table 7-2 lists values for Ω_{nm} for a few representative forbidden-line transitions promi-

Table 7-2 Atomic collision parameters.

Species	Transition	A_{mn} (sec^{-1})	Ω_{mn}[a]	N_e^c (cm^{-3})
O II	$^2D_{5/2} \longrightarrow {}^4S_{3/2}$	4×10^{-5}	1.5	$10^{2.8}$
	$^2D_{3/2} \longrightarrow {}^4S_{3/2}$	2×10^{-4}	1.5	$10^{3.5}$
O III	$^1S_0 \longrightarrow {}^1D_2$	1.6	0.6	10^8
	$^1D_2 \longrightarrow {}^3P_2$	2×10^{-2}	2.5	$10^{5.8}$
	$^1D_2 \longrightarrow {}^3P_1$	7×10^{-3}	2.5	$10^{5.8}$
	$^1S_0 \longrightarrow {}^3P_1$	2×10^{-1}	0.3	10^7
N II	$^1S_0 \longrightarrow {}^1D_2$	1.1	0.4	10^8
	$^1D_2 \longrightarrow {}^3P_2$	3×10^{-3}	3.0	$10^{4.9}$
	$^1D_2 \longrightarrow {}^3P_1$	1×10^{-3}	3.0	$10^{4.9}$
	$^1S_0 \longrightarrow {}^3P_1$	3×10^{-2}	0.4	10^6
S II	$^2P_{3/2} \longrightarrow {}^4S_{3/2}$	3×10^{-1}	2.7	10^7
	$^2D_{5/2} \longrightarrow {}^4S_{3/2}$	5×10^{-4}	5.7	$10^{3.5}$
	$^2D_{3/2} \longrightarrow {}^4S_{3/2}$	2×10^{-3}	5.7	$10^{4.5}$

[a] Calculated for $T_e = 10^4$ K.

nent in the visual spectrum of gaseous nebulae. The final column denotes that value of N_e^c, sometimes known as the critical density, much above which the second choice of equation (7-33) should be used to calculate Λ_c. This density, derived from detailed multilevel solutions, obviously represents the point where collisional de-excitation becomes appreciable, lessening the magnitude of Λ_c.

In reality, many of the important coolant transitions occur in the spectroscopically unobserved ultraviolet and far infrared regimes. For example, the $^3P_1 \rightarrow {}^3P_0$ and $^3P_2 \rightarrow {}^3P_1$ transitions occur in O^{++} at λ 88 μ and λ 52 μ, while the $^1S_0 \rightarrow {}^3P_1$ and $^1S_0 \rightarrow {}^3P_2$ transitions occur in O^{++} at λ 2321 and λ 2331. All the appropriate coolant transitions must be considered when calculating the net cooling rate Λ_c.

Figure 7-10 shows a summary of bremsstrahlung and fine-structure cooling considerations compared with the heating rate discussed earlier. Oxygen, nitrogen, and neon are assumed to have their normal cosmic abundances and are taken to be 80% singly ionized and 20% doubly ionized. Individual contributions to the total radiative cooling for reasonable densities, $N_e \simeq 10^3$ cm^{-3}, are shown by dashed lines.

Having specified the principal heating and cooling mechanisms, we return to evaluate equation (7-23). Figure 7-10 shows that the net photoionization-recombination heating rate $(\Gamma - \Lambda_R)$ and the dominant fine-structure cooling rate Λ_c have comparable values of about 10^{-24} erg cm^{-3} sec^{-1}. On the other hand, the right side of equation (7-23) is of the order

$$p/t = RTv/\mu Vx \simeq 10^{-30.5}T \tag{7-37}$$

for typical nebular motions having a scale length $x \simeq 1$ pc, a velocity $v \simeq 10$ km/sec and a mass density $m(\text{H}^+)N(\text{H}^+) = (\mu V)^{-1} \simeq 10^{-24}$ kg/cm^3. Thus, for any temperature $10^2 < T < 10^5$ K, the right side of equation (7-23) vanishes, and the temperature at any nebular location is determined completely by the thermal equilibrium between the heating and cooling rates,

$$\Gamma = \Lambda_R + \Lambda_{ff} + \Lambda_c. \tag{7-38}$$

Re-expressed as

$$\Gamma - \Lambda_R = \Lambda_{ff} + \Lambda_c, \tag{7-39}$$

the nebular temperature then becomes uniquely determined at the point of intersection in Figure 7-10. For $N_e < 10^4$ cm^{-3}, the resulting nebular temperature, $T_e \simeq 7500$–8500 K, is independent of the total density, since all the above rates are proportional to the relative abundance of the various ions and their ionization states. Hence, higher T_e can be achieved for the same N_e either with hotter exciting stars or by depletion of the trace elements. T_e can also be increased if $N_e > N_e^c \simeq 10^4$ cm^{-3}, at which point some of the collisionally excited fine-structure levels become depopulated by collisional de-excitation, and not by a radiative process that serves to cool nebulae. Consequently, the cooling curve of Figure 7-10 decreases, and the net heating and cooling curves intersect in equilibrium at nebular temperatures $T_e \simeq 9000$–$10,500$ K.

Nevertheless, despite ways to increase T_e, it is unlikely that nebular temperatures ever exceed about 15,000 K. At such high temperatures, electron collisions with *neutral* hydrogen, comparable in abundance to the trace ions, sufficiently populate the first few excited hydrogen states, and the resulting radiative de-excitations in turn can strongly and rapidly cool the nebula.

Continuous Nebular Radiation

In the absence of an appreciable magnetic field, the ensemble of nebular electrons interacts with ions to emit weak continuous radiation by free-free and free-bound transitions. The free-bound continuum is strongest in the optical regime, whereas the free-free continuum dominates in the radio and infrared regimes.

Radio-Frequency Regime

We have already considered nebular cooling by thermal bremsstrahlung when free electrons are decelerated in the Coulomb field of positive ions. The resulting photons that successfully escape the nebula can be detected by instruments primarily efficient in the radio-frequency domain.

Returning to our previous development of the transfer of radiation from the nebula to the earth, we note that the photon intensity for nebular thermal bremsstrahlung in the absence of appreciable background radiation—i.e., $I_0 = 0$ in equation (7-11)—is ultimately a function of the absorptivity, since

$$I_\nu = \int B_\nu(T) \exp\left(-\int \kappa_\nu \, dx\right) d\tau. \qquad (7\text{-}40)$$

In the Rayleigh-Jeans radio-frequency approximation ($h\nu \ll kT_e$) to the

Planck function, the absorptivity is given by Kirchhoff's law,

$$\kappa_{ff} = 0.5 j_{ff} c^2 (kT_e)^{-1} \nu^{-2}, \tag{7-41}$$

provided that a Maxwellian velocity distribution and thermodynamic equilibrium prevail for the electron gas at temperature T_e. Substitution of equation (7-29) into equation (7-41) and evaluation of the atomic and physical constants yields an absorption coefficient for thermal bremsstrahlung,

$$\kappa_{ff} = 0.01 T_e^{-1.5} \nu^{-2} N_e N(H^+ + He^+) \ \ln \ (5 \times 10^7 \ T_e^{1.5} \nu^{-1}), \tag{7-42}$$

where T_e, ν, and N_e are expressed in the operational units K, Hz, and cm^{-3}, respectively. Because of the weak temperature and frequency dependence of the logarithmic term, commonly known as the Gaunt factor, this expression can be simplified to the more customary form, valid for a range of frequencies $0.1 < \nu < 50$ GHz and a range of nebular temperatures $6000 < T_e < 18,000$ K, to wit,

$$\kappa_{ff} = 0.21 \ a(\nu, T_e) T_e^{-1.35} \nu^{-2.1} N_e N(H^+ + He^+); \tag{7-43}$$

or, from equation (7-7)

$$\tau_{ff} = 6.5 \times 10^{17} \ a(\nu, T_e) T_e^{-1.35} \nu^{-2.1} E_{ff}. \tag{7-44}$$

Here, E_{ff} (pc/cm^6) is the continuum emission measure, which, for a homogeneous region of geometrical thickness $d = 2R$, is defined as $E_{ff} \equiv N_e N(H^+ + He^+) \ d$, where d is in pc ($= 3.1 \times 10^{18}$ cm). And, $a(\nu, T_e)$ is a slowly varying factor, no more than 10% different from unity, that relates the precise form of κ_{ff} (equation 7-42) to its approximate form (equation 7-43).

Unfortunately, radio astronomers do not speak of photon intensity but rather of brightness temperature, a quantity that can be readily related to the signal strength detected by radio telescopes. Specifically, the brightness temperature T^B is defined as that temperature to which a black body must be raised in order to emit the observed flux density. (Actually, there are further complications since radio telescopes are not perfectly efficient and do not detect T^B directly; specified instead is an antenna temperature which is related to T^B, given the characteristics of the instrument and of the observed source.)

Invoking again the Rayleigh-Jeans approximation, easily satisfied for all radio-frequency studies of gaseous nebulae, we can re-express equation

(7-40) as

$$T_{ff}{}^B = \int_0^{\tau_{ff}} T_e e^{-\tau'} d\tau' \qquad (7\text{-}45)$$

and, for an isothermal nebula, distinguish two limiting cases,

$$T_{ff}{}^B = \begin{cases} T_e & (\text{for } \tau_{ff} \gg 1), \\ T_e \tau_{ff} & (\text{for } \tau_{ff} \ll 1). \end{cases} \qquad (7\text{-}46)$$

Figure 7-13a shows the classical continuum spectrum of several hypothetical nebulae with gas temperature T_e and emission measure E_{ff}. At sufficiently low frequency, $T_{ff}{}^B$ is independent of ν and the nebula is considered optically thick; conversely, at high frequencies (generally $\nu > 1$ GHz), $T_{ff}{}^B$ varies approximately as $\nu^{-2.1}$ and the nebula is optically thin.

More meaningful is the total nebular energy budget or integrated brightness across the nebular extent Ω, defined as the flux density

$$F_\nu = \int I_\nu d\Omega, \qquad (7\text{-}47)$$

where F_ν is expressed in units of W m^{-2} Hz^{-1} for which 10^{-26} W m^{-2} Hz^{-1} is known as the jansky (jy). Thus, with $B_\nu(T) \propto \nu^2$, the theoretical variation of flux density as a function of frequency is shown in Figure 7-13b for two representative cases. By comparison, Figure 7-13c illustrates the observed flux density spectra for the Orion and Trifid nebulae. As can be seen, these Trifid observations are insufficient to delineate any low-frequency turnover produced by optical depth, but Orion's spectrum clearly turns over at $\nu \simeq 1$ GHz. For $\nu \gtrsim 1$ GHz, each object shows a relationship $F_\nu \propto \nu^{-0.1}$, close to that expected from the above mathematics. Measurement of the continuum spectrum is then an easy way to identify the true nature of a thermal ($F_\nu \propto \nu^{-0.1}$) source such as a gaseous nebula or a nonthermal source such as a supernova, external galaxy, or quasar for which F_ν typically varies as $\nu^{-0.7}$.

But what can be learned from studies of continuous radio-frequency radiation emitted by gaseous nebulae? The most obvious advantage is that, because of their inability to scatter appreciably, nebular radio waves completely and genuinely delineate the distribution of nebular gas. This is true not only for many well-known local nebulae that have substantial debris along their lines of sight, but also for those more distant galactic nebulae obscured from optical study altogether. Figures 7-14, 7-15, and 7-16 illus-

Figure 7-13. Frequency distribution of thermal bremsstrahlung. *a*, Models of brightness temperature for nebular $T_e = 1.2 \times 10^4$, 10^4, and 8×10^3 K, and emission measure $E_{ff} = 10^6$ pc/cm^6 (dashed lines) and 10^7 pc/cm^6 (solid lines). *b*, Models of integrated brightness for $T_e = 9000$ K and $E_{ff} = 10^6$ pc/cm^6 (dashed line) and 10^7 pc/cm^6 (solid line). *c*, Distribution of observed flux density for the Orion and Trifid nebulae.

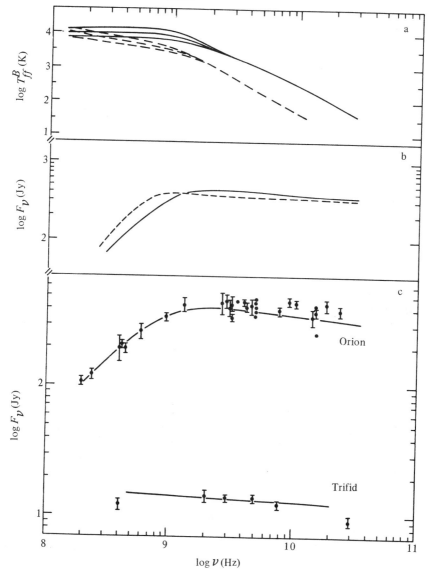

Figure 7-14.   Distribution of 24-GHz thermal bremsstrahlung radiation superposed on the optical image of the Orion Nebula, reproduced here with the permission of the Kitt Peak National Observatory. The isothermal contours increase inward in uniform increments. The cross at the upper right designates the angular response of the radio telescope.

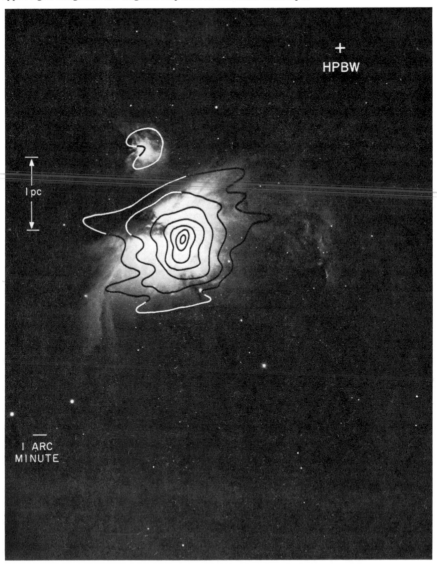

trate maps of T_{ff}^B contours distributed across the Orion and Trifid nebulae. The Orion radiation was sampled at the reasonably high microwave frequency of 24 GHz (λ 1.2 cm) for which the telescope beam or effective spatial resolution was about 1 arcmin. The intensity distribution observed toward the Trifid Nebula is less detailed because, at the 8-GHz (λ 3.8 cm) operational frequency, the spatial response was nearly 4 arcmin. The real power of radio-frequency mapping, however, is demonstrated toward objects completely hidden to optical astronomers. Figure 7-17 shows a 15-GHz (λ 2 cm) radio map made over a larger piece of interstellar real estate. This particular map delineates the visually obscured galactic-center region, containing numerous types of celestial objects including some of the Milky Way's most fascinating gaseous nebulae.

Multifrequency radio maps specify the continuous spectrum from which nebular temperature and density can, in principle, be extracted. Figures 7-13a–13c have already demonstrated how different models specifying T_e

Figure 7-15. Distribution of Orion's radiation viewed from an elevated 30° perspective.

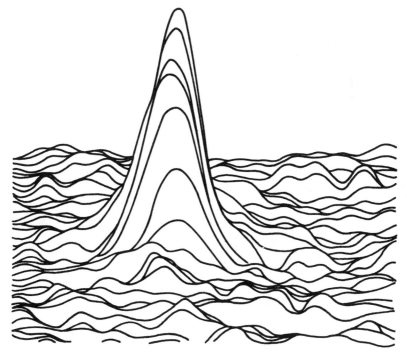

Figure 7-16.   Same as Figure 7-14, but for the Trifid Nebula at 8 GHz. The optical image is reproduced with the permission of Kitt Peak National Observatory.

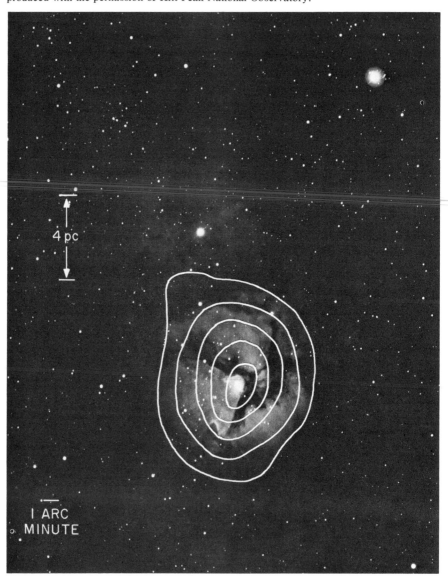

and E_{ff} can be varied appropriately to match an observed nebular spectrum. But fitting the continuum spectrum is not as easy as it looks. For example, since the emergent intensity from an optically thick nebula is actually the same as that from a black body, the measured T_{ff}^B should equal T_e for suffi-

Figure 7-17.   Distribution of 15-GHz radio-frequency radiation in the direction of the totally obscured galactic center. The dotted line denotes a common intensity level, which increases inward thereafter. The dashed line indicates the orientation of the galactic plane. Several prominent sources are noted, including Sagittarius A, the nonthermal galactic nucleus, and Sagittarius B$_2$, the largest, most massive, most turbulent and most excited gaseous nebula known (adapted from the research of Kapitzky and Dent, 1974).

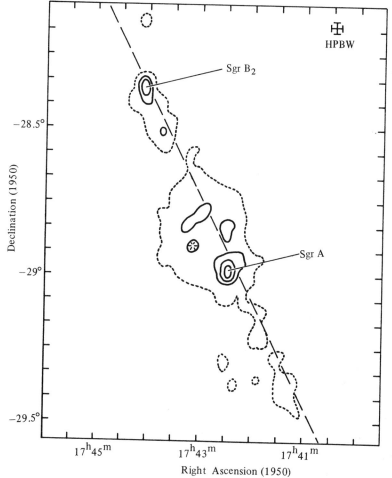

ciently low ν. In practice, however, there are numerous difficulties, chiefly because, at low ν, even the largest radio telescopes have spatial resolutions comparable to or larger than the angular diameters of typical nebulae. Whenever the nebula fails to fill the telescope beam, it is nearly impossible to convert accurately from the instrumentally measured antenna temperature to the physically meaningful brightness temperature. Two other difficulties are that an optically thick nebula necessarily emits radiation only from its foreground periphery and that the galactic background radiation becomes significant at small ν. For these reasons, radio mapping techniques often measure T_e values considerably lower (3000–8000 K) than those found by all other radio and optical techniques. Nonetheless, careful measurements recently made at 400 MHz (λ 75 cm), utilizing modern knowledge of source sizes and of the low-frequency galactic background, have furnished T_e values closer to the predicted 8000 K range. Column 2 of Table 7-3 lists a few values of T_e determined by this technique. That these values are still slightly lower than the 9000–10000 K found by other methods could be explained by the functional weighting of the measured bremsstrahlung intensity toward low temperature regions, that is, $j_{ff} \propto T_e^{-0.5}$.

Another practical problem concerns the derivation of N_e from the continuum spectrum. In principle, the turnover frequency ν_{to} is sufficient to specify N_e since, from equation (7-44),

$$E_{ff} = 1.5 \times 10^{-18} \tau_{ff} a^{-1} T_e^{1.35} \nu_{to}^{2.1}. \tag{7-48}$$

Consequently, if the nebular size and temperature are well determined, N_e follows directly for unit τ_{ff}. But the operational limitation here concerns uncertainties in the estimation of ν_{to}.

Thus few nebulae have an accurately known continuum spectrum. Experimental difficulties abound, notably the inability to secure accurate low-frequency measurements of the total flux density. Fortunately, radio maps made at a sufficiently high frequency where the nebula is optically thin and when the source extent exceeds the telescope beam can yield a wealth of information concerning the bulk properties of the nebular gas.

Firstly, a mean angular size $\theta_g (\propto \Omega^{-1/2})$ of the source follows directly from radio maps like those of Figures 7-14 and 7-16. A deconvolution of the apparent size θ_a of the source with the response pattern θ_b of the telescope is usually necessary to delineate the true-source spatial dimension, $\theta_g = (\theta_a^2 - \theta_b^2)^{1/2}$. θ_g, usually taken to be the half-power width of the radio

Table 7-3 Nebular parameters derived from radio continuum observations.

Nebula	T_e (K)	D (kpc)	$2R$ (pc)	N_e (cm^{-3})	E_{ff} (pc/cm^6)	U (pc/cm^2)	M ($M_\odot$)	$N(Ly_c)$ (photons/sec)	Number O7 stars
Orion (NGC 1976)	8550	0.5	0.5	2000	$10^{6.5}$	50	10	10^{49}	1
Trifid (M 20)	7850	2.1	2.5	100	$10^{4.7}$	60	150	$10^{48.8}$	1
Omega (M 17S)		2.4	2.1	800	$10^{6.3}$	130	300	$10^{49.8}$	7
Lagoon (M 8)		1.1	1.1	400	$10^{5.5}$	40	20	$10^{48.2}$	<1
NGC 2024		0.5	0.5	1300	$10^{5.9}$	30	2	$10^{48.2}$	<1
W 49A[a]		14	11	400	$10^{6.5}$	280	$10^{3.5}$	$10^{50.8}$	60
W 51A[a]		6.5	5.5	1000	$10^{6.5}$	200	900	$10^{50.5}$	30
W 3A (IC 1795)		2.5	2.2	400	$10^{5.5}$	60	50	$10^{49.5}$	2
Sgr B2 (W 24)[a]		10	14	200	$10^{5.7}$	280	10^4	10^{51}	140

[a] These regions are totally obscured from optical view.

map, can be converted into a physical dimension equal to twice the nebular radius R (pc) provided the nebular distance D (kpc) is known. (The distance is often found from a velocity-distance relation derived from a rotation model of the Milky Way Galaxy. For those objects that have identified central stars, a photometric distance can be derived from stellar colors.) Column 4 of Table 7-3 lists $2R$ in parsecs for a wide range of nebulae.

Secondly, nebular density, mass, and emission measure can be determined for a variety of models. Substitution of equation (7-44) into the second choice of equation (7-46) yields

$$T_{ff}^B(x,y) = 6.5 \times 10^{17} a T_e^{-0.35} \nu^{-2.1} \int_0^{2R} N_e N(H^+ + He^+) dx', \quad (7-49)$$

which, when converted from Cartesian to polar coordinates specified by the relations,

$$\theta = (x^2 + y^2)^{1/2} D^{-1}; \qquad \phi = 2RD^{-1}, \quad (7-50)$$

yields

$$T_{ff}^B(\theta,\phi) = 6.5 \times 10^{17} a T_e^{-0.35} \nu^{-2.1} D \int N_e^2 dr. \quad (7-51)$$

Here, we have let $N_e = N(H^+ + He^+)$ since for every hydrogen and helium ion there is a free electron. But regardless of the coordinate system in which this equation is cast, it is impossible to escape the fact that the average density and mass depend sensitively on the density-distribution model used to evaluate the observations. Suppose, for example, that we adopt a gaussian model for which the density distribution is exponentially tapered to a central peak:

$$N_e \simeq N_0 \exp\left[-(\theta^2 + \phi^2)/2(0.6\theta_g)^2\right]. \quad (7-52)$$

Inversion of equation (7-51), substitution of equations (7-47) and (7-52), and some manipulation eventually yields an expression for the density (cm^{-3})

$$N_e \simeq 35 a^{-0.5} \nu^{0.05} T_e^{0.175} F_\nu^{0.5} D^{-0.5} \theta_g^{-1.5}. \quad (7-53)$$

The emission measure (pc/cm^6) follows in a straightforward manner,

$$E_{ff} = 2R N_e \simeq 503 a^{-1} \nu^{0.1} T_e^{0.35} F_\nu \theta_g^2, \quad (7-54)$$

as does the nebular excitation parameter (pc/cm^2),

$$u = R N_e^{2/3} \simeq 2.5 (a^{-1} \nu^{0.1} T_e^{0.35} F_\nu D^2)^{1/3}. \quad (7-55)$$

Note that E_{ff} and u are independent of frequency since $F_\nu \propto \nu^{-0.1}$. Finally, the computation for the total mass of ionized hydrogen in units of solar mass ($M_\odot$) follows directly from equation (7-53) but must include a correction for the electrons produced by helium ionization,

$$M(H^+) \simeq 0.03 a^{-0.5} \nu^{0.05} T_e^{0.175} F_\nu^{0.5} D^{2.5} \theta_g^{1.5} [1 + N(He^+)N(H^+)^{-1}]^{-1}. \quad (7\text{-}56)$$

Note again that the units associated with all the equations of this section are those most useful on an operational basis, namely, ν(Hz), T_e(K), F_ν(jy), D(kpc), R(pc), and θ_g(arcmin).

Table 7-3 compiles values of N_e(cm^{-3}), E_{ff}(pc/cm^6), u(pc/cm^2), and $M(H^+)$ ($M_\odot$) for several well-known nebulae. These values are considered representative of the gross physical properties of extended gaseous nebulae. If the relative ionized-helium number abundance $N(He^+)/N(H^+)$ is taken to be 10%, then the total mass of ionized matter is some 30% larger than the values listed in column 8.

Of course, the nebular quantities just mentioned are model-dependent. If, instead of a gaussian distribution of density, we had chosen a spherical distribution whereby N_e is constant within a sphere of size θ_s and zero elsewhere, then the values of N_e, E_{ff}, and $M(H^+)$ listed in Table 7-3 would be multiplied by factors of 0.8, 1.4, and 0.4, respectively. Values of N_e and E_{ff} are nearly independent of the model, but $N(H^+)$ is considerably larger for a gaussian distribution of N_e, than for a constant distribution since larger amounts of ionized gas are needed to produce a given flux density.

Sampling thermal bremsstrahlung radiation with a radio telescope is therefore a useful means to elucidate several important nebular parameters. However, the single-dish instrument is diffraction-limited in its resolution and is unable to determine the distribution of nebular parameters on a truly fine scale, as can optical studies.

A radio-frequency interferometer can provide finer detail. Radio astronomers are now able to synthesize the response of a partially filled aperture of large diameter and thus to map variations in nebular brightness. Maps, currently made over kilometer dish spacings and with angular resolutions on the order of a few arc seconds, show strong variations in radio emission, called fine structure or clumps, across angular extents comparable to the instrumental resolution. The observed T_{ff}^B variations occur primarily in the inner part of nebulae and imply changes in the physical parameters describing the thermal plasma. Since observations at different frequencies show the clumps to be optically thin, the second choice of equation (7-46) demands that changes in $T_{ff}{}^B$ directly relate to changes in $N_e{}^2 T_e{}^{-0.35}$; furthermore, since T_e is restricted to a relatively narrow range, as discussed earlier, it follows that brightness variations noted by interferometers are primarily the result of N_e inhomogeneities. Two barely resolved clumps recently uncovered in the Orion Nebula require $N_e \simeq 10^{4.5}$ cm^{-3} and $E_{ff} \simeq 10^7$ pc/cm^6, considerably larger than those parameters describing the more extended gas. Their individual masses do not exceed 0.05 $M_\odot$ and their structure typically extends over path lengths $x < 0.04$ pc, appearing to be intrinsically related to remnants of the protostellar cloud from which the young, hot stars of the Orion Nebula formed.

The existence of fine structure is puzzling, however. With T_e invariant for the most part, pressure gradients resulting from such N_e variations should destroy the fine structure on a time scale given by the sound travel time, $x/s \simeq 10^{3.5}$ years. Here, the isothermal sound speed s is taken to be $\simeq 11$ km/sec, a typical value for a gas at 10^4 K. Since this is considerably less than typical nebular lifetimes ($\simeq 10^6$ years), the very existence of high-density clumps is enigmatic unless there exist mechanisms of density resupply or gravitational confinement. Consequently, the observed fine structure must be considered to be the manifestation of new, but not yet understood, physical processes within the nebular plasma. Models currently favored include regions of ionized gas that disperse from knots of high-density matter associated with imbedded neutral hydrogen complexes, and protostellar regions of ionized gas in the interior of compact, gravitationally stable, neutral hydrogen cocoons.

One final caveat: Although optical photographs have long ago demonstrated the existence of density inhomogeneities, there is no evidence at present that nebulae are massively clumped. The two regions of fine structure noted above in Orion contribute no more than 5% of the total bremsstrahlung radiation. Furthermore, interferometers operating over very long baselines (VLBI) have failed to reveal clumped structure on the order of fractions of arc seconds. Thus, it seems that the preponderance of nebular gas can be described adequately by the bulk physical parameters listed in Table 7-3.

Continuous Optical Radiation

Essentially two types of continuous radiation dominate in the optical regime. The first of these is continuous emission at frequency ν, resulting from free-bound transitions when free electrons with velocity v recombine to an atomic level having the principal quantum number n,

$$v = \frac{m_e v^2}{2h} + \frac{v_1}{n^2}. \qquad (7\text{-}57)$$

In analogy with the absorption coefficient for free-free continuous radiation, equation (7-42), we can write an expression for the free-bound absorptivity,

$$\kappa_{fb} = 3.6 \times 10^8 T_e^{-0.5} v^{-3} N_e N(H^+ + He^+)[\exp(hv/kT_e) - 1], \qquad (7\text{-}58)$$

where the parametric units are identical to those of equation (7-42). Figure 7-18 shows a comparison of the two principal continuum mechanisms involving a free electron for a 10^4 K gas. At radio frequencies, the free-free

Figure 7-18. Comparison of the free-free and free-bound continuous absorption coefficients throughout the electromagnetic spectrum. Note how the two-photon emission process confuses measurement of the visible free-bound radiation.

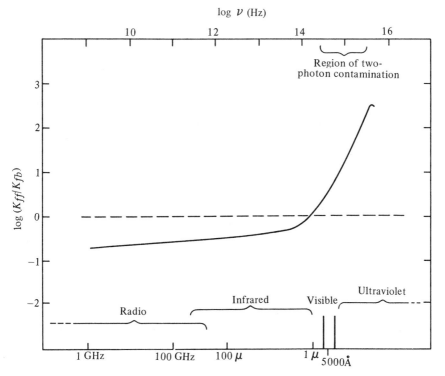

mechanism dominates, ensuring that the previous analyses of the nebular continuous radio radiation are not contaminated by other atomic processes; in the optical regime, the free-bound mechanism is stronger.

However, analysis of optical continuous radiation is not as useful as that at radio frequencies. Two problems arise. The first concerns scattering, which varies as λ^{-4}, making the optical continuum difficult to measure. The second concerns an additional emission process, contaminating the free-bound spectrum. Recall that atoms in the $2^2S_{1/2}$ state, populated by direct recombinations and by cascades from higher levels, are forbidden to relax rapidly to the ground level. Of course, the $2^2S_{1/2}$ level would eventually become depopulated either by forbidden-line emission or by collisional de-excitation. But in low-density regions, another decay process, the so-called two-photon emission mechanism, depopulates the $2^2S_{1/2}$ level with a relatively rapid rate $A(2^2S_{1/2} \rightarrow 1^2S_{1/2}) = 8 \ \text{sec}^{-1}$. The individual energies possessed by each of the two photons are unrestricted, provided that together they conserve the $h\nu = 10.2$ eV for hydrogen. The emission probability distribution centers about $h\nu/2$ and thus peaks at λ 2431 Å (cf. Figure 7-18). The process is further complicated in denser nebular regions ($> 10^4 \ \text{cm}^{-3}$) by collisions, which often alter the angular momentum of a hydrogen atom, shifting it from the $2^2S_{1/2}$ to the $2^2P_{1/2}$ level, which is then permitted to relax radiatively with the emission of precisely 10.2 eV. Consequently, the population of the $2^2S_{1/2}$ level is never known well enough for rigorous theoretical prediction of the two-photon intensity. We discuss this process no further here, except to note that its intensity can become comparable to that of the free-bound process, especially shortward of ~ 6000 Å.

Nebular Energetics

By now, it is obvious that measurement of continuous radiation, particularly in the radio-frequency domain, can provide substantial information about the nebular gas. But this radiation can also yield useful information about the source(s) of nebular excitation, namely the central stars. Consider the following.

The total amount of thermal bremsstrahlung radiation emitted from an optically thin nebula is naturally proportional to the absorption rate of Lyman continuum photons ($h\nu > 13.6$ eV). Provided that none of the ultraviolet radiation generated by the exciting star(s) escapes the nebula (ionization bounded) and that dust does not compete appreciably with the gas in absorbing photons with $h\nu \geq 13.6$ eV, then the total ultraviolet emission rate

$N(\text{Ly}_c)$ (photons/sec) at frequency ν (Hz) from a nebula of temperature T_e (K) depends primarily on the product of the distance D (kpc) squared and the radio flux density F_ν (jy);

$$N(\text{Ly}_c) = 6 \times 10^{47} F_\nu D^2 \nu^{0.1} T_e^{-0.45}. \qquad (7\text{-}59)$$

The rate of Lyman-continuum photon emission can then be calculated by simply measuring the radio flux from a nebula whose distance and temperature are known. The relationship plotted in Figure 7-19 between stellar type and ultraviolet energy output in the Lyman continuum as measured by the

Figure 7-19.   The Lyman-continuum photon emission rate plotted as a function of effective stellar surface temperature, according to the Harvard spectral type classification.

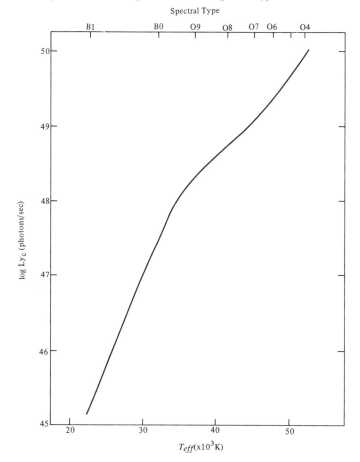

stellar surface temperature T_{eff} can then be used to estimate the type and number of stars necessary to account for the nebular ionization structure. For example, the 24-GHz bremsstrahlung map of the Orion Nebula shown in Figure 7-14 yields $F_\nu = 380$ jy, for which $N(Ly_c) \simeq 10^{49}$ photons/sec for $D = 500$ pc and $T_e = 10^4$ K. This rate of Ly_c photon emission agrees (Figure 7-19), with the central O6-type star known to excite Orion. Similarly, the 8-GHz map of the Trifid Nebula's continuous radiation shown in Figure 7-16 implies an O7-type star, in agreement with the known spectral type of a star centrally located and positively identified as the Trifid's source of excitation.

The exciting stars for most nebulae, however, are unknown, especially for those distant regions totally obscured from optical view. Yet even for some nearby nebulae, like the large M 17 region, the central stars are not yet identified. Thus, the integrated radio brightness provides us with a reasonably good estimate of the effective properties of some nebular exciting stars we can never hope to see. The last two columns of Table 7-3 list the $N(Ly_c)$ photon-emission rate and the implied effective number of O7-type stars needed to excite sufficiently several prominent nebulae.

Nebular Spectroscopy

Discrete bound-bound transitions within some of the previously noted species give rise to narrow features called spectral lines, superposed on the spectrum of continuous radiation. The study of these spectral features constitutes the science of nebular spectroscopy, a powerful tool enabling experimentalists to study any single nebular species to the exclusion of all others. Year after year, spectroscopy provides for us more accurate experimental data with which to build more realistic models of nebulae and their interstellar environment.

Having achieved considerable understanding of many bulk nebular properties from analysis of the radio-frequency continuum, we begin our study of nebular spectroscopy in the radio-frequency domain.

Radio-Frequency Regime

Recombination lines are the only spectral features emitted by gaseous nebulae at very low-energy radio frequencies. To date, only five species—hydrogen, helium, ionized helium, carbon, and a heavier emitter, probably sulfur—are known to produce detectable radio lines. Of these, only hydrogen and helium are directly associated with gaseous nebulae, as carbon and the heavier emitter arise in cool, predominantly neutral gas in the direction of nebulae. The recently discovered ionized helium radio recombination line has so far been found only toward planetary nebulae, interstellar regions often excited by exceedingly hot stars ($> 10^5$ K) capable of emitting copious quantities of the 54-eV photons necessary to doubly ionize He. Planetary nebulae, however, are thought to be in an advanced stage of stellar evolution and thus differ considerably from the younger gaseous nebulae considered in this chapter. Consequently, this section is restricted to the mathematical methods used to extract physical, chemical, and kinematic information from the measurement of H and He spectral features.

Figure 7-2 has already displayed the energy differences between adjacent ($\Delta n = 1$) levels of the H atom as a function of terminal quantum number. Provided that levels $n > 30$ are sufficiently populated, transitions between Δn levels will yield H spectral features in the radio domain ($\nu < 300$ GHz) at the precise frequencies (cf. equation 7-4)

$$\nu = RcZ^2[n^{-2} - (n + \Delta n)^{-2}], \tag{7-60}$$

or if $n \gg \Delta n$,

$$\nu \simeq 2RcZ^2 \Delta n n^{-3}(1 - 3\Delta n/2n). \tag{7-61}$$

Here, the Rydberg constant R can be generalized to include more massive emitters,

$$R = R_\infty(1 - m_e/M), \tag{7-62}$$

where R_∞ is the Rydberg constant for infinite mass and M is the mass of the entire atomic system, including the electron mass m_e. Consequently, for a given n, emission lines of elements heavier than hydrogen are shifted toward higher frequencies. Because this mass shift is a linear function of ν, the separation between elements of different M is frequency invariant when expressed in radial velocity v_x space according to the radio astronomical Doppler convention:

$$\nu_{obs} = \nu_{rest}(1 - v_x/c). \tag{7-63}$$

The top of Figure 7-20 shows a schematic diagram of the relative separation of emitters for three consecutive values of n, provided $\Delta n = 1$. Note that, although the mass shift is highly nonlinear, emission from all members of the periodic table are condensed into a velocity interval of 163 km/sec for each n. The effects of fine structure, caused by spin-orbit interaction, and of non-Coulomb potentials for emitters heavier than hydrogen decrease rapidly with increasing azimuthal quantum number, producing entirely negligible shifts of the line frequencies. In other words, the recombining electron sees an effective nuclear charge $Z = 1$ for any singly-ionized element, provided that $n > 30$.

The lower half of Figure 7-20 shows a brief recombination-line observation toward the Orion Nebula for $n = 92$, $\Delta n = 1$. Recombination lines are mathematically specified by their intensity, full-width at half intensity, and velocity centroid, and are conventionally labeled by their chemical symbol, terminal level of the transition, and change in n. The strongest line displayed in Figure 7-20 is then called H 92α. (Another example would be the $n = 156 \rightarrow 153$ transition in helium, for which the line is designated He 153γ.)

Spectral lines can yield a wealth of information about the emitting or absorbing gas, whether it be laboratory or cosmic. The starting point in the analysis of any spectral line is always the equation of transfer. Generalizing

Figure 7-20. Schematic diagram showing frequency-invariant velocity displacements among adjacent α sets and among elements of different mass within those α sets. Also shown is an actual observation of the 92α set toward the Orion Nebula, showing evidence for H 92α and He 92α.

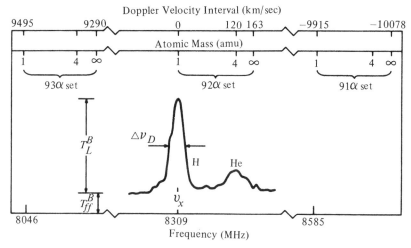

equation (7-6) to include emission and absorption for both line and continuum radiation, we have for the alteration in total intensity at any ν,

$$\frac{dI_\nu}{dx} = -\kappa_L I_\nu - \kappa_{ff} I_\nu + j_L + j_{ff}. \tag{7-64}$$

Here the subscripts ff and L refer to the continuum and spectral-line quantities. Integration, manipulation, and utilization of the Rayleigh-Jeans approximation yields a general expression for the brightness temperature of a spectral line in the absence of a background radiation field:

$$T_L{}^B = T_e[1 - e^{-(\tau_L + \tau_{ff})}] - T_e(1 - e^{-\tau_{ff}}), \tag{7-65}$$

where T_e is a homogeneous measure of the nebular gas or electron temperature. If local thermodynamic equilibrium (LTE) prevails, then T_e becomes a true measure of the Maxwellian velocity distribution of the free electrons. Since spectral lines are superposed on the nebular continuous radiation field, it is pleasing that the form of the above expression, $(T_L{}^B + T_{ff}{}^B) - T_{ff}{}^B$, is as expected heuristically and drawn schematically in Figure 7-20.

If the line optical depth is vanishingly small, as will be subsequently proved for a 10^4 K gas, then equation (7-45) and a Taylor series expansion of equation (7-65) about τ_L combine to form a ratio of line and continuum brightness temperatures,

$$\frac{T_L{}^B}{T_{ff}{}^B} = \frac{\tau_L e^{-\tau_{ff}}}{(1 - e^{-\tau_{ff}})} \qquad \text{(for } \tau_L \ll 1\text{)}, \tag{7-66}$$

for which we can distinguish two limiting cases, in analogy with equation (7-46),

$$\frac{T_L{}^B}{T_{ff}{}^B} = \begin{cases} \tau_L/\tau_{ff} & \text{(for } \tau_L, \tau_{ff} \ll 1\text{)}, \\ \tau_L e^{-\tau_{ff}} & \text{(for } \tau_L \ll 1, \tau_{ff} \gg 1\text{)}. \end{cases} \tag{7-67}$$

Immediately it becomes apparent that the detectability of recombination lines improves at high ν, since below the turnover point ($\nu_{to} < 1$ GHz) the lines merge progressively into the continuum. But instrumental difficulties at present restrict accurate observations to $\nu < 25$ GHz.

A mathematical understanding of τ_{ff} has already been achieved. If similar appreciation can now be developed for τ_L, then the dynamic observables $T_L{}^B$ and $T_{ff}{}^B$ could yield considerable information about nebulae, especially gas temperature, perhaps the most meaningful nebular parameter. But no doubt a question persists at this point: Why not analyze simply the line radiation without recourse to the continuum? The reason is that ratios of different brightness temperatures at a given ν, a line to another line, or a line to its adjacent continuum, alleviate the difficulty of experimentally converting from the directly observed antenna temperature to the physically understood brightness temperature; a ratio of brightness temperatures approximates very well a ratio of antenna temperatures, provided the antenna responses are similar. Hereafter, then, the superscript B is omitted from the analysis.

To find a mathematical expression for τ_L, we seek to specify κ_L, the absorption coefficient in the line, given generally by the net difference between absorption and emission as electrons transit from an upper level m to a lower level n, to wit

$$\kappa_L = h\nu(4\pi)^{-1}\psi(N_n B_{nm} - N_m B_{mn}).   \tag{7-68}$$

Evaluation of this expression is not easy. Essentially three items must be specified; the line-shape factor ψ, the volume number density N_n in level n, and the rates of absorption B_{nm} and stimulated emission B_{mn}.

Firstly, the chief source of line shape is Doppler broadening, provided the nebular gas is characterized by a Maxwell-Boltzmann distribution at some kinetic temperature T_k, not necessarily equal to T_e unless, of course, LTE prevails;

$$\psi = 2\Delta\nu_D^{-1}(\ln 2/\pi)^{0.5} \exp(-M v_x^2/2kT_k)   \tag{7-69}$$

or

$$\psi = 2\Delta\nu_D^{-1}(\ln 2/\pi)^{1/2} \exp[-4 \ln 2\Delta\nu_D^{-2}(\nu - \nu_{mn})^2]   \tag{7-70}$$

where $\Delta\nu_D$ is the spectral line's full width at half intensity. Of course, equation (7-68) implicitly assumes that this line-shape factor applies identically for absorption and emission.

Secondly, for collisionally excited states populated according to LTE at some temperature T_e, the Saha-Boltzmann equation specifies the number density of atoms in some level having a statistical weight $g_n = 2n^2$, to wit,

$$N_n = N_e N_i (h^2/2\pi m_e k T_e)^{3/2}(g_n/2P_i) \exp(h\nu_{n\infty}/kT_e),   \tag{7-71}$$

where, fortunately, for a 10^4 gas, the ionic partition function P_i and the exponential term for high n transitions are unity. Note that for line emission, each species has its own Saha-Boltzmann equation; N_i here is not the sum of all ions in the thermal gas, as was earlier the case for the interaction of all ions giving rise to the bremsstrahlung continuum.

And thirdly, the radiative rates are related to the spontaneous decay rate A_{mn} according to conventional quantum mechanics,

$$A_{mn} = \frac{2h\nu^3}{c^2} B_{mn} = \frac{2h\nu^3}{c^2} \frac{g_n}{g_m} B_{nm}. \tag{7-72}$$

Here, A_{mn}, B_{mn}, and B_{nm} are the Einstein coefficients for spontaneous emission, stimulated emission, and absorption, respectively. Expressed alternatively,

$$A_{mn} = 8\pi e^2 \nu^2 g_n m_e^{-1} c^{-3} g_m^{-1} f_{nm}, \tag{7-73}$$

where f_{nm}, the oscillator strength or f-value, is a measure of the probability that an atom will undergo a given transition; f-values depend only on the atom's structure and have been computed from theory or measured in the laboratory for a variety of possible transitions. The results show that, for $50 \lesssim n \lesssim 900$, the values of f_{nm} diminish rapidly with Δn: $(f_{nm}n^{-1}) \simeq 0.20$, 0.03, and 0.01 for $n = 1(\alpha)$, $2(\beta)$, and $3(\gamma)$ transitions, respectively. Consequently, the less probable $\beta, \gamma, \cdots$ transitions will produce lines only 28, 13, $\cdots$ % as intense as α transitions since, as we shall see below, $T_L \propto \Delta n f_{nm} n^{-1}$.

Actually, the oscillator strength can be estimated for any level, provided the Einstein A coefficient is known. To do this, consider for a moment the fact that the atomic transitions discussed here obey the electric-dipole selection rules. Classical electrodynamics then predicts that the power radiated as the dipole oscillates along some displacement y in time t will be

$$dE/dt = h\nu_{mn}A_{mn} = \frac{2e^2}{3c^3} \left\langle \left| \frac{\partial^2 y}{\partial t^2} \right| \right\rangle^2. \tag{7-74}$$

Consequently, if the oscillatory displacement at some distance r from the nucleus is

$$y = re^{i\omega t},$$

then differentiation and substitution yield an evaluated expression for the radiative rate

$$A_{mn} \simeq 10^9 n^{-5}. \tag{7-75}$$

For typical radio lines around $n \simeq 100$, the lifetime against spontaneous decay is then on the order of a second.

Substitution of equations (7-70), (7-71), and (7-73) into equation (7-68) finally furnishes a general expression for the absorption coefficient which, upon evaluation at the line center, becomes

$$\kappa_L = 3.3 \times 10^{-12} N_e N_i T_e^{-2.5} \Delta\nu_D^{-1} f_{nm} n^{-1} \Delta n Z^2 (1 - 3\Delta n/2n)(1 - m_e/M),$$
(7-76)

or, from equation (7-7),

$$\tau_L = 10^7 T_e^{-2.5} \Delta\nu_D^{-1} E_L f_{nm} n^{-1} \Delta n Z^2 (1 - 3\Delta n/2n)(1 - m_e/M), \quad (7-77)$$

where $E_L = \int N_e N_i dx$ (pc/cm^6) is the line emission measure averaged along the line of sight. Substitution of reasonable quantities shows that $\tau_L \simeq 10^{-3} < 1$, verifying the Taylor series expansion that led to equation (7-66).

Having specified τ_L under the special conditions of LTE assumed above, we are now in a position to evaluate the so-called line-to-continuum ratio. Substitution of equations (7-44) and (7-77) into equation (7-67) for the optically thin regime (τ_L, $\tau_{ff} \ll 1$), where most experiments are currently conducted ($\nu > 1$ GHz), yields

$$T_L T_{ff}^{-1} = 1.5 \times 10^{-11} (\Delta\nu_D n a E_{ff})^{-1} T_e^{-1.15} f_{nm} \Delta n Z^2 E_L \nu^{2.1}$$
$$\times (1 - 3\Delta n/2n)(1 - m_e/M) \quad (7-78)$$

where, contrary to most equations appearing in the literature, $\Delta\nu_D$ and ν are expressed in Hz. For an H ($Z = 1$) recombination line arising from transitions between adjacent levels ($\Delta n = 1$), this equation reduces considerably to an operational expression which depends on the observables T_L, T_{ff}, and $\Delta\nu_D$:

$$\Delta\nu_D T_L T_{ff}^{-1} = 2.5 \times 10^{-12} a^{-1} T_e^{-1.15} \nu^{2.1} (6f_{nm} n^{-1})(E_L E_{ff}^{-1}). \quad (7-79)$$

When H recombination lines were first detected about ten years ago, both a^{-1} and the parenthetical terms in this equation were taken to be unity. However, more accurate calculations of the oscillator strengths now suggest that $(6f_{nm} n^{-1}) \simeq 1.16$ for any α-transition in the range $60 < n < 170$. In the same range, a can be taken to be 0.98, which is accurate within 5%. The final term

$$E_L E_{ff}^{-1} = \int N_e N_i \, dx \Big/ \int N_e N(\text{H}^+ + \text{He}^+) \, dx \quad (7-80)$$

reduces for H to

$$E_L E_{ff}^{-1} = N(\text{H}^+) N(\text{H}^+ + \text{He}^+)^{-1}, \quad (7-81)$$

provided the path lengths for continuous and line radiation are similar. Measurement of He $n\alpha$ lines can provide the necessary estimate of E_L/E_{ff}, as described below.

Numerous observations have employed equation (7-79) to derive the nebular electron temperature, based on LTE considerations, for a wide variety of gaseous nebulae. The pioneer workers almost invariably found $T_e \simeq 5000$–7500 K, considerably lower than the 9000–10,000 K values expected theoretically from studies of nebular thermal balance or determined observationally by several other techniques. However, measurements of nebular line and continuum radiation made recently by a number of observers have furnished the value $T_e \simeq 9000$ K, which agrees reasonably well with expected values, but disagrees considerably with pioneering derivations of T_L/T_{ff}. For example, recent observations of hydrogen recombination lines toward the Trifid and Orion nebulae yield $T_e \simeq 8200$ and 8900 K. Table 7-4 compiles average values of T_e derived from some recent measurements of the hydrogen line and its associated continuous radiation for several representative nebulae. Almost invariably, they show $T_e = 8000$–$10,000$ K. Furthermore, reasonable values of T_e ($\simeq 9000$ K) are now derived from hydrogen recombination-line observations over a wide range of n, as shown for the Orion Nebula in Figure 7-21.

One can then ask: Why did the pioneering radio-line researchers derive values of $T_e \simeq 6500$ K, whereas modern measurements clearly show $T_e \simeq 9000$ K? The answer probably relates to improvements in receiving equipment and observational techniques. Radio observers now compare the

Table 7-4   Nebular parameters derived from radio spectroscopy.

Nebula	T_e (K)	v_t (km/sec)	T_k (K)[a]	He+/H+
Orion (NGC 1976)	9000	10	10,500	0.08
Trifid (M 20)	8200	7	—	0.10
Omega (M 17S)	8700	17	7,500	0.10
Lagoon (M 8)	7500	9	—	0.09
NGC 2024	8000	5	—	0.02[b]
W 49A	9800	12	—	0.06
W 51A	7800	16	—	0.08
W 3A (IC 1795)	9500	10	12,000	0.08
Sgr B2 (W 24)	9000	24	—	$\begin{cases} <0.02 \text{ for } n > 85 \\ \phantom{<}0.09 \text{ for } n = 76 \end{cases}$

[a] T_k cannot be specified for most nebulae because of Δv_D (He) uncertainties.
[b] Antenna beamwidth $\gg R$ (He) dilutes He signal.

observed nebular signal with cosmic or laboratory standards more frequently than typical instrumental fluctuations can occur, thus calibrating
line intensities better and preserving unprecedented baseline stability. Also,
continuum measurements T_{ff} are now made concurrently with line measurements T_L, to yield an accurate T_L/T_{ff} ratio. Yet, despite this current work,
some researchers still regard nebular temperatures derived from radio recombination lines as considerably lower than those derived from other techniques. This is a classical problem that exists throughout many areas of science. In the present case, the process goes like this: Pioneering observers
make an incorrect measurement, finding $T_e \simeq 6000$ K. Theorists build new
models to incorporate the low temperature. Other researchers, respecting
the pioneering and theoretical work, also report $T_e \simeq 6000$ K. A bandwagon
effect ensues and publication after publication repeats the same theme, that

Figure 7-21. All H recombination-line data acquired toward the Orion Nebula, provided that
$\tau_{ff} \ll 1$. Dashed curves are LTE models for: *a*, 8000 K, *b*, 9000 K, and *c*, 10,000 K homogeneous nebular temperatures.

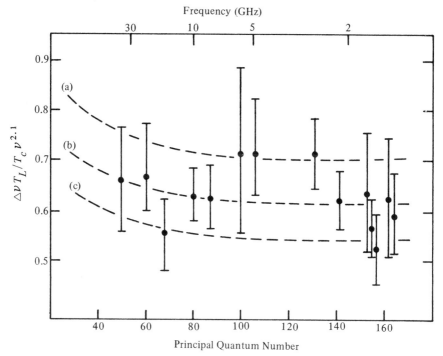

radio-line nebular temperatures are lower than expected. A myth thus evolves into a "truth"—at least for a while.

At any rate, the attempt to account for the earlier 6000 K nebular temperatures led to the development of non-LTE theory. We examine this briefly here because, although non-LTE processes may not have consequences as great as once thought, they may still alter recombination-line intensities to some degree.

Generally, departures from thermal equilibrium modify LTE line-intensities by making the spontaneous transition rate proportional to the level departure coefficient b_n and by altering the magnitude of the stimulated emission in proportion to the differential variation of b_n with n; that is, $d \ln b_n/dn$. Let the true population of the nth level be

$$N_n = b_n N_n^*, \tag{7-82}$$

where N_n^* denotes the corresponding population under LTE conditions. Alteration of equation (7-68) to include these departures then leads to a new expression for τ_L which, when compared with τ_{ff}, yields a general non-LTE expression

$$\left(\frac{\Delta \nu_D T_L}{T_{ff} \nu^{2.1}}\right) = \left(\frac{\Delta \nu_D T_L}{T_{ff} \nu^{2.1}}\right)^* b_n \left(1 + \frac{\tau_{ff}}{2} \frac{kT_e}{h\nu} \frac{d \ln b_n}{dn} \Delta n\right). \tag{7-83}$$

General solutions for the b_n coefficients are calculated under a steady-state assumption that governs the radiative and collisional rates in and out of level n. In principle, the problem is impossible to solve because an infinite number of equations need be solved simultaneously. However, a solution for several hundred levels, like that shown in Figure 7-22, has been found by using the best electron-ion collision cross-sections available. The resultant shape of the b_n curve is as expected; the farther a bound electron is from its parent nucleus, the more susceptible it becomes to increased collisions. Collisional effects, which redistribute the populations toward LTE, grow progressively weaker with decreasing n until radiative effects completely control the level population. The b_n curves furthermore show how, for a given n, larger N_e causes the population to approach LTE through more frequent collisions. A reasonable choice of N_e and n then guarantees the validity of the Saha-Boltzmann equation used previously for collisionally excited levels.

Thus, the first term on the right of equation (7-83) demonstrates how the spontaneous emission is controlled by the b_n coefficients, and the second

term represents the stimulated emission contribution. For radio applications, then, it appears that if non-LTE processes are going to alter the radio-
line intensity at all, the stimulated emission term must play an important
role, since $1 \geqslant b_n > 0.75$ for all $n > 60$. Early theorists, attempting to en-
hance T_L by such non-LTE processes in order to account for the earlier low
T_e values, were often forced to conclude that $N_e > 10^{4.5}$ cm^{-3} for the regions
responsible for radio-line emission. That this is so can be seen from equation

Figure 7-22. Population departure coefficient b_n and its differential variation $d \ln b_n/dn$ plotted
against principal quantum number for $T_e = 10^4$ K and a variety of N_e (from Brocklehurst, 1970).

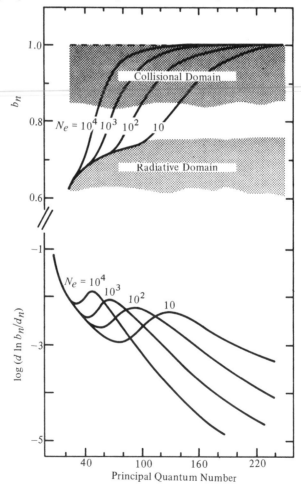

(7-83) where the second term, proportional to $\tau_{ff} \, d \ln b_n/dn$, requires large $\tau_{ff} \propto N_e^2$, regardless of the magnitude of $d \ln b_n/dn$.

However, as explained above, such line enhancements are no longer needed to explain the observed values of T_e. In fact, there is no way whatever that models incorporating constant $T_e = 10^4$ K and constant $N_e = 10^{4.5}$ cm^{-3} can fit the Orion data of Figure 7-21.

The question remains: Are nebulae out of LTE? The answer assuredly is yes, for the b_n coefficients do in fact depart from unity. But it is not at present clear, especially in view of the data of Figure 7-21, that non-LTE processes have any measurable effect on the radio-line formation mechanism. Indeed, there are no compelling observational data demonstrating that nebulae are removed from LTE. The final answer may require elaborate models incorporating radial variations of N_e and T_e, as some researchers have recently proposed but, in view of the data in Figure 7-21, it seems ironic indeed that non-LTE models will require the ratio T_L/T_{ff} to vary with n in nearly the same way as LTE theory predicts!

Enough discussion of line intensity. What affects the line width? Theoretically, it takes only a few minutes for the electron gas to redistribute itself according to a Maxwellian velocity function. Consequently, observed spectral lines will be Doppler broadened, producing a gaussian profile having, from equations (7-63) and (7-69), a full width at half intensity,

$$\Delta \nu_D = \frac{2\nu_{mn}}{c} \left(\frac{2kT_k}{M} \ln 2 \right)^{1/2}. \tag{7-84}$$

But previously discussed thermodynamic knowledge dictates that thermal Doppler motions corresponding to 10^4 K can account for no more than about 50% of the observed line widths (≈ 30 km/sec). In reality, nonthermal motions of bulk nebular matter are also likely to contribute to the broadening of an observed feature. Microturbulent motions having a most probable velocity v_t will not cause departures from a gaussian line shape, provided there are many turbulent cells randomly distributed in the telescope beam. By microturbulence, we mean turbulence that is small compared with the instrumental beam size (of order arc minutes), but not necessarily small compared with the mean free path of an ultraviolet photon. With this caveat, equation (7-84) can be generalized:

$$\Delta \nu_D = \frac{2\nu_{mn}}{c} \left[\left(\frac{2kT_k}{M} + v_t^2 \right) \ln 2 \right]^{1/2}. \tag{7-85}$$

The magnitude of the turbulence then follows directly from the observed line width if the nebular gas temperature is known. Column 3 of Table 7-4 lists values of v_t, generally in excess of the 11 km/sec sound speed of a 10^4 K gas. But some nebulae, notably those with small excitation requirements, have v_t measures less than Mach 1. For example, Figure 7-23 displays the distribution of v_t as a function of stellar $N(\mathrm{Ly}_c)$ emission rate. The relation between $N(\mathrm{Ly}_c)$ and the effective number of O-type stars (cf. Figure 7-19) then suggests that the often observed supersonic turbulence may be the accumulated product of individual subsonic cells maintained by high-velocity stellar winds.

The Stark effect presents a potential source of additional line broadening. Bound electrons, in levels of high n and hence far removed from their parent nuclei, are substantially influenced by inelastic collisions with the surrounding plasma. Although a complete development of Stark or, to be more precise, impact broadening is beyond the scope of this chapter, we can

Figure 7-23. Distribution of most probable nebular turbulence as a function of stellar ultraviolet emission rate. The dashed line denotes a least-squares solution.

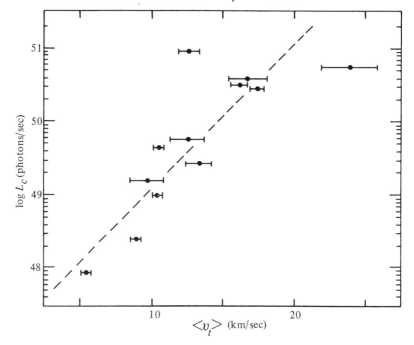

sketch the principal result, expressed as a ratio of Stark half-width $\Delta\nu_s$ to Doppler half-width $\Delta\nu_D$:

$$\frac{\Delta\nu_s}{\Delta\nu_D} \simeq \frac{9.5 \times 10^{-16}N_e n^7[\ln(nT_e) - 11.5]}{(T_e^2 + 4 \times 10^{-5}T_e v_t^2)^{0.5}}. \tag{7-86}$$

The net observed profile is actually a convolution of a Doppler-broadened gaussian function and a Stark-broadened Lorentzian function. Ultimately, the effect of Stark broadening is a sharp deviation of the observed line width from the linear dependence on ν_{mn} specified for pure Doppler effect. This is particularly true at large n (cf. equation 7-86) as the electron-atom collision cross-section increases dramatically. The solid curves of Figure 7-24 show the result of the convolution for different values of N_e. These calculations assume $T_e = 9000$ K and are normalized to the lowest n (= 39) currently ob-

Figure 7-24.   H recombination line width observed toward the Orion Nebula when $\tau_{ff} \ll 1$ superposed on a theoretical convolution of a Doppler Gaussian (dashed) and a Stark Lorentzian (solid) curve for different N_e.

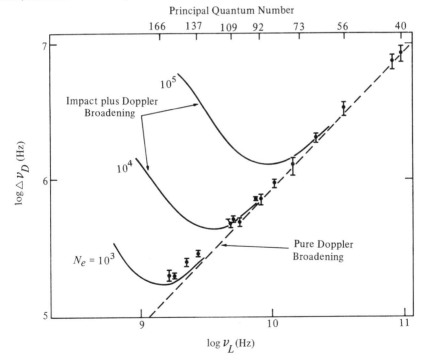

served. The data, compiled from observations made toward the Orion Nebula, suggest a departure from pure Doppler broadening at $n \simeq 150$, consistent with $N_e \simeq 10^{3.2}$ cm^{-3}. Interestingly enough, this is about the same N_e
derived earlier from studies of Orion's radio-frequency continuum. Better
observations, especially around $n \simeq 160$, will be necessary to confirm this
suggestion as genuine Stark effect or refute it as perhaps an effect of increasing size of the telescope beam. Unfortunately, observations at $n > 180$ will
be of little value since $\tau_{ff} \rightarrow 1$. At any rate, there is no evidence for emission
from clumps with $N_e > 10^4$ cm^{-3}.

The line width can also be used to estimate the nebular temperature, independent of the level populations. This method, however, requires detection
of two reasonably well-mixed nebular elements. Helium $n\alpha$ lines, detected
now toward many nebulae, can be used provided the helium line profile is
measured well. Describing each hydrogen and helium width by the general
form of equation (7-85) and solving these equations simultaneously under the
assumption that the hydrogen and helium species are reasonably well mixed,
we derive the nebular kinetic temperature:

$$T_k = \frac{c^2}{8k \ln 2}\left[\frac{M(\mathrm{H})M(\mathrm{He})}{M(\mathrm{He}) - M(\mathrm{H})}\right]\left\{\left[\frac{\Delta\nu_D(\mathrm{H})}{\nu_{mn}(\mathrm{H})}\right]^2 - \left[\frac{\Delta\nu_D(\mathrm{He})}{\nu_{mn}(\mathrm{He})}\right]^2\right\}. \quad (7\text{-}87)$$

Column 4 of Table 7-4 lists mean values of T_k determined in this way from
low-n observations uninfluenced by Stark effect. These values are similar to
the modern T_e values found from line and continuum measurements and,
being independent of the level populations, suggest once again that non-LTE
processes have a negligible effect on the radio line formation mechanism.
Unfortunately, values of T_k are generally uncertain since T_k depends on the
difference between the squares of the line widths. Specification of $\Delta\nu_D$ (He)
is further complicated by the presence of the carbon $n\alpha$ recombination line
that arises external to the nebula. Reasonably accurate observations of helium line widths, like that shown in Figure 7-25 for $n = 76$ toward the Orion
Nebula, have been made toward only a few nebulae to date.

Observations of helium radio recombination lines also permit an estimate of relative helium abundance in gaseous nebulae. This estimate is easier to obtain if hydrogen and helium lines are observed for the same n value,
to avoid differences in oscillator strength, departure from LTE if any, and
instrumental beam response. Furthermore, if the regions of ionized hydrogen and ionized helium spatially coexist, then equation (7-79) yields a

ratio of (directly observed) integrated line energies,

$$\frac{T_L(\text{He})\Delta\nu_D(\text{He})}{T_L(\text{H})\Delta\nu_D(\text{H})} = \frac{N(\text{He}^+)}{N(\text{H}^+)} . \tag{7-88}$$

An average of several individual values listed in column 5 of Table 7-4 yields the ratio $N(\text{He}^+)/N(\text{H}^+) \simeq 0.09$, which in turn equals $N(\text{He})/N(\text{H})$ if there are no appreciable amounts of neutral or doubly-ionized helium. Actually, this is regarded as a fairly good assumption for most gaseous nebulae, though there are some low-excitation objects like NGC 2024 that have Strömgren spheres $R(\text{He}^+) < R(\text{H}^+)$ and thus artificially low values of $N(\text{He}^+)/N(\text{H}^+)$, as noted in Table 7-4. Other objects like Sagittarius B2 are mystifying, appearing to emit helium $n\alpha$ lines at some frequencies but not at others.

Helium abundance determinations are of considerable importance not only for gaseous nebulae but for all astrophysics as well. Since the proton-proton cycle that fuses helium in stellar interiors generally accounts for $N(\text{He})/N(\text{H}) < 0.01$ given the lifetime of our Galaxy, cosmologists generally seek to account for the remaining 8–9% He in the early stages of the big bang. Objects for which $N(\text{He})/N(\text{H})$ is substantially less than the 10% cosmic abundance therefore present potential difficulties for many cosmol-

Figure 7-25. Long-integration observations of the 76α set toward the Orion Nebula demonstrating the accuracy to which $\Delta\nu_D(\text{He})$ must be specified for an accurate determination of $\langle T \rangle$. The H intensity is ten times that of He. A C 76α line may be noted to the high-frequency side of the He 76α profile. The length of the horizontal arrow is a measure of the instrumental resolution.

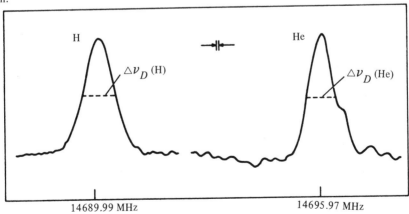

ogies, there being no known way to destroy inert He once formed. Thus, He abundance measurements made with unscattered radio waves toward a large number of galactic objects may ultimately have profound implications for our knowledge of the nature of the universe itself.

Finally, in addition to intensity and width, spectral lines are characterized by their central frequency or, by equation (7-63), their radial velocity centroid v_x. Together with a specific rotation model of the Milky Way Galaxy, observed values of v_x can be converted into distance. Large surveys of hundreds of nebulae in the northern and southern hemispheres, most of them totally obscured, have now furnished a radial distribution of H II gas in much the same way that λ 21-cm line studies have delineated the spatial distribution of galactic H I gas. Figure 7-26 shows the spatial distribution of gaseous nebulae, mostly concentrated within a zone 4–6 kpc from the galac-

Figure 7-26. Galactic distribution of gaseous nebulae derived from measurement of radial velocity and the Schmidt model of galactic rotation. Notable nebulae tabulated throughout this chapter are labeled. Galactic latitude of a polar coordinate system centered on our sun is indicated, along with distances (kpc) from the galactic center. (Adapted from Reifenstein, 1968, and Wilson, 1970.)

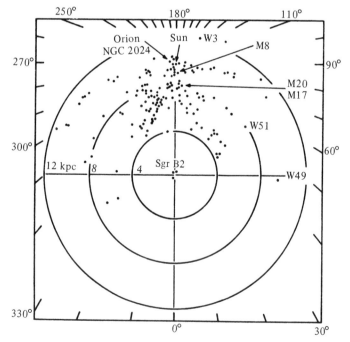

tic center. Some hint of spiral structure can be seen, especially for the objects in the 50–180° quadrant. If radio recombination lines observed toward the so-called diffuse interstellar gas regions, devoid of well-known nebulae, nevertheless eventually come to be interpreted as emission from nebulae, as now seems probable, then the statistics in Figure 7-26 will improve considerably.

All the above discussion has dealt with spectroscopy at nebular centers. One of the principal advantages of spectroscopy, however, is its ability to yield spatial distributions of nebular parameters. Profile intensities, widths, and centroids have been mapped extensively across several nebulae. Variations in temperature, turbulence, abundance, and kinematics have provided substantial information concerning individual gaseous nebulae. For example, Figure 7-27 shows the distribution of radial velocity across the Trifid Nebula, the result of mapping the H 110 α and H 94 α lines at some twenty locations. Many nebulae, like parts of Trifid, appear to be rotating with periods on the order of 10^6 years. However, differential expansion or shearing of the hot gas against the cold surrounding environment could give rise to similar radial-velocity distributions. Some nebulae have recently been shown to have helium Strömgren spheres smaller than those of hydrogen, implying that dust may effectively compete with the gas for 24-eV photons. Still other nebulae have displayed suggestive evidence for a temperature gradient with radial distance.

Optical Spectroscopy

Recombination lines occurring in the optical regime are restricted to the visible Balmer series of hydrogen and helium. Some photographic plates are sensitive to the Paschen and higher emission-line series but generally the Hα and Hβ lines are the most useful recombination lines. These optical lines give virtually no information on T_e, since the relative strengths of hydrogen recombination lines are very insensitive to temperature. However, as for the radio recombination lines, the intensity ratio of a line and its adjacent continuum varies more rapidly with temperature. Still, unlike the radio continuum, the optical continuum is seriously affected by scattering, weak lines, and two-photon emission. Separation of the free-bound continuum from the total measured optical continuum is tricky indeed.

Consider the Hβ line at λ 4861. The emissivity of this line is simply

$$j(\text{H}\beta) = N_4 A_{42} h\nu_{42}, \tag{7-89}$$

Figure 7-27.   Distribution of radial velocities derived from H 94α and H 110α lines observed at 20 locations across the Trifid Nebula. The contours refer to the magnitude of the radial velocity (km/sec). The optical image is reproduced with the permission of Kitt Peak National Observatory.

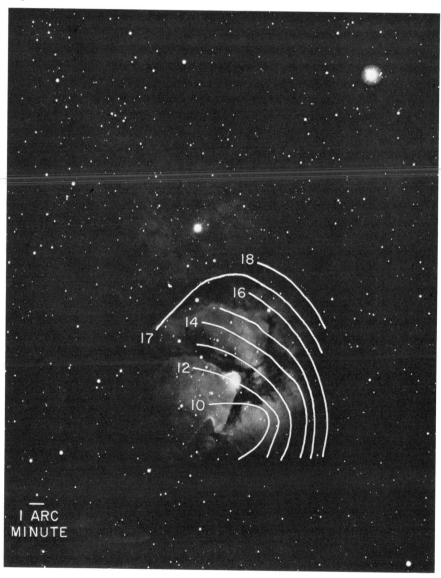

where, from equation (7-75), $A_{42} \simeq 10^7 \text{ sec}^{-1}$. The number density of atoms follows directly from equation (7-71), suitably altered for departures from LTE that undoubtedly affect the optical recombination lines (cf. Figure 7-22 for low n), so that

$$j(\text{H}\beta) = b_4 N_e N(\text{H}^+) n^2 A_{42} h\nu_{42} \left(\frac{h^2}{2\pi m_e kT_e}\right)^{1.5} \exp{(h\nu_{4\infty}/kT_e)}, \quad (7\text{-}90)$$

which when approximated becomes

$$j(\text{H}\beta) \simeq 2.8 \times 10^{-22} N_e N(\text{H}^+) T_e^{-0.84}. \quad (7\text{-}91)$$

Here, the value $b_4 \simeq 0.2$ was adopted, corresponding to the so-called Menzel case B for $T_e = 10^4$ K, a nebula optically thick in all the Lyman lines.

Some researchers either have made theoretical calculations of the λ 4861 continuum, including corrections for all the difficulties noted above, or have estimated the Balmer continuum by measuring $[I(\lambda\ 3646^-) - I(\lambda\ 3646^+)]$ at the series limit. They subsequently derive a mean $T_e \simeq 7000$ K for several nebulae. However, the considerable disagreement in individual values derived by numerous researchers for a single nebula reflects the large uncertainties of this method.

Better estimates of T_e and N_e can be extracted from analyses of forbidden lines. Fortunately, the formalism sketched earlier for estimating the nebular cooling rate Λ_c can be used directly to calculate the strengths of these collisionally excited spectral lines. Obviously, comparison of emission lines arising from ions like O^{++} and N^+ that have upper levels of considerably different excitation energies will be most useful in deriving T_e. Conversely, ions like O^+ and S^+ have closely spaced upper energy levels relatively insensitive to T_e but capable of yielding an estimate of N_e.

Consider first the O^{++} ion whose low-lying energy-level structure has already been illustrated in Figure 7-4. In the low N_e limit of negligible collisional de-excitations, every collisional excitation to the $^1D_2$ level produces either a λ 4959 or a λ 5007 photon, according to the ratio of their respective Einstein A coefficients (cf. Table 7-2). (The λ 4931 photon emitted from the electric-quadrupole $^1D_2 \rightarrow {}^3P_0$ transition is ignored here since its Einstein $A \simeq 2 \times 10^{-6} \text{ sec}^{-1}$ is negligibly small in comparison.) Similarly, photons at either λ 4363 or λ 2321 emit from the $^1S_0$ level according to their relative radiative probabilities. Thus, a ratio of line emissivities yields

$$\frac{j(\lambda\ 4959) + j(\lambda\ 5007)}{j(\lambda\ 4363)} = \frac{\Omega(^3P, {}^1D_2)}{\Omega(^3P, {}^1S_0)} \left[\frac{A({}^1S_0, {}^1D_2) + A({}^1S_0, {}^3P)}{A({}^1S_0, {}^1D_2)\nu({}^1S_0, {}^1D_2)}\right] \times$$

$$\exp\left[h\nu({}^1D_2, {}^1S_0)/kT_e\right]\left[\frac{A({}^3P_2, {}^1D_2) + A({}^3P_1, {}^1D_2)}{A({}^3P_2, {}^1D_2)\nu({}^1D_2, {}^3P_2) + A({}^3P_1, {}^1D_2)\nu({}^1D_2, {}^3P_1)}\right],$$

$$(7\text{-}92)$$

provided that the ${}^1D_2$ level is populated collisionally from below and not radiatively from above. When numerically evaluated with values of A and Ω given in Table 7-2, this expression becomes

$$\frac{j(\lambda\ 4959) + j(\lambda\ 5007)}{j(\lambda\ 4363)} = 8.3\ \exp\ (3.3 \times 10^4 T_e^{-1}). \qquad (7\text{-}93)$$

A similar treatment for the N^+ ion yields an analogous equation,

$$\frac{j(\lambda\ 6548) + j(\lambda\ 6583)}{j(\lambda\ 5755)} = 7.5\ \exp\ (2.5 \times 10^4 T_e^{-1}). \qquad (7\text{-}94)$$

When integrated over the nebular path length, these emissivity ratios become ratios of emergent intensity. Consequently, measurement of the relative intensities of forbidden lines directly yields a measure of the gas temperature, as shown in Figure 7-28.

Collisional de-excitation does, however, have some effect for $N_e > 10^{4.5}$ cm^{-3}. Because the radiative lifetime $A^{-1}({}^1D_2) \gg A^{-1}({}^1S_0)$, the ${}^1D_2$ level collisionally de-excites at lower N_e than does the ${}^1S_0$ level. The λ 4959 and λ 5007 lines of [O III], and the λ 6548 and λ 6583 lines of [N II] are consequently weakened. The λ 4363 [O III] and the λ 5755 [N II] line are simultaneously strengthened by collisional excitation of ${}^1S_0$ at the expense of the ${}^1D_2$ level. Figure 7-28 shows the numerical results of a complete statistical-equilibrium solution for large N_e. Few gaseous nebulae, however, have a large enough N_e to warrant this additional complication.

In practice, observations of the λ 4363 [O III] line are often contaminated by the Hg I λ 4358 feature emitted by street lights, making photographic comparison of $I(\lambda\ 4959 + \lambda\ 5007)/I(\lambda\ 4363)$ virtually impossible. Values of T_e derived from a few recent photoelectric measurements are listed in columns 2 and 3 of Table 7-5. The [N II] lines are weak but, when measured, tend to predominate at the nebular periphery where the ionization is lower. Indeed, the T_e discrepancies in columns 2 and 3 of Table 7-5 may reflect T_e in different nebular zones, as the ionization potentials of O III and N II are considerably different.

Analysis of the degree of collisional de-excitation can provide an estimate of N_e. Ions like O^+ or S^+ giving rise to two lines that originate from closely spaced energy levels are especially useful. In the lower N_e limit, we once again have the result that every collisional excitation is followed by photon emission. The line intensity is then simply proportional to the level popula-

Figure 7-28. [O III] and [N II] intensity ratios as a function of T_e. Solid lines refer to the low-density limit ($N_e \lesssim 10^4$ cm^{-3}), whereas the dashed lines refer to $N_e \simeq 10^5$ cm^{-3}.

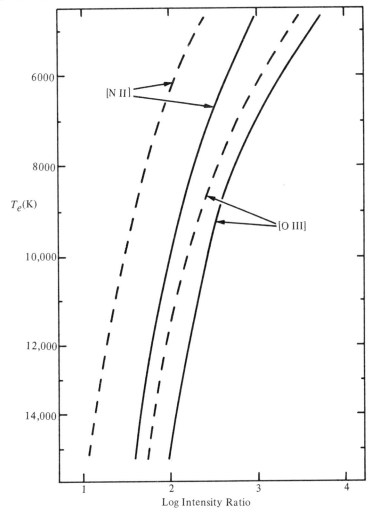

Table 7-5 Nebular parameters derived from optical spectroscopy.

Nebula	T_e (K) O III	N II	T_k (K)	N_e (cm^{-3}) O II	S II
Orion (NGC 1976)	8600	10,200	7000–12,000	10^2–$10^{3.5}$	10^3
Trifid (M 20)	—	—	8000–10,000	10^2–$10^{2.3}$	$10^{2.5}$–$10^{3[a]}$
Lagoon (M 8)	8300	8100	7000–10,000	10^2–10^3	10^3–$10^{3.5}$
Omega (M 17)	9000	7000	—	—	—

[a] Reaches $10^{3.8}$ cm^{-3} near bright rim structure and imbedded dust lanes.

tion of statistical weight $(2J + 1)$, all other factors canceling since the levels of excitation energies are virtually identical. For either the [O II] or [S II] doublet,

$$\frac{j(\lambda\ 3729)}{j(\lambda\ 3726)} = \frac{j(\lambda\ 6716)}{j(\lambda\ 6731)} = \frac{2J(^2D_{5/2}) + 1}{2J(^2D_{3/2}) + 1} = 1.5. \qquad (7\text{-}95)$$

On the other hand, a Boltzmann population is established for the high N_e limit as collisional excitations and de-excitations dominate. Consequently, the relative intensities of the doublets become

$$\frac{j(\lambda\ 3729)}{j(\lambda\ 3726)} = \frac{j(\lambda\ 6716)}{j(\lambda\ 6731)} = \frac{2J(^2D_{5/2})A(^2D_{5/2},\ ^4S_{3/2})}{2J(^2D_{3/2})A(^2D_{3/2},\ ^4S_{3/2})}, \qquad (7\text{-}96)$$

which equals 0.35 for both [O II] and [S II]. Figure 7-29 shows the results of a complete statistical equilibrium solution for $T_e = 10^4$ K, including population of the ^2D levels by cascade from the ^3P levels. The transition between the extreme cases of high and low N_e occurs for the N_e^c values listed in Table 7-2; the [S II] curve is shifted to the right since N_e^c [S II] $> N_e^c$ [O II]. The shape of the curves demonstrates that this technique, for these ions or others like [Cl III], [Ar IV], and [K V] having similar atomic structure, is most sensitive for $N_e \simeq 10^3$ cm^{-3}.

Operationally, observation of the closely spaced [O II] doublet is often hampered by lack of sufficient spectral resolution, and calculations of the [S II] collisional cross sections by researchers are not in agreement.

Columns 5 and 6 of Table 7-5 list some typical N_e values derived from these doublets. Often, N_e is found to be several thousand cm^{-3} at the nebular core, and a few hundred cm^{-3} at the periphery. However, since this technique is weighted somewhat toward regions of high N_e, simply because these locations are brighter, such implied N_e gradients may only reflect the distribution of the high-density clumps known to be prevalent near the nebular core. These studies are not sufficiently complete to demonstrate any substantial gradient in the lower-density ($10^{2.5}$ cm^{-3}) gas that characterizes the bulk matter in gaseous nebulae.

The optical determinations of T_e and N_e just discussed have for years required only intensity measurements at the line centers. Only recently have astronomers, using more accurate and stable spectrometers, begun to mea-

Figure 7-29.   [O II] and [S II] intensity ratios as functions of N_e for $T_e = 10^4$ K (taken from Osterbrock, 1974).

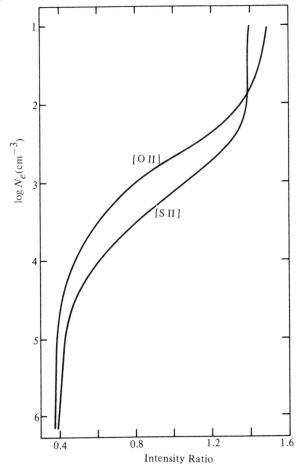

sure entire line profiles. Consequently, respective widths of ions with differing mass can now be used to extract thermal and turbulent information in the manner that led to the development of equation (7-87). Column 4 of Table 7-5 lists a few representative values of T_k derived from a width analysis of the Hα and [N II] lines. These values, as well as the turbulence $v_t \simeq 10$ km/sec values subsequently derived, agree reasonably well with those found from H-He radio-line observations (cf. Table 7-4), though some studies have recently suggested that comparison of Hβ and [O III] widths may yield more accurate temperatures.

Until recently, there was no compelling evidence for a clear-cut nebular temperature gradient, though numerous researchers reported differing values of T_e throughout nebulae. Two recent studies, however, one using [O III] intensity ratios at optical wavelengths, and the other using hydrogen and helium 76α line-width comparisons at radio wavelengths (cf. Figure 7-25), have found definite evidence for increasing T_e with nebular radius. Toward Orion's core and about ± 3 arcmin to the north and south, these studies show the following:

	Optical temperature	Radio temperature
Orion (0,0)	9,200 K	10,500 K
(0, 3.2 N)	11,200	12,900
(0, 3.5 S)	12,200	13,200

These results, in remarkably good agreement, argue that ultraviolet radiation does indeed obey the $(v_1/v)^3$ frequency dependence of the photoionization cross section discussed earlier. Consequently, the radiation is said to "harden," since all the low-energy photons are selectively absorbed in the nebular interior.

Nebular Environments

Until recently, gaseous nebulae were thought to be among the most massive objects in our Galaxy. But within the past several years, radio astronomers have uncovered evidence for large, dense molecular clouds that have considerably greater mass than any other interstellar constituent. Whereas many of these molecular clouds reside in galactic spiral-arm locations rich in gas and dust but devoid of nebulae, several of the more prominent clouds are known to be directly associated with gaseous nebulae. In fact, it is now widely recognized that molecular clouds play an integral role in the prestellar evolutionary sequence of interstellar matter. Consequently, a complete understanding of nebulae and of interstellar evolution necessarily includes a knowledge of their very complex and inhomogeneous galactic environment.

No attempt is made here to review the extraordinarily fast-developing field of interstellar molecules. Instead, as with the previous sections on nebulae, we offer a representative view of molecular regions now thought to engulf most gaseous nebulae. The Orion and Trifid nebulae are once again used as examples.

Interstellar Debris Near Nebulae

Qualitative evidence for the presence of interstellar dust either along the line of sight or within gaseous nebulae has been recognized for years. Optical photographs like those of figures 7-14 and 7-16 clearly show regions of obscuration, often superposed across the face of the optical image. The tongue of totally obscuring matter to the east of the Orion Nebula's centrally excited region is a good example, though the thin dark lane to the northeast is probably the result of insufficient excitation. Similarly, lanes of obscuring material trisect the Trifid Nebula, hence its name; these lanes are known from optical studies of the [S II] doublet to be interacting directly with the ionized gas, since nebular N_e increases markedly toward the lane edges.

Additional evidence for interstellar debris can be appreciated by visually scanning the nonuniform star fields adjacent to nebulae. Numerous dark areas partially or totally obscure an otherwise uniform distribution of background stars. For example, a zone of low star count nearly surrounds the Trifid Nebula, particularly toward the south and west. Such apparent inhomogeneities in the stellar distribution, including the appearance of several small blotches of complete obscuration, become even more obvious on photographs of longer exposure (cf. *Palomar Sky Atlas*). Finally, recent infrared studies of nebulae clearly imply the presence of dust, although it is not known whether the dust is mixed with or external to the nebular gas.

It is nearly impossible to estimate accurately the amount of extinction for those dark interstellar regions angularly removed from gaseous nebulae; laborious star counts yield uncertain values of optical depth that range from zero to as high as a hundred. However, the extent of obscuration can be made more quantitative for regions along the nebular line of sight. The easi-

est way to accomplish this is to measure the intensity of some optical spectral line and of the optically-thin radio continuum. The optical radiation of
wavelength comparable to the typical size of a dust particle will be scattered
and therefore effectively attenuated. But the radio radiation is completely
unattenuated by the intervening dust and can be scaled several orders of
magnitude in frequency to predict the true value of the optical line intensity.
The difference between the predicted (true) and observed values of the optical intensity yields the extent of obscuration. For example, the emissivity
ratio of the 24-GHz radio bremsstrahlung (equation 7-29) to that of the Hβ
optical line (equation 7-91) yields, for an optically thin nebula, the ratio of
flux densities

$$\frac{F_{\nu}(24\text{GHz})d\nu}{F_{\nu}(\text{H}\beta)_{\text{true}}} = 1.3 \times 10^{-16} T_e^{0.34} \frac{N(\text{H}^+ + \text{He}^+)}{N(\text{H}^+)} \ln (2.1 \times 10^{-3} T_e^{1.5}).$$

$$(7\text{-}97)$$

The amount of extinction can then be calculated provided the He abundance
and T_e are known:

$$\tau(\text{H}\beta) = \ln [F_{\nu}(\text{H}\beta)_{\text{true}}/F_{\nu}(\text{H}\beta)_{\text{obs}}]. \qquad (7\text{-}98)$$

This extinction optical depth is often expressed in magnitudes according to
the relation $A(\text{H}\beta) \simeq 1.1\tau(\text{H}\beta)$ mag or, more commonly for the center of
the visual band (5500 Å), $A_v \simeq \tau(\text{H}\beta)$ mag.

Figure 7-30 shows a map of $A(\text{H}\beta)$ distributed across the Orion Nebula.
This map was derived from the 24-GHz bremsstrahlung contours of Figure
7-14 and from an Hβ photograph whose angular response was convolved to
match the radio beam shape. As expected, the dark tongue to the east of
Orion's core can be recognized as the region of heaviest obscuration.

The distributions of $A(\text{H}\beta)$ toward the Trifid Nebula, derived from a comparison of the 8-GHz radio map (Figure 7-16) and Hβ observations, show the
obscuration to increase from about 1 mag at the nebular center to about 2–4
mag near the nebular periphery. Observations have not yet been made with
fine enough radio resolution to delineate quantitatively the extent of extinction along each of Trifid's dark lanes.

Similar studies have shown that major parts of some nebulae, notably
M 17 and W 3, are completely invisible. Yet despite improvements in the
quantitative estimates of Hβ and of other optical line intensities and in the
angular resolution at radio frequencies, it is generally impossible to deter-

Figure 7-30.   Obscuration measured toward the Orion Nebula in units of $A(H\beta)$ mag. The optical image is reproduced with the permission of Kitt Peak National Observatory.

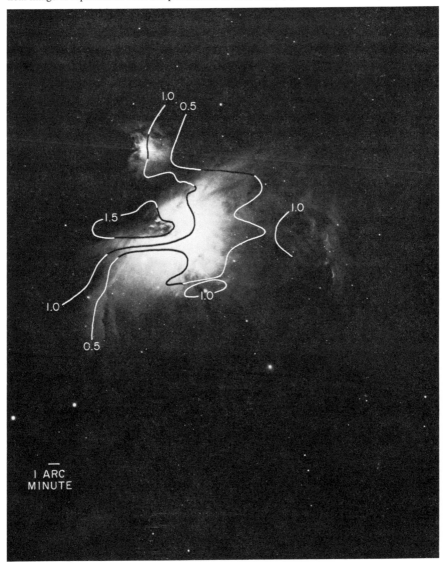

mine the precise physical relationship of the dusty regions of obscuring matter to the ionized regions of nebular gas. It appears that detailed spectroscopy holds the key to a better understanding of nebular environments.

Spectroscopy of the Obscuring Regions

The passage of optical and ultraviolet radiation through interstellar clouds can provide considerable information concerning the physics and chemistry of rather low-density $[N(H_2) < 10^{2.5}$ cm$^{-3}]$ interstellar regions along the line of sight toward hot O- and B-type stars (see Chapter 8.) But optical and ultraviolet studies of denser regions, often associated with nebulae, or of nearby locations angularly removed from nebulae, cannot provide meaningful information because these high-frequency photons are badly scattered and because there are few O-B type stars adjacent to nebulae. Infrared studies also suffer, since spectrometers capable of dispersing lower-frequency radiation are just beginning to become operational. Thus, radio waves are the only electromagnetic phenomena that can be sampled at present with sufficient spectral resolution and with little interstellar attenuation.

The first suggestion that gaseous nebulae represent only the ionized segment of larger interstellar clouds came from studies of the λ 21-cm line of neutral hydrogen. Recall that this transition is a forbidden line arising from levels split by an electron-proton magnetic-moment interaction. Several regions have been surveyed, but the inability to distinguish background H I gas from that associated with nebulae often produces unconvincing results. However, there are a few incontrovertible cases for which the H I gas has been shown to engulf the ionized gas, being at least 25 times more massive. The difficulty, as will be demonstrated, is that most of the neutral hydrogen in dusty clouds near nebulae is in the molecular form, H_2, which has no net nuclear moment and thus no radio-frequency hyperfine spectrum.

The carbon radio-recombination line provides additional observational evidence that partially ionized gas (and presumably dust) exists near nebulae. Figure 7-31a shows the C 92α line observed toward the Orion Nebula. The carbon feature was called the anomalous recombination line for several years because its large intensity, narrow width, and sometimes strange velocity made its identification uncertain. However, theoretical abundance considerations and observations of an equally narrow hydrogen line that is always displaced by 12 atomic mass units provides support for identification with the element carbon (cf. Figure 31b). These same reasons suggest strongly that the carbon line, now observed toward several nebulae and at several radio frequencies, arises in a predominantly neutral region immediately adjacent to and possibly surrounding gaseous nebulae. Out beyond the zone of H II gas there is likely to be a zone of C II into which leaks radiation of energy less than 13.6 eV but larger than 11.1 eV, the first ionization po-

Figure 7-31. Radio-frequency spectral lines observed with instrumental resolution indicated by the horizontal arrows. The vertical scale differs for each line. a, 92α spectrum of the Orion Nebula showing features arising from He, C, and heavier element X. b, 158α spectrum observed toward Orion showing the 12 *amu* separation between C and its associated narrow H feature. c, Kinematic comparison of C 92α recombination line with the OH ($J = \frac{3}{2}$, F = 2 → 2), H$_2$CO ($J = 1_{10} \rightarrow 1_{11}$) and λ 21-cm H (^{2}1S$_{1/2}$) spectral lines observed toward NGC 2024. The vertical arrow at the bottom denotes the gaseous nebular velocity. d, Similar observations of a dark cloud devoid of nebulae, called Rho Ophiuchi, for which the heavier feature labeled X 158α can be tentatively identified with the element S.

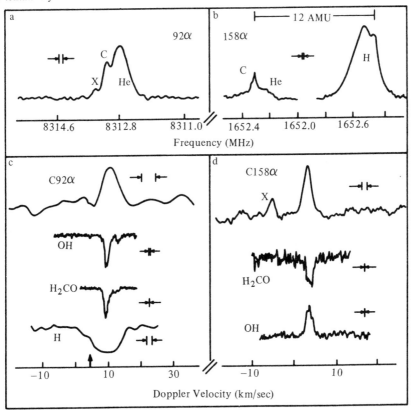

tential of carbon. It is unclear whether the carbon line and its associated narrow H feature arise by spontaneous emission from a dense $[N(H_2) > 10^4 \text{ cm}^{-3}]$ region that can be either in front or in back of the nebula, or by stimulated emission from an exclusively foreground region of lower density. In any case, the carbon-line velocity is always similar to the associated nebular velocity and often in precise agreement with prominent molecular features. Figure 7-31c shows this kinematic correlation most clearly toward NGC 2024, a nebula having a weak He recombination line. Although weak and difficult to measure, the carbon line has now been detected toward about a dozen nebulae. It is thought to be closely associated with nebular regions of large obscuration, primarily because several observations suggest that the carbon line is prominent in the Orion Nebula's eastern tongue. A better observational and theoretical understanding of radio recombination lines of carbon and of heavier elements (cf. Figure 7-31d) having ionization potentials < 13.6 eV is necessary, since the abundance and degree of ionization of these species will be important in cooling the cloud and possibly exciting the molecules.

Foremost among the techniques used to study dense interstellar regions is molecular radio astronomy, a relatively new and rapidly developing interdisciplinary science. Within the past half-dozen years, nearly 40 new species and more than 100 new spectral lines have been detected by microwave and millimeter-wave spectroscopy. Some of these molecules, such as OH and H_2CO, absorb radiation from intense background radio sources; others, such as CO, SiO, CS, and CH_3OH, emit characteristic radiation on their own accord, and still others, such as OH and H_2O, emit intense lines by way of maser-like processes.

Molecules rotate and vibrate by virtue of their heat. Vibrational modes usually produce characteristic spectral lines in the relatively inaccessible infrared. However, a large variety of molecules, from simple inorganic diatomics to complex organic polyatomics, rotate at quantized frequencies that can be detected in the radio-frequency spectrum. The basic physical and chemical properties of more than a million molecules have been accumulating for years in laboratory studies, but not until recently did astrophysicists recognize that molecules are widespread throughout our Galaxy. Unfortunately, most of the chemical processes that occur under terrestrial conditions are inapplicable since the densities of interstellar space ($\approx 10^{3-7} \text{ cm}^{-3}$) are usually orders of magnitude smaller than the best laboratory vacuums ($\simeq 10^{11}$ cm^{-3}). (For a discussion of interstellar chemistry, see Chapter 8.) Nonetheless, the previous notion that even simple molecules would be easily destroyed by interstellar radiation has now perished, and we recognize that large amounts of dust not only provide useful sites for molecule formation but also serve to protect molecules from the harsh interstellar environment. Figure 7-32 shows the results of a theoretical study emphasizing the degree of protection necessary for the survival of several representative molecules. Within the past few years, it has become an observa-

Figure 7-32. Molecule lifetimes τ against photodissociation for clouds having obscuration A_v. The strong double bond of C$=$O is primarily responsible for the widespread abundance of carbon monoxide (from Stief, 1973).

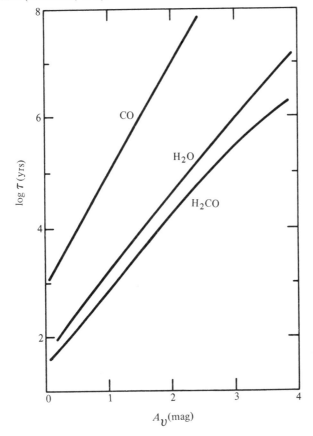

tional fact that the abundance of all molecules is strongly correlated with interstellar regions of obscuration.

Consider a simple diatomic rotor having an angular velocity ω and a moment of inertia I. Let r be the interaction separation between the two atoms of masses m and M. The rotational energy is then

$$E = \tfrac{1}{2}I\omega^2 = \tfrac{1}{2}\mu r^2\omega^2, \tag{7-99}$$

where μ is the reduced mass of the system, $mM/(m + M)$. If, as was done earlier for atoms, we invoke quantization of angular momentum, then

$$I\omega = \hbar J, \tag{7-100}$$

where J is the total angular-momentum quantum number. Combining these two equations yields a general expression for the rotational energy,

$$E = \hbar^2 J(J + 1)/2\mu r^2, \tag{7-101}$$

where the classical operator $J \cdot J$ has been transformed into its quantum mechanical analog $J(J + 1)$. Note how the energy levels of a simple diatomic molecule are evenly spaced. Transitions among them obey the selection rules $\Delta J = 0, \pm 1$. Equation (7-101) is often expressed as $E = hBJ(J + 1)$, where the rotational constant B, strictly dependent on the molecule's physical disposition, has been measured in the laboratory for a large variety of species.

For hydrides, the most common type of molecule expected *a priori* in interstellar space, μ will be small, making E large. For example, $B = 5.5 \times 10^{11}$ Hz for hydroxyl (OH), so that the frequency of the lowest rotational transition ($J = 1 \rightarrow 0$) is about 3000 GHz. This line falls in the far-infrared portion of the spectrum for which there are at present no astronomical spectrometers. All hydrides present this problem.

But a heavier molecule like carbon monoxide (CO) has a small B value, 5.8×10^{10} Hz, and hence produces spectral lines at lower frequencies. Figure 7-33 shows the CO energy-level diagram, and notes several transitions observable in the millimeter-wave regime. Figure 7-33 also illustrates actual observations of the $J = 1 \rightarrow 0$ transition in CO toward both the Orion and Trifid nebulae. The slightly larger value of μ for the ^{13}CO species results in a smaller spacing between adjacent J and thus a feature that is isotope-shifted toward lower frequencies, as also shown in Figure 7-33.

Triatomic and larger molecules have more complex energy-level diagrams. Consider the polyatomic molecule formaldehyde (H_2CO), an asymmetric rotor with two moments of inertia nearly equal but substantially larger than the third. The small asymmetry about the principal axis causes a splitting of the energy levels; Figure 7-34 shows this K-type doubling for the first few energy levels labeled according to the nomenclature $J_{K^-K^+}$, the subscripts K^- and K^+ being the angular momenta about the symmetry axis for the corresponding levels of a prolate and an oblate symmetric top. Because the nuclear spins of the two hydrogen atoms can be opposed or parallel, H_2CO or any other H_2-like molecules occur in para and ortho states between which there are no allowed transitions. Figure 7-34 shows representative spectra of the $J = 1_{11} \rightarrow 1_{10}$ H_2CO absorption lines observed toward the Orion and Trifid nebulae.

Energy-level diagrams for some molecules can become exceedingly complicated even for simple species. Consider OH again, a radical with one unpaired electron. The odd number of electrons will always distribute their

Figure 7-33. Carbon monoxide energy-level diagram and spectra observed toward the molecular cloud associated with the Orion and Trifid nebulae. The horizontal arrow indicates the instrumental response.

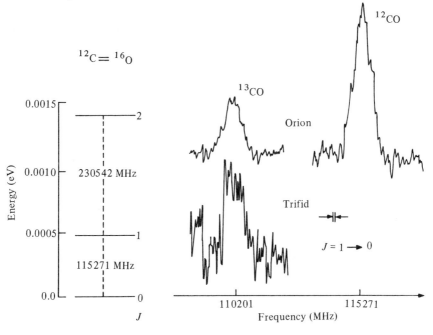

angular momenta unevenly about the principal axis of rotation. Consequently, the rigid-rotator energy levels previously discussed will be split, as
shown in Figure 7-35 for the $J = 3/2$ ground state. The mean value of a transition across this so-called Λ-doublet would occur at 1666 MHz if it were not
for hyperfine structure which further complicates the distribution of energy.
The hydrogen atom's nuclear spin of net angular momentum I interacts with
spin of the unpaired electron to split the levels further, as shown in Figure
7-35. Four spectral lines corresponding to the indicated microwave frequencies arise from transitions between levels of different F ($= J + I$), the
grand total angular momentum. Lines from the ground state of this particular
molecule have been observed in more than a thousand individual interstellar
clouds.

Figure 7-35 also shows an extremely complex OH spectrum observed
toward the Orion Nebula. Narrow features, probably the result of maser
emission, appear at several velocities. Some of the 1612-MHz ($J = 3/2$,

Figure 7-34.   Same as for Figure 7-33, but for the asymmetric-top formaldehyde molecule.

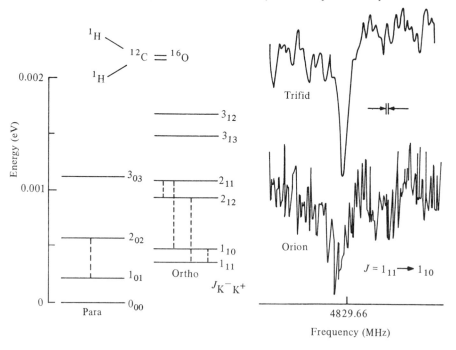

$F = 1 \rightarrow 2$) and 1665-MHz ($J = 3/2$, $F = 1 \rightarrow 1$) lines are circularly polarized, possibly forming a Zeeman pattern consistent with an interstellar
magnetic field of milligauss strength. Such an effect is expected to occur because, subsequent to the proton-electron spin interaction producing hyperfine structure, the levels are still $2F + 1$ degenerate. A weak magnetic field
can remove this degeneracy by additionally splitting the energy levels,
shifting, splitting, and polarizing the observed spectral features resulting
from allowed transitions. All the energy levels throughout the ladder of OH
excitation are similarly split, making the observed spectrum of OH complex
indeed. (Further details of OH and its apparent maser-like properties can be
found in Chapter 9.)

The first three columns of Table 7-6 list the laboratory rest frequencies for
some of the CO, H_2CO, and OH transitions noted in Figures 7-33–7-35.

Figure 7-35. Hydroxyl energy-level diagram of only the ground state under the influence of a
magnetic field, and the OH spectra observed toward the Orion molecular cloud. The dashed and
solid lines represent right- and left-circularly polarized radiation for a logitudinal magnetic field
directed toward earth. The energy interval of the entire splitting ($F \rightarrow F'$) = $(2 \rightarrow 1)$ is less than
10 μ eV.

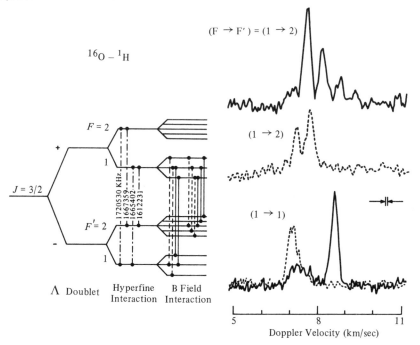

Table 7-6 Physical properties of some representative molecules.

Molecule	Transition	Rest frequency (MHz)	A (sec^{-1})	N_t (cm^{-2})	N (H$_2$) (cm^{-3})
$^{12}C^{16}O$	$J = 1 \longrightarrow 0$	115271.2	6×10^{-8}	10^{18}–10^{20}	10^3
	$2 \longrightarrow 1$	230542.4	—	—	—
$^{13}C^{16}O$	$1 \longrightarrow 0$	110201.4	—	10^{16}–10^{18}	—
$^1H_2{}^{12}C^{16}O$	$J_{k^-k^+} = 1_{10} \longrightarrow 1_{11}$	4829.66	3×10^{-8}	10^{14}–10^{16}	10^3
	$2_{11} \longrightarrow 2_{12}$	14488.65	3×10^{-7}	—	—
	$2_{02} \longrightarrow 1_{01}$	145602.97	—	—	—
	$2_{12} \longrightarrow 1_{11}$	140839.53	5×10^{-5}	—	$10^{5.5}$
	$2_{11} \longrightarrow 1_{10}$	150498.36	—	—	—
$^{16}O^1H$	$J = 2/3,\ F = 1 \longrightarrow 2$	1612.231	1.3×10^{-11}	—	$10^{1.5}$
	$1 \longrightarrow 1$	1665.402	7.7×10^{-11}	—	—
	$2 \longrightarrow 2$	1667.359	7.7×10^{-11}	10^{14}–10^{17}	—
	$2 \longrightarrow 1$	1720.530	9×10^{-12}	—	—

How are the molecular spectral lines analyzed for astrophysical information? Returning to the equation of transfer used to describe the atomic nebular radiation, and restricting the analysis for the moment to a two-level molecule, we have

$$T^B = (T_{mn} - T_{bg})(1 - e^{-\tau}), \qquad (7\text{-}102)$$

where T_{mn} is the rotational excitation temperature characterizing the population of the upper m and lower n energy levels. T_{bg} is the continuum background (thermal and/or nonthermal) temperature including the 3 K universal microwave background neglected earlier for nebulae since 3 K $\ll T_e$. The spectral line appears in emission if $T_{mn} > T_{bg}$, and conversely. At millimeter-wave frequencies ($\nu \simeq 100$ GHz), $T_{bg} \simeq 3$ K, there being no appreciable galactic continuous radiation. This is not necessarily true, however, at microwave frequencies ($\nu \simeq 1$ GHz).

Actually, equation (7-102) incorporates the Rayleigh-Jeans approximation, $h\nu_{mn} \ll kT_{mn}$, valid only for molecules that rotate in the microwave regime; some higher-frequency millimeter-wave spectral lines that arise in exceptionally cool regions ($T_{mn} < 10$ K so that $h\nu_{mn} \simeq kT_{mn}$) require equation (7-102) to be generalized to include the complete Planck function:

$$T^B = \{[\exp(h\nu_{mn}/kT_{mn}) - 1]^{-1} - [\exp(h\nu_{mn}/kT_{bg}) - 1]^{-1}\}h\nu_{mn}k^{-1}(1 - e^{-\tau}).$$
$$(7\text{-}103)$$

But equation (7-102) is obviously preferred for pedagogical reasons and will be used henceforth.

In analogy with our previous development of atomic line radiation, a two-level system has population densities (cm^{-3}) characterized by a Boltzmann distribution,

$$\frac{N_m}{N_n} = \frac{g_m}{g_n} \exp(-h\nu_{mn}/kT_{mn}),$$ (7-104)

and a line absorption coefficient

$$\kappa = \frac{g_m A_{mn} c^2}{g_n 8\pi\nu^2} \psi \left(N_n - N_m \frac{g_n}{g_m}\right).$$ (7-105)

All symbols have been defined previously. Substitution of equation (7-104) and integration over the path length yields the optical depth in the line,

$$\tau = \int_0^\infty \frac{g_m A_{mn} c^2}{g_n 8\pi\nu^2} \psi N_n [1 - \exp(-h\nu_{mn}/kT_{mn})] dx,$$ (7-106)

so that equation (7-102) reduces to

$$T^B = \begin{cases} (T_{mn} - T_{bg}) & (if\ \tau \gg 1) \\ const\ \mathcal{N}_n \Delta\nu^{-1} \left(\dfrac{T_{mn} - T_{bg}}{T_{mn}}\right) & (if\ \tau \ll 1). \end{cases}$$ (7-107)

Here, $\mathcal{N}_n$ is the column density (cm^{-2}) of lower-level molecules having velocities within the frequency interval $\Delta\nu$, commonly known as the line's full width at half maximum intensity. Thus saturated line profiles ($\tau \gg 1$) immediately yield $(T_{mn} - T_{bg})$ which, for millimeter-wave lines ($T_{bg} \rightarrow 3$ K), can nearly equal the gas kinetic temperature T_k, provided there are sufficient collisions (see below), but give no other molecular abundance or even kinematic information. On the other hand, a single, optically thin spectral line for which $T_{mn} \gg T_{bg}$ cannot be used to find T_{mn} but can furnish $\mathcal{N}_n$ and kinematic information from measurement of its integrated-line energy and line centroid respectively. The constant in equation (7-107) is a function of the Einstein spontaneous decay rate (sec^{-1})

$$A_{mn} = \frac{(J_n + 1)64\pi^4\nu_{mn}^3\mu_{mn}^2}{(2J_n + 3)3hc^3},$$ (7-108)

tabulated in column 4 of Table 7-6 for several representative transitions. The factor μ_{mn} is known as the dipole-moment matrix element, the inverse strength of which partly determines, in conjunction with abundance, the detectability of an astrophysical molecule.

But it is not enough to know the ground state line-of-sight concentration $\mathcal{N}_n$; the molecular column density of all levels must be computed. In the simplest approximation, all levels can be assumed to be populated according to a Maxwell-Boltzmann distribution at the temperature T_{mn}. For example, for a typical value $T_{mn} \simeq 20$ K characterizing the CO molecule, each excited level is populated according to its statistical weight (or multiplicity) $2J + 1$, so that, upon summing, the total column density $\mathcal{N}_t$ is approximately $7\mathcal{N}_n$. Larger T_{mn} obviously implies more CO molecules in higher states, according to the Boltzmann convention. Similar computations, made for some representative molecules, are listed in column 5 of Table 7-6.

Of course, with improved instrumental sensitivity, spectral lines will eventually be observed from all the important low-lying levels of many molecules, thus yielding $\mathcal{N}_t$ directly by summing $\mathcal{N}$ for observations of different transitions. Similarly, multilevel solutions, beyond the scope of this chapter but now solvable by the use of computers, can yield T_{mn} precisely for any pair of levels, even if the spectral lines are optically thin. Likewise, departures from an LTE Boltzmann distribution, if any, will some day be measured directly by means of multifrequency observations of different states of a single molecule.

Let us sketch the several competing processes that serve to excite molecular levels. The current scenario goes as follows: Local sources of ultraviolet radiation (e.g., embedded stars) heat the dust, which then collisionally transfers energy to H_2, the most abundant neutral-particle species, which in turn excites the molecules by collision. The heat input is then primarily stellar radiation, while the dominant cooling mechanism is thought to result from radiation from high rotational levels of CO. However, millimeter-wave radiation becomes opaque and cannot escape or cool the cloud for $N(H_2) > 10^3$ cm^{-3}, at which point increasing density generally means increasing temperature.

Consider now the collisional C and radiative A rates for a two-level molecule in the absence of a strong radiation field. In equilibrium, the excitation and de-excitation rates are equal,

$$N_n C_{nm} = N_m (C_{mn} + A_{mn}). \qquad (7\text{-}109)$$

In the limit of collisional dominance where $A \ll C$, the levels become populated according to the kinetic temperature T_k of the ambient gas. The collisionally dominated form of equation (7-109) can then be combined with the T_k analogue of equation (7-104) to yield

$$C_{nm} = C_{mn} \frac{g_m}{g_n} \exp(-h\nu_{mn}/kT_k). \tag{7-110}$$

This relationship is independent of A_{mn} and can be used in equation (7-109). Thus

$$\frac{N_m}{N_n} = \left[C_{mn} \frac{g_m}{g_n} \exp\left(-h\nu_{mn}/kT_k\right) \right] \bigg/ (C_{mn} + A_{mn}), \tag{7-111}$$

which when combined with equation (7-104) yields a general relation between T_{mn} and T_k, namely

$$T_{mn} = \frac{h\nu_{mn}}{k} \left[\frac{h\nu_{mn}}{kT_k} + \ln\left(1 + \frac{A_{mn}}{C_{mn}}\right) \right]^{-1}. \tag{7-112}$$

Clearly, $T_{mn} \rightarrow T_k$ until collisional effects dominate, at which point $T_{mn} = T_k$. In other words, for sufficiently large C_{mn}, the excited levels are populated according to a Maxwell-Boltzmann distribution characteristic of the ambient gas temperature T_k; in this case we say that the molecular levels are thermalized.

Just how large C_{mn} must be for thermalization to prevail is a function of the A_{mn} for a particular transition and of the collision cross section σ. The collision rate is given by

$$C_{mn} = N(\text{H}_2)\sigma v, \tag{7-113}$$

where neutral particles of density $N(\text{H}_2)$ and kinetic velocity v are implicitly assumed to be the only important collisional mechanisms. The cross section is difficult to calculate or measure, but is known for CO in the ground state, $\sigma \simeq 2 \times 10^{-15}$ cm^2. Since A_{10} (CO) $= 6 \times 10^{-8}$ sec^{-1} (cf. Table 7-6), $N(\text{H}_2) \simeq 10^3$ cm^{-3} for $A_{10} \simeq C_{10}$ and any reasonable T_k. Similar cross sections are currently used for most other molecules under the assumption that σ is not a strong function of a molecule's individual properties.

Figure 7-36 shows how T_{mn} approaches T_k as a function of neutral-particle density for the $J = 1, 0$ level pair of CO. Had a radiative field of intensity T_{bg} been included, T_{mn} would approach T_{bg} in the limit of small $N(\text{H}_2)$, as shown.

A potentially important caveat is that the electron-molecule collision cross-section is considerably larger than its neutral-particle counterpart, owing to the longer-range effect of a charged particle. An appreciable electron-molecule collision rate shifts the curve of Figure 7-36 to the left, requiring smaller $N(H_2)$ for thermalization. However, the electron density in typical molecular clouds is regarded as too small (< 1 cm^{-3}) to excite molecules appreciably.

Thus molecules having larger values of A require larger values of $N(H_2)$ to populate their upper levels. Column 6 of Table 7-6 lists the inferred values of $N(H_2)$ for which $A_{mn} \simeq C_{mn}$, and therefore the minimum values necessary to populate appreciably excited molecular levels. Of course, densities considerably larger than those of column 6 are necessary for complete thermalization.

The densities of molecular clouds quoted in the literature are determined by this method. The mere detection of a spectral line from a species like CO ($J = 1 \rightarrow 0$) or from any number of several other species like HC$_3$N

Figure 7-36. Excitation temperature of the $J = 1,0$ level pair in CO as a function of total particle density for a 50 K kinetic temperature.

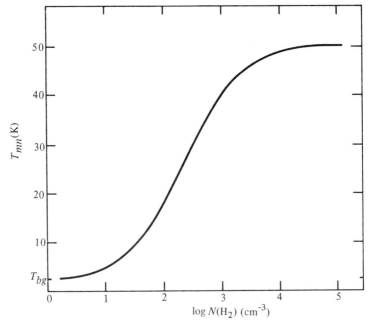

($J = 1 \rightarrow 0$) or NH_3 (1, 1) having comparable Einstein A values is enough to place a lower limit on the density $N(H_2) > 10^3$ cm^{-3}. Other observed molecules like CS ($J = 1 \rightarrow 0$) or SiO ($J = 3 \rightarrow 2$) have larger Einstein A's ($\simeq 10^{-5}$–10^{-4} sec^{-1}) and require $N(H_2) \simeq 10^{5.5}$ cm^{-3} for sufficient excitation.

We emphasize that the total density $N(H_2)$ is not measured directly, but is inferred from excitation considerations for those molecules observed in interstellar space. Actually, there is another method capable of yielding estimates of $N(H_2)$, independent of collisional arguments. Division of the measured column density $\mathcal{N}_t$ derived from an optically thin molecular line by the cloud's diameter yields a volume density of that molecule. With its apparently overwhelming abundance relative to other carbon-containing species, CO probably contains most of the atomic carbon. Consequently, the terrestrial ratio $H/^{13}CO \simeq 2.7 \times 10^5$ increases the observed $\mathcal{N}_t(^{13}CO) \simeq 10^{17}$ cm^{-2} to $\mathcal{N}_t(H_2) \simeq 10^{22}$ cm^{-2} which, for reasonable cloud sizes, implies $N(H_2) \simeq 10^{3-4}$ cm^{-3}. However, as stated, this calculation assumes negligible abundances of other carbon-containing molecules, a point that appears valid on face value but may be confused by non-LTE subtleties. Collisional and/or radiative processes can contrive to create severe non-LTE conditions, and consequently unusual energy-level populations in some molecules. In fact, population inversion causes the line absorption coefficient, given by equation (7-105), to become negative, thus amplifying exponentially the line intensity by means of equation (7-102). Molecules radiating under such weird conditions are known as interstellar masers (cf. Chapter 9).

In any case, estimates of the total particle density in turn furnish the total cloud mass, provided the cloud diameter can be estimated from molecular maps. Columns 2 and 3 of Table 7-7 give mean sizes $2R$ (pc) and total masses $M(H_2)$ ($M_\odot$) determined from CO observations for a variety of molecular clouds associated with well-known galactic nebulae. Large as they appear, these masses are surely lower limits since the implied $N(H_2)$ required for sufficient excitation of CO is only a lower limit and because other molecules requiring larger $N(H_2)$ doubtlessly reside in the cloud's core. Lack of complete knowledge of the size and shape of most clouds, however, makes even these lower limits very uncertain at present. Probably the only significant point to be made here is that the mass of a molecular cloud universally dwarfs that of its apparently associated gaseous nebula (cf. Table 7-3).

Column 4 of Table 7-7 lists observed values of T^B for the ^{12}CO species which, from isotopic studies, is regarded optically thick. Consequently, with

Table 7-7   Physical characteristics of nebular clouds near nebulae.

Nebula	$2R$ (pc)	M ($M_\odot$)	T^B (K)
Orion (NGC 1976)	2	—	50
Trifid (M 20)	8	$10^{3.5}$	20
Omega (M 17)	5	$10^{3.5}$	50
Lagoon (M 8)	2.5	$10^{2.5}$	20
NGC 2024	—[a]	—	30
W 49A	20	$10^{4.8}$	20
W 51A	15	$10^{4.5}$	20
W 3A (IC 1795)	8	$10^{3.5}$	20
Sgr B2 (W 24)	30	$>10^5$	20

[a] Undelineated extent.

$T_{bg} \simeq 3$ K, we find that $T^B \simeq T_{mn} \simeq T_k$ via equation (7-107) and the above collisional arguments. Virtually all galactic molecular clouds are cool, $5 < T_k < 100$ K.

Observationally, our Galaxy is richly populated with a widespread distribution of molecules, particularly CO, OH, CS, SO, NH_3, HCN, H_2O, H_2S, and H_2CO. These species are mainly associated with gaseous nebulae, though some infrared objects and dark clouds devoid of nebulae contain substantial abundances. Small kinematic differences observed in the atomic (nebular) and molecular (interstellar) velocities are not surprising because if nebulae indeed evolve from the collapse of a molecular condensation, they will do so predominantly at the cloud's periphery, 90% of the mass of a uniformly dense sphere being concentrated in the outer 50% of its radius. Thus it seems likely that most visible nebulae, like Orion and Trifid, represent ionized foreground fragments of larger molecular clouds.

On the other hand, some nebulae appear to be completely engulfed by exceedingly large molecular clouds. Recent studies of very distant and totally obscured regions like W 49 and W 51 suggest that these nebulae are surrounded by molecular clouds that peak near the location of the continuum source, and have similar atomic and molecular velocity variations. Sgr B2, the richest and most massive molecular cloud in our Galaxy, also appears to surround what is, not surprisingly, the largest and most massive gaseous nebula. The velocity variations of the H 92α and $J = 1_{11} \rightarrow 1_{10}$ H_2CO spectral lines, shown in Figure 7-37, demonstrate remarkable similarity in orientation and gradient, suggesting rotation of the cloud-nebula as a cohesive unit. The spatial distribution of H_2CO, however, bears little resemblance to that of other molecules, which appear to concentrate into two separate regions that may ultimately be shown to be related to a foreground-background shell surrounding the nebula.

What about the relative distribution of different molecules? As might be expected from the previous discussion, the observational picture is that those molecules requiring $N(H_2) \geq 10^4$ cm^{-3} for excitation are found near the relatively shielded core of an otherwise lower-density cloud. Figure 7-38 shows the angular distribution of several species observed toward the Orion Nebula, but centered on the infrared source about 1 arcmin northwest of the nebular source of ultraviolet excitation; indeed, high-excitation molecules like H_2CO ($J = 2_{12} \rightarrow 1_{11}$) or CS ($J = 1 \rightarrow 0$) are found in an approximately 0.3 pc core within a much larger molecular cloud (>1.5 pc) delineated by low-excitation molecules like H_2CO ($J = 1_{10} \rightarrow 1_{11}$) and CO ($J = 1 \rightarrow 0$). Nearby point-source ($\approx 10^{-3}$ pc) sites of OH and H_2O maser emission are found within the core, presumably requiring total densities approaching those of protostellar objects ($>10^8$ cm^{-3}) for sufficient excitation.

Actually, the distribution of CO toward Orion is very large, possibly extending over several arc degrees (or tens of parsecs). CO is more widespread than any other molecule, with the presumed exception of H_2, partly because of its uncharacteristically small dipole moment but also because its strong

Figure 7-37. Distribution of H 92α (left) and $J = 1_{11} \rightarrow 1_{10}$ H_2CO across the Sgr B$_2$ complex. The contours refer to the magnitude of radial velocity (km/sec) and the dashed line denotes the orientation of the galactic plane.

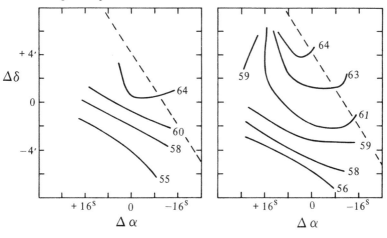

double bond ($\approx$ 11-eV binding energy) enhances its abundance relative to other, more easily destroyed species. In fact, it is not uncommon for molecular radio astronomers to fail to find the periphery of a typical CO cloud within a normal amount of telescope-access time.

Figure 7-38. Approximate extent of several molecules observed toward the Orion region. (Adapted in part from Buhl and Snyder, 1973.) The optical image is reproduced with the permission of Kitt Peak National Observatory.

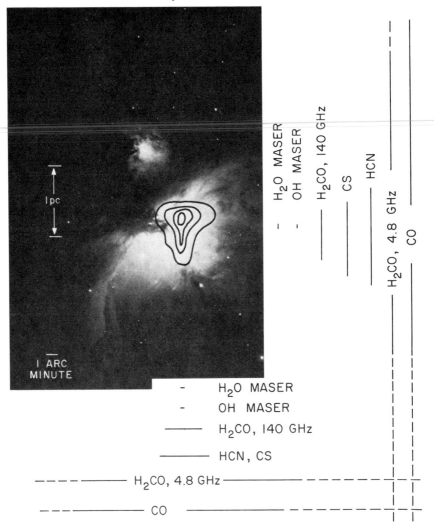

The rapidly developing science of molecular radio astronomy is not without numerous difficulties. In particular, the above considerations represent a simplified version of an exceedingly difficult radiative transfer problem. A complete solution incorporates many additional subtleties. One of these, a multilevel solution, has already been mentioned. Another is that net radiative de-excitations do not cool the cloud when an emitted photon is absorbed by another molecule. Trapped photons actually undergo τ scatterings so that the effective radiative decay rate should be reduced from A_{mn} to $A_{mn}(1 - e^{-\tau})/\tau$ for large τ. In other words, equation (7-112) is strictly correct only for small τ. More generalized solutions to the equations of radiative transfer, incorporating many molecular levels, are a topic of frontier research. A third important area of current observational and theoretical research is the rather remote possibility that N_e may be large enough to cause excitation by electron-molecule collisions. The inferred total density and mass listed for the clouds of Table 7-7 could be reduced considerably, particularly if the apparently close association of ionized carbon (via recombination-line studies) and molecular species is substantiated by further observations.

Interstellar Cloud Evolution

Molecular radio astronomy has thus found direct observational evidence for the widespread existence of exceedingly massive interstellar clouds throughout the Milky Way Galaxy. Most regions studied to date display the same type of density stratification as that inferred above for the Orion cloud, including the presence of ultrahigh density sites of maser emission. Without adequate retardation by rotation, turbulence, heat, and/or magnetism, these clouds are gravitationally unstable. Collapse is likely to set in rapidly, terminating in 10^4–10^5 years.

The scenario for an interstellar evolutionary sequence, from a dense galactic cloud to a newborn star and its associated nebula, is well formulated in the minds of most astrophysicists. It goes something like that outlined in Figure 7-39. Until recently, however, virtually no observational evidence existed for such a scenario. Now, molecular radio astronomy is beginning to provide at least suggestive evidence that large, dense clouds are indeed collapsing to form stars.

Consider the Trifid Nebula complex as a representative example of a region rich in star formation. Figure 7-40 displays a contour map of $\mathcal{N}_t$

(H₂CO) and $\mathcal{N}_t(^{13}CO)$ for the reasonable set of assumptions noted in the foregoing section. These molecules appear to engulf the gaseous nebula from the south and west, mimicking the distribution of dust obscuration noted in Figure 7-16. In fact, the molecules peak on a very dark blotch, which we shall refer to here as Trifid SW. There are fewer molecules toward the east and northeast where the background stellar density appears relatively un-obscured—interestingly enough, the reflection nebula to the northeast ap-parently contains enough dust to scatter starlight, but not enough to protect the molecules from ultraviolet destruction. As for most interstellar regions, there seems little doubt that molecular abundance is strongly correlated with regions of apparent dust.

Figure 7-40 also shows the line width and velocity distribution for the op-tically thin H₂CO species, both of which also peak at Trifid SW. The ¹³CO profiles yield similar results. The common symmetry of the kinematic distri-butions with those of $\mathcal{N}_t$ suggests that the Trifid SW molecular cloud may be in a state of gravitational collapse—at the cloud periphery, infall motion is primarily transverse to the line of sight, yielding more narrow molecular fea-tures as observed. If so, then the Trifid region provides one of the first pieces of observational evidence for the evolutionary scenario sketched in Figure 7-39.

Figure 7-41 schematically illustrates some tentative observational evi-dence for the three principal phases of star formation. In the northwest, the nebular velocity of about 18 km/sec (cf. Figure 7-27) meshes nicely with that of the adjacent molecular cloud (cf. Figure 7-40), thus establishing the initial

Figure 7-39.   Theoretical scenario of interstellar cloud evolution.

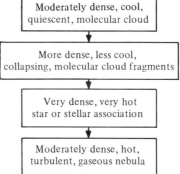

or quiescent phase for which the total density is probably not very high ($\approx 10^2$ cm^{-3}). The velocity of 18 km/sec can be viewed as the kinematic motion of the interstellar cloud prior to the formation of Trifid's exciting star. More sensitive mapping of these or other molecules may show that the 18 km/sec quiescent-cloud phase completely surrounds the Trifid complex, with the exception of the southwest region. There, at Trifid SW, the molecular cloud has probably reached sufficiently high density ($> 10^3$ cm^{-3}) and mass ($> 2000\ M_\odot$) to initiate gravitational collapse, the intermediate phase of star formation.

The Trifid Nebula itself represents the final phase of star formation. The outer contour of 8-GHz nebular emission shows two "kinks" that could be real (cf. Figure 7-16). The first represents an outer bulge toward the northeast, where optical photographs show some ionization and where, we note in retrospect, the molecular cloud is apparently absent or at least of tenuous density. And a second, inner bulge occurs at the nebular position closest to Trifid SW. It is easy to visualize an ionization front created by the action of stellar ultraviolet radiation, propagating into the dense molecular region near Trifid SW. The curved arrows of Figure 7-41 are meant to depict the rapid flow of nebular gas away from the neutral-cloud/ionized-gas interface, a view supported by the anomalous velocities of the H 110α recombination line in the southwest part of the nebula (cf. Figure 7-27). The shock front

Figure 7-40. Distribution of $\mathcal{N}_t(H_2CO)$ and $\mathcal{N}_t(CO)$ (left), H_2CO and ^{13}CO velocity (center), and $\Delta\nu_D(H_2CO)$ (right) toward the Trifid SW molecular cloud. The dashed curve is the outer 8-GHz thermal contour.

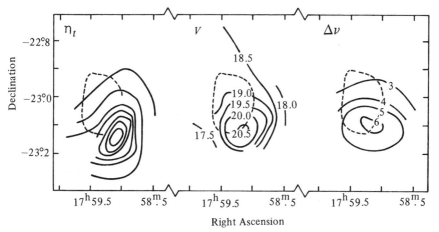

Figure 7-41. Suggestive observational evidence for an evolutionary scenario depicting a cold, dense cloud proceeding to a young, hot star. (Adapted from the research of Chaisson and Willson, 1975.) The optical image is reproduced with the permission of Kitt Peak National Observatory.

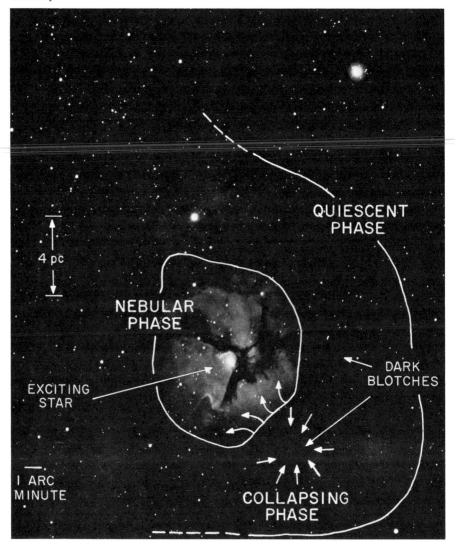

preceding the ionization of material will certainly compress the gas and probably contribute to the enhanced density in the molecular cloud near Trifid SW. It is even conceivable that such a shock could act as a triggering mechanism for rapid star formation at that location.

Future atomic and molecular line observations with better spatial resolution may allow quantitative estimates of the degree of shock-like interaction of hot, tenuous nebular gas against cold, dense molecular matter, and may possibly delineate an additional phase intermediate to the collapsing and nebular regions, namely, small-scale fragmentation into individual sites of protostellar formation.

It is a pleasure to acknowledge the contributions of those Harvard College undergraduate students with whom I have worked during the past few years: Charles Beichman, Daniel Jaffe, Donald Luyre, Matthew Malkan, Jean Turner, and Robert Willson. I have learned much from them. I am also grateful to Professor A. E. Lilley for providing the stimulating (terrestrial) environment in which this type of research can be undertaken.

References

References cited in text

Brocklehurst, M. 1970. Level populations of hydrogen in gaseous nebulae, *Monthly Not. Roy. Astron. Soc. 148,* 417–434.

Buhl, D., and Snyder, L. E. 1973. The problem of X-ogen, *Astrophys. J. 180,* 791–800.

Chaisson, E. J., and Willson, R. F. 1975. A microwave investigation of the Trifid Nebula and its surrounding environment, *Astrophys. J. 199,* 647–659.

Churchwell, E. B., and Walmsley, C. M. 1973. Observations of optical nebulae at 2695 MHz, *Astron. Astrophys. 23,* 117–124.

Kapitzky, J. E., and Dent, W. A. 1974. A high-resolution map of the galactic-center region, *Astrophys. J. 188,* 27–32.

Osterbrock, D. E. 1974. *Astrophysics of Gaseous Nebulae,* W. H. Freeman and Co., San Francisco.

Reifenstein, E. C. 1968. The structure of the Galaxy as determined by studies of hydrogen recombination line emission, Ph.D. thesis, Massachusetts Institute of Technology.

Rubin, R. H. 1969. Helium abundances and the sizes of He II and H II regions, *Astron. J. 74,* 994–998.

Stief, L. J. 1973. Photochemistry of interstellar molecules, in M. A. Gordon and L. E. Snyder, eds., *Molecules in the Galactic Environment,* pp. 313–328, John Wiley and Sons, New York.

Wilson, T. L. 1970. The study of H II regions, Ph.D. thesis, Massachusetts Institute of Technology.

General references. The most recent and complete reference describing atomic processes in nebulae is the book by Osterbrock cited above; there is no comparably complete publication for molecular processes in interstellar clouds.

Aller, L. H. 1956. *Gaseous Nebulae,* Chapman and Hall, London.

Aller, L. H. 1971. *Atoms, Stars and Nebulae,* Harvard University Press, Cambridge, Mass.

Dufay, J. 1968. *Galactic Nebulae and Interstellar Matter,* Dover Publications, New York.

Herzberg, G. 1944. *Atomic Spectra and Atomic Structure,* Dover Publications, New York.

Kaplan, S. A., and Pikelner, S. B. 1970. *The Interstellar Medium,* Harvard University Press, Cambridge, Mass.

Menzel, D. H., ed. 1962. *Selected Papers on Physical Processes in Ionized Plasmas,* Dover Publications, New York.

Mezger, P. G., and Hendersen, A. P. 1967. Galactic H II regions. I. Observations of the continuum radiation at the frequency 5 GHz, *Astrophys. J. 147*, 471–489.

Middlehurst, B. M., and Aller, L. H., eds. 1968. *Nebulae and Interstellar Matter*, volume 7 of *Stars and Stellar Systems*, University of Chicago Press, Chicago.

Spitzer, L. Jr. 1968. *Diffuse Matter in Space*, Interscience Publishers, New York.

Recent reviews

Carson, T. R., and Roberts, M. J., eds. 1972. *Atoms and Molecules in Astrophysics*, Academic Press, New York.

Dupree, A. K., and Goldberg, L. 1970. Radiofrequency recombination lines, *Annu. Rev. Astron. Astrophys. 8*, 231–264.

Gordon, M. A. 1974. The radio characteristics of H II regions and the diffuse thermal background, in G. L. Verschuur and K. I. Kellerman, eds., *Galactic and Extragalactic Radio Astronomy*, pp. 51–81, Springer-Verlag, New York.

Gordon, M. A., and Synder, L. E., eds. 1973. *Molecules in the Galactic Environment*, John Wiley and Sons, New York.

Heiles, C. E. 1971. Physical conditions and chemical constitution of dark clouds, *Annu. Rev. Astron. Astrophys. 9*, 293–322.

Hummer, D. G., and Rybicki, G. 1971. Formation of spectral lines, *Annu. Rev. Astron. Astrophys. 9*, 237–270.

Matthews, W. G., and O'Dell, C. R. 1969. Evolution of diffuse nebulae, *Annu. Rev. Astron. Astrophys. 7*, 67–98.

Mezger, P. G. 1972. Interstellar matter: An observer's view, in N. C. Wickramasinghe, F. D. Kahn, and P. G. Mezger, eds., *Interstellar Matter*, pp. 1–191. Geneva Observatory, Geneva.

Miller, J. S. 1974. Planetary nebulae, *Annu. Rev. Astron. Astrophys. 12*, 331–358.

Neugebauer, G., Becklin, E., and Hyland, A. R. 1971. Infrared sources of radiation, *Annu. Rev. Astron. Astrophys. 9*, 67–102.

Pinkau, K., ed. 1974. *The Interstellar Medium*, Reidel, Dordrecht and Boston.

Turner, B. E. 1974. Interstellar molecules, in G. I. Verschuur and K. I. Kellerman, eds., *Galactic and Extragalactic Astronomy*, pp. 199–255, Springer-Verlag, New York.

Zuckerman, B., and Palmer, P. 1974. Radio radiation from interstellar molecules, *Annu. Rev. Astron. Astrophys. 12*, 279–314.

8

Chemistry of the Interstellar Medium

A. Dalgarno

The discovery of molecules in the interstellar medium of our Galaxy, and recently of other galaxies, has profound implications. Most of the more complicated molecular species are found in dark cloud complexes containing regions of relatively high density in which star formation is apparently taking place. These dense interstellar clouds are opaque to visible and ultraviolet radiation and the molecular emissions in the radio frequency range of the electromagnetic spectrum provide a unique and powerful diagnostic tool for exploring the physical conditions in the interior of gravitationally collapsing regions.

Molecules are found also in more diffuse clouds, through which radiation can penetrate, where they can be studied by their absorption of the starlight. Because there are no hidden internal sources of energy the interpretation of the molecular observations is much simpler than for dense clouds. With a careful theoretical analysis various physical parameters such as the gas density, temperature, and pressure, and the ultraviolet and ionizing radiation fluxes can be derived. By observing in the direction of many stars, the distribution of these physical parameters in the interstellar gas can be obtained and we can learn much about the interactions of the individual stars with the interstellar medium surrounding them. The interpretation of the observations in dense clouds is attended by many uncertainties but often it is possible to infer the gas density, temperature, and pressure.

Molecules are, however, much more than diagnostic tools. Energy is lost from the clouds by infrared and millimeter emissions from molecules and the thermal structure of the clouds is profoundly influenced by the molecular composition. It is probable that the formation of molecules leads to dynamic and thermal instabilities which cause the cloud to break into fragments that collapse to produce stars. The molecular composition of the cloud will be modified by the subsequent interaction of the newly formed stars with the remnant interstellar cloud material out of which they were formed. Thus if we are to understand how an interstellar cloud evolves with time from some initial condensation to the dissipation of the cloud by stellar winds and radiation pressure from the stars formed within it, we must first identify the mechanisms for the formation and destruction of the interstellar molecules. We discuss such mechanisms in this chapter.

A list of the interstellar molecules that have been discovered so far is given in Table 8-1. The simple diatomic species CH, CH^+, CN, CO, H_2, and OH have been detected by their absorption of visible or ultraviolet radiation from stars. With the exception of CH^+ and of H_2, all the other interstellar molecules have been detected by emission or absorption at radio wavelengths. Other molecules with either absent or weak radio-frequency transitions such as carbon dioxide, CO_2, are undoubtedly present.

Molecules containing various nuclear isotopes, such as $C^{12}O^{16}$, $C^{13}O^{16}$ and $C^{12}O^{18}$, are also found. Their detection is of particular interest not only to the elucidation of the molecular chemistry of the interstellar medium but also to the study of the nuclear chemistry of the Galaxy.

Molecular Structure

Because of their larger masses, the nuclei in a molecule move slowly compared to the electrons and an accurate description of a molecular system can be obtained by first determining the electronic energy with the nuclei held fixed.

Table 8-1 Interstellar molecules.

CH^+	CH	CN	CO	H_2	OH
CS	SiO	SO	NS	SiS	
HDO	HCN	H_2S	OCS	SO_2	N_2H^+
C_2H	HCO^+	HNC			
NH_3	H_2CO	H_2CS	HNCO		
HCOOH	HC_3N	NH_2CN			
CH_3OH	CH_3CN	NH_2HCO			
CH_3C_2H	CH_3HCO	CH_3NH_2	CH_3C_2N	C_2HC_3N	
HCOOCH$_3$					
$(CH_3)_2O$	CH_3CH_2OH				

As for an atom, the distinct discrete electronic energy levels are separated typically by energies of several electron volts. The electronic energy levels are functions of the set of nuclear separation vectors $\{R\}$ and they can be regarded as potential energy surfaces $V(R)$ on which the nuclei move. Associated with each electronic potential energy surface are the energy levels of nuclear motion, which consist of a series of vibrational levels, separated typically by energies of about 0.2 eV and labelled by vibrational quantum numbers v. They are similar to the energy levels of a harmonic oscillator. Superimposed upon each of the vibrational levels are the rotational energy levels whose separation is proportional to $2J/I$, where J is the quantum number that labels the rotational level and I is the moment of inertia of the molecule. Most observations in the radio region are of rotational transitions of heavy molecules for which the rotational level separations are about $5 \times 10^{-3} J$ eV and it is transitions between these rotational energy levels that control the thermal structure in dense clouds.

The energy level structure is further complicated by the interactions between nuclear and electronic motion and by the interactions between spin and orbital angular momenta. Although the resulting structural detail greatly complicates the molecular energy level diagrams, observations of the many transitions that are made possible provide a wealth of information about the molecular environment. A full description of the energy level structure is given in the book *Microwave Spectroscopy*, by Townes and Schawlow (1955); a summary description is given in an article by Foley (1972).

Diffuse Interstellar Clouds

Molecular Absorption Lines

A visible or ultraviolet photon emitted by a star may be absorbed by interstellar molecules which undergo a radiative transition from the initial ground electronic state to an excited electronic state. Figure 8-1 reproduces spectra in the region of 4232 Å observed in the direction of several stars. The depressions in the residual intensity are due to absorptions by the molecule CH^+ in its lowest rotational and vibrational level. At zero velocity the line is located at 4232.54 Å, but because the diffuse clouds containing the CH^+ molecules are moving with respect to the observer, the lines are shifted in wavelength by amounts determined by the velocities along the line of sight.

The strength of the absorption is related to the number of absorbing molecules that lie between the observer and the star that is being observed. The amount of absorption at a particular wavelength depends upon the width of the line over which the molecular absorption cross section is distributed, and the derived number of absorbing molecules depends upon the absorption cross section and the velocities of the molecules. If several absorbing lines are detected that originate in a common initial level, an empirical curve of growth relating the absorption strengths to the number of absorbers can be constructed, which reproduces the measured absorptions; then both the mean velocity dispersion and the number of absorbing species can be derived. The methods by which such estimates are made are described by Spitzer (1968). The molecules CH and CN have also been detected in the interstellar gas by the absorption of visible radiation. CH is found only in the lowest rotational level of its lowest vibrational level but in the case of the heavier molecule CN, the rotational levels lie close together and absorption lines have been detected that originate in the lowest two rotational levels, $J = 0$ and 1, of the lowest vibrational level. From the measured absorptions, the column densities $N(J = 0)$ and $N(J = 1)$ can be derived. The population ratio $N(J = 1)/N(J = 0)$ can be conveniently expressed by an excitation temperature T, defined by

$$\frac{N(J = 1)}{N(J = 0)} = \frac{g_1}{g_0} \exp\left(-\Delta E/kT\right) = 3 \exp\left(-\Delta E/kT\right),$$

Figure 8-1.   Measured residual intensities of the interstellar line λ 4232 of CH^+ towards the stars 20 Taurus, 23 Taurus and ξ Persei, from Hobbs (1973). The velocity shifts show that the CH^+ is moving away from us.

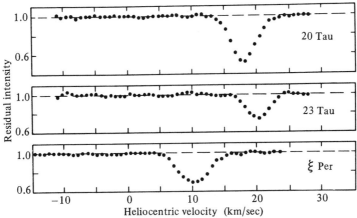

where ΔE is the energy separation of the two levels and $g_J = (2J + 1)$ is the statistical weight of the level. The derived excitation temperature is 2.7 K.

Because the particle volume densities are low (as we shall later argue) and the radiative depopulation of level $J = 1$ in a downward transition to level $J = 0$ is rapid, the upper level cannot be populated significantly by collisions. It is now generally accepted that the population of the $J = 1$ level is due to the equilibration of the rotational levels with the universal black-body background radiation that is the relict of the primordial fireball that heralded a beginning of the universe some twenty billion years ago (see the last two sections of Chapter 11). The relative population of the two rotational levels of CN measures the temperature of that radiation.

Most molecules absorb strongly, not in the visible, but in the ultraviolet region of the spectrum at wavelengths shorter than the ozone cutoff of the atmosphere at 3000 Å, with the consequence that they cannot be detected by ground-based observations. The detection of carbon monoxide CO and of molecular hydrogen H_2 and its isotopic form HD by ultraviolet absorption had to await the advent of rocket-borne and satellite-borne instrumentation. The molecule OH has also been detected by ultraviolet absorption and upper limits to the column abundances of other molecules have been obtained. CO and OH were discovered earlier by observations in the radio region of the spectrum.

The Rotational Populations of H_2

Molecular hydrogen has no radio frequency spectrum but it has been detected by ultraviolet absorption in diffuse clouds. The presence of H_2 in diffuse clouds where it can be destroyed by interstellar photons provides valuable support for the belief that molecular hydrogen is the major constituent in dense clouds from which interstellar radiation is excluded. The discovery that H_2 exists in detectable amounts in several rotational levels of the lowest vibrational state is important for our understanding of diffuse clouds. In Figure 8-2 we reproduce a sample spectrum obtained with the ultraviolet spectrometer on the satellite Copernicus.

The potential-energy curves of the two electronic states involved are illustrated in Figure 8-3. The ground electronic state is labelled $X^1\Sigma_g^+$ and the excited electronic state is labelled $B^1\Sigma_u^+$. Allowed absorptions are possible to any vibrational level v' of the upper electronic state subject to the rotational selection rule $J' = J'' \pm 1$. An increase by unity corresponds to the R branch

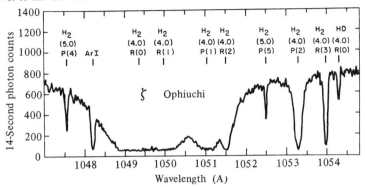

Figure 8-2. A high-resolution scan of the spectrum towards ζ Ophiuchi, from Spitzer and Jenkins (1975). The absorption lines originating in several rotational levels of H₂ and one rotational level of HD are indicated.

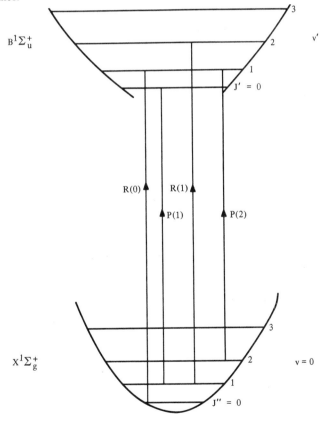

Figure 8-3. A schematic diagram of the potential energy curves of H₂ illustrating the observed absorption lines.

and a decrease by unity corresponds to the P branch. Thus $P(5)$ in Figure 8-2 labels the line originating in rotational level $J'' = 5$ of the $v'' = 0$ vibrational level of the ground electronic state and terminating in the rotational level $J' = 4$ of some vibrational level (actually $v' = 5$ for the line near 1052.5 A) of the upper electronic state.

The populations $N(J)$ of the individual rotational levels can be derived from the measured absorption strengths by a curve-of-growth analysis. It is instructive to compare them by plotting $\ln[N(J)/g_J]$ against E_J where g_J is the statistical weight of level J and E_J is its rotational energy. If the points lie on a straight line, the slope of this straight line gives an effective temperature.

Figure 8-4 reproduces the rotational populations obtained with the Copernicus spectrometer towards the stars ζ Ophiuchi, ϵ Persei, and ζ Puppis. Typically there occur two effective temperatures, one appropriate for the populations of the low-lying rotational levels and the other, higher, temperature appropriate for the populations of the higher rotational levels; for some stars, such as ζ Puppis, there is only one temperature and it is high, of the order of 1000 K.

There are three distinct mechanisms through which the rotational levels of H_2 can be populated. In the first, a shock wave propagating rapidly through the interstellar cloud containing the molecular hydrogen creates a warm high-density region behind the shock front and collisions populate the levels, which then decay by radiative processes. In the second mechanism, ultraviolet photons are absorbed (as is observed) which populate the excited electronic states. The electronic excited states then decay by spontaneous emission and produce an enhanced vibrational and rotational population of the ground electronic state. Molecular hydrogen has no dipole moment but the excited vibrational levels of the ground electronic state can radiate within a time, τ_r, of between 10^5 and 10^6 sec by the emission of electric quadrupole photons. The cascading process ultimately leaves the molecule in various rotational levels of the $v'' = 0$ vibrational state. The ultraviolet pumping process, in which photons are absorbed into the excited $B^1\Sigma_u$ and $C^1\Pi_u$ states of H_2, is illustrated schematically in Figure 8-5. The net effect is an enhanced rotational population.

The third possible mechanism for producing an enhanced rotational population is the formation mechanism. At the low kinetic temperatures characteristic of the majority of interstellar clouds, molecular hydrogen is formed only by association on interstellar grains. The details of the formation are uncertain. If the molecule remains on the grain surface long enough to reach thermal equilibrium with the grain, it is released with little development of its internal rotational and vibrational modes. If it is detached immediately after it is formed, the molecule will have a substantial population of excited

Figure 8-4. The rotational populations of H_2 measured towards ζ Ophiuchi, ϵ Persei and ζ Puppis. For each star, the value of $N(J)/g_J$ is plotted on a logarithmic scale against the energy of excitation E_J. The symbol T_{ex} denotes the excitation temperature obtained from the slope of the high J populations. The symbol T is the excitation temperature derived from the slope of the low J populations. I indicates the error. From Spitzer and Cochran (1973).

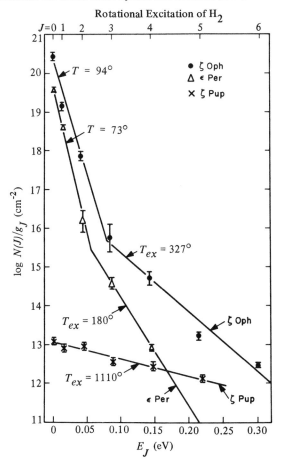

vibrational and rotational levels which soon radiate down to the various rotational levels of the lowest vibrational level.

The rotational levels of the ground vibrational level may be depopulated
by radiation. Because it is electric quadrupole radiation, the transition probability varies as the fifth power of the transition frequency v^5 (in contrast to
electric dipole transitions which vary as v^3). The frequency v is proportional
to the rotational quantum number J so that the lifetime of level J to radiative decay varies as J^{-5}. For $J = 2$, $\tau_r = 3.4 \times 10^{10}$ sec and for $J = 6$,
$\tau_r = 3.8 \times 10^7$ sec.

A redistribution of the population of the rotational levels can also occur
through collision processes, but the collision cross section σ_J varies slowly
with J. Then if v is the velocity, the mean time for collisions is $\tau_c = (n \overline{v \sigma_J})$
where the bar denotes an average over the velocity distribution. It is nearly
independent of J and inversely proportional to the density n. The value of
$\overline{v \sigma_J}$ is of the order of 5×10^{-11} cm³/sec so that with a density of 100 cm⁻³, τ_c

Figure 8-5. A schematic diagram of the ultraviolet pumping mechanism for enhancing the rotational populations of the ground vibrational state $v = 0$.

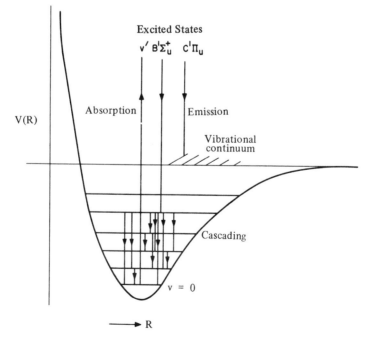

is of the order of 2×10^8 sec. Then τ_c and τ_r are approximately equal at $J = 4$. Then because $\tau_c < \tau_r$ at lower J, it follows that the populations of the low J levels are controlled by collision processes so that they are characterised by the kinetic temperature of the interstellar cloud, whereas, because $\tau_c > \tau_r$ at high J, the high J levels are unaffected by collisions. Because we know the rate of their depopulation by quadrupole emission, the steady-state populations of the high J levels provide a measure of the efficiency of the mechanisms that populate them.

If the rapid shock-wave mechanism were operative, the excited molecules would have a substantial translational velocity. The longer-lived molecules in low J states would be slowed by collisions and velocity differences should exist between the low and high J level populations. Such differences are in general not observed, and although shock waves are probably responsible for the creation of many of the dense regions in which the hydrogen molecules are found, they are usually not the source of the enhanced rotational populations.

The ultraviolet pumping mechanism is surely a source of enhanced rotational population, the initiating step being just those absorptions that are the observational features which establish the presence of rotationally excited hydrogen. The efficiency with which the mechanism occurs is of course directly proportional to the ultraviolet flux incident upon the molecules.

The formation mechanism may also lead to rotational enhancement, and we may easily argue that in the steady state its efficiency is also directly proportional to the incident ultraviolet flux. Thus, in equilibrium, molecular formation proceeds at the same rate as molecular destruction and interstellar hydrogen molecules are destroyed by the absorption of the same ultraviolet photons that participate in the ultraviolet pumping. Interstellar molecular hydrogen cannot be destroyed by the direct continuum absorption of a photon in a transition to an excited electronic state because this process requires photons of wavelength shorter than 845 Å, and stellar photons with wavelengths shorter than the Lyman limit at 912 Å are consumed in the creation of the H II regions surrounding the stars. Even if some do escape into the interstellar medium, they will be excluded from the regions containing the hydrogen molecules by the hydrogen atoms which are also present. However, molecular hydrogen can be destroyed by the discrete absorption of photons with wavelengths between 912 Å and 1108 Å in a fluorescent process that differs from the ultraviolet pumping sequences only in that the spontaneous radiative decays out of the intermediate excited electronic states terminate not in the bound rotation vibration levels of the ground electronic state but in the continuum rotation-vibration levels above the dissociative threshold. Calculation of the efficiencies of the spontaneous line and continuum emission probabilities shows that fluorescent destruction occurs about once for every ten interstellar ultraviolet photons absorbed.

Because the ultraviolet pumping mechanism and the formation mechanism for populating the rotational levels are both directly proportional to the incident ultraviolet flux and because the high rotational levels decay by the known process of quadrupole emission, the intensity of the incident interstellar ultraviolet flux $\bar{I}_\lambda$ between the Lyman limit, 912 Å, and the threshold wavelength for the pumping, 1100 Å, can be inferred from the measurements of the populations of the high J levels.

The H/H$_2$ Equilibrium

It is known from observations of optical obscuration, polarization, and reddening of starlight that the interstellar medium contains dust grains, and it is widely believed that in cold clouds the grain surfaces are the sites for the formation of molecular hydrogen and possibly of other interstellar molecules. When hydrogen atoms strike a grain, a fraction γ of these atoms is assumed to leave the grain in the form of hydrogen molecules; the value of γ is assumed to be about 0.3. The rate of collisions of hydrogen atoms with grains is given by $n(g)v_H\sigma_g$ where $n(g)$ is the grain number density, v_H is the mean velocity of the hydrogen atoms and σ_g is the grain cross section. The product $n(g)\sigma_g$ can be related to the density of hydrogen nuclei $n_H = n(H) + n(H^+) + 2n(H_2)$ by using measures of the extinction due to the grains (Spitzer, 1968) and it appears that for a kinetic temperature of 100 K, the formation rate of molecular hydrogen is of the order $10^{-17}n(H)n_H$ cm^{-3} sec^{-1}. It may well be larger because of the presence of very small grains that are not revealed by extinction in the visible and we adopt a rate of $3 \times 10^{-17}n(H)n_H$.

The destruction rate is given by $\Sigma k_J\beta_J n_J(H_2)$ cm^{-3} sec^{-1}, where k_J is the fraction of absorptions from level J that lead to dissociation, β_J is the rate of absorption of photons and is proportional to the ultraviolet radiation flux, and $n_J(H_2)$ is the number density of the Jth rotational level of molecular hydrogen. In a steady state,

$$\sum_J k_J\beta_J n_J(H_2) = 3 \times 10^{-17}n(H)n_H.$$

In the average unshielded interstellar radiation field β_J is almost independent of J and has a value of the order 10^{-10} sec^{-1}. The parameter k_J is determined essentially by the relative efficiency of the spontaneous radiative dissociation process and is about 0.1, independent of J. Thus exterior to an interstellar cloud

$$n(H_2) = \sum_J n_J(H_2) = 3 \times 10^{-6} [n(H)]^2,$$

and the hydrogen is almost entirely in atomic form. With increasing depth, the absorption rates β_J decrease exponentially as the available photons are used up in dissociating H_2. The hydrogen molecules are self-shielding and in the inner core of the cloud the hydrogen will be almost entirely in molecular form.

The observational data on molecular hydrogen are summarized in Figure 8-6. The figure includes the results of theoretical calculations for total hy-

Figure 8-6. The observed H_2 column densities (Spitzer and Jenkins, 1975) as a function of the color excess $E(B - V)$. The full curves are theoretical predictions of J. H. Black for total hydrogen densities of 10, 100, and 1000 cm^{-3}.

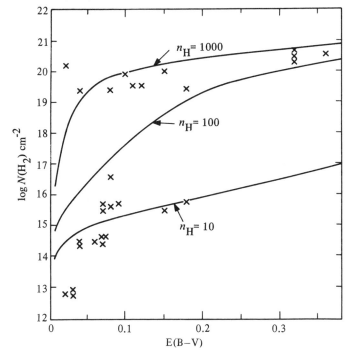

drogen densities n_H = 10, 100, and 1000 cm^{-3} and for a specific incident ultraviolet flux. The column abundances of H$_2$, n(H$_2$), obtained by integrating n(H$_2$) through the cloud, are shown as a function of the color excess $E(B - V)$. The color excess is a measure of the dependence on wavelength of interstellar extinction and it is proportional to the column density of interstellar grains. By assuming that the gas and the grains are uniformly mixed, the approximate relationship $N_H = 7.5 \times 10^{21} E(B - V)$ cm^{-2} mag^{-1} can be established (cf. Spitzer and Jenkins, 1975). Figure 8-6 demonstrates that the fraction $f = 2N(H_2)/N_H$ of hydrogen in molecular form exceeds 0.1 for total column densities $N_H > 10^{21}$ cm^{-2} but is generally less than 2×10^{-4} for small color excesses.

There are no cases in which the molecular fraction f lies between 2×10^{-4} and 5×10^{-2} and it may be there are two distinct classes of clouds.

Ortho- and Para-Populations

Molecular hydrogen exists in two forms. In para-hydrogen, the nuclear spins of the protons are antiparallel and the Pauli principle requires that only even-numbered J levels can exist. The opposite is true for ortho-hydrogen, for which the proton spins are parallel and only odd-numbered J levels exist. Because a nuclear spin exchange is required, there is no purely radiative mechanism which can change an odd J level into an even J level at other than a negligible rate, and yet the observational data show clearly that the $J = 0$ and $J = 1$ levels are usually in an equilibrium determined by the same temperature that characterizes the relative populations of the $J = 0$ and $J = 2$ levels. A collision mechanism that interchanges the proton spins must occur in the interstellar cloud.

Chemical reactions are often slowed by the presence of a potential barrier and do not proceed rapidly until the collision is energetic enough to surmount the barrier. The rate coefficient is characterized by an activation energy ΔE and its value at temperature T is proportional to $\exp(-\Delta E/kT)$. As the nuclear spin-labeling demonstrates, the exchange reaction

$$\overset{\downarrow}{H} + \overset{\uparrow\uparrow}{H_2}(\text{ortho}) \longrightarrow \overset{\downarrow\uparrow}{H_2}(\text{para}) + \overset{\uparrow}{H} \tag{8-1}$$

can transform ortho-H$_2$ into para-H$_2$. However, the reaction has an activation energy of about 0.5 eV equivalent to kT for $T \sim 5000$ K, and is consequently very slow at interstellar cloud temperatures. It seems, then, that the equilibration reaction must be the proton interchange reaction

$$\overset{\downarrow}{H^+} + \overset{\uparrow\uparrow}{H_2}(\text{ortho}) \rightleftarrows \overset{\downarrow\uparrow}{H_2}(\text{para}) + \overset{\uparrow}{H^+}. \tag{8-2}$$

The long-range polarization force between H^+ and H_2 overcomes any activation energy and the reaction is predicted to be rapid. In equation (8-2), the H^+ ion acts as a catalyst for the equilibration. We infer that there is a source of hydrogen ionization in the clouds other than ultraviolet radiation, presumably cosmic rays or X-rays, and we conclude that the $J = 1$ to $J = 0$ population ratio is a reliable indicator of the cloud temperature. For clouds exceeding the column density $N(H_2) = 10^{19}$ cm^{-2}, the average temperature derived from the ortho- and para-populations is 80 K, in good agreement with temperatures derived from atomic hydrogen 21-cm data.

Cloud Models

Figure 8-6 presents the theoretical distributions for a specific choice of the incident radiation field. We can be more precise. Using our earlier arguments, we can determine the actual radiation field from the measured populations of high J levels. We can determine the temperature from the $J = 1$ and $J = 0$ population ratio and we can estimate the density from the dependence of the population on intermediate values of J as we move from a physical regime dominated by collisions at low J where $\tau_r > \tau_c$ to a regime dominated by radiation processes at high J where $\tau_c > \tau_r$. Then we can derive a value of the efficiency of the rate of formation of hydrogen molecules from the relative abundances of H and H_2.

The analysis is straightforward but complicated. When it is carried through for the clouds observed by the Copernicus satellite, the formation rates that are obtained are within the expected range near $3 \times 10^{-17} n_H n(H)$ cm^{-3} sec^{-1}. The derived densities n_H range from 20 cm^{-3} for δ Orionis to more than 10^3 cm^{-3} for ζ Ophiuchi. The derived radiation fields range from about the mean interstellar field with an energy density of 8×10^{-15} erg/cm^3 between 912 Å and 1100 Å, to a value 100 times greater.

Several clouds are associated with a large ultraviolet intensity. They presumably lie close to the parent star and are physically connected with it. In some instances, N_H is small but n_H is large and the cloud configurations are thin, dense, and sheet-like. They may be manifestations of interstellar bubbles, caused by stellar winds emanating from early-type stars. Such winds cause the propagation of shock waves which sweep up interstellar material into shells around the star. In other instances, the thin sheets may be condensations created during the cooling and recombination of a hot ionized plasma produced perhaps by a supernova explosion.

If indeed the temperature for ζ Puppis indicated by Figure 8-4 is a kinetic temperature the grains may be too hot to permit the formation of H_2 on their surfaces. In such physical conditions, the mechanism of associative detachment,

$$H + H^- \longrightarrow H_2 + e, \tag{8-3}$$

may have produced the molecular hydrogen. The process is rapid, occurring at a rate of about $10^{-9} \, n(H)n(H^-)$ cm^{-3} sec^{-1} but the efficiency is limited by the photodetachment of H^- and by the process that forms the negative ion H^-. That process is radiative attachment,

$$H + e \longrightarrow H^- + h\nu, \tag{8-4}$$

which demands a high temperature and a high fractional ionization to be effective.

The Molecule HD

The spectrum of Figure 8-2 contains in addition to several absorption lines of H_2 an absorption line $R(0)$ of the 4-0 band of the molecule HD.

The ratio $n(D)/n(H)$ has been derived in the direction of several stars by measurements of the interstellar absorption in the atomic absorption lines of D and of H. For those directions where no molecules are detected, the ratio gives the relative abundance of deuterium n_D in all its forms to hydrogen n_H in all its forms. The value is 1.4×10^{-5} and it is apparently independent of position though the error bars of the measurements are larger for the more distant stars. For the direction towards the nearby star β Centauri, the value is $(1.4 \pm 0.2) \times 10^{-5}$.

The deuterium abundance is of cosmological significance. With the assumption that all the deuterium was produced at early times in the formation of the universe, the abundance can be used to derive a value for the present mass density of the universe. The value so derived, 1.5×10^{-31} g/cm^3, is much less than the closure density so that the universe is open and will continue to expand. The conclusion will be modified if deuterium is produced in the Galaxy by other mechanisms.

The molecule HD is destroyed in a manner analogous to the destruction of

H_2. Because the destruction is initiated by line absorption, and the abundance of interstellar HD is not enough to achieve self-shielding of HD, its lifetime towards photodestruction is much less than that of H_2. If it were not for the different lifetimes in the clouds, the measured ratio of $n(HD)$ to $n(H_2)$ would exceed 10^{-3}. Since $n_D/n_H = 1.4 \times 10^{-5}$, there must be a source of interstellar HD in addition to the grain formation and associative detachment processes that produce H_2.

The source is in fact already identified. The analogue of equation (8-2) is

$$D^+ + H_2 \rightleftharpoons HD + H^+. \tag{8-5}$$

It can be a large source of HD, the D^+ being provided by the near-resonance charge-transfer reaction

$$H^+ + D \rightleftharpoons H + D^+. \tag{8-6}$$

A quantitative interpretation of the measured ratio of $n(HD)$ to $n(H_2)$ leads therefore to the determination of the density of D^+ and of H^+ and ultimately, if the chemistry of the H^+ removal processes is understood, to a determination of the magnitude of the ionizing flux that produces the H^+ ions.

The Cloud Chemistry

Before considering the possible chemical sequences, we recall that the efficiency of a chemical reaction is described by its rate coefficient k measured in cm^3/sec, which is the product of the cross section σ and velocity v averaged over the velocity distribution. Then the number of reactions occurring in unit volume in unit time is given by the product of k and the number densities of the reactants. Thus in the reaction

$$X + YZ \longrightarrow XY + Z \tag{8-7}$$

the production rate of XY molecules is given by $kn(X)n(YZ)$ cm^{-3} sec^{-1}. The efficiency of the process in the gas can be characterized by the mean lifetime $\tau_X(YZ)$ of the molecule YZ towards destruction by X, defined by $\tau_X(YZ) = 1/kn(X)$ measured in seconds.

Rate coefficients vary widely and are often sensitive to the kinetic temperature. However, many exothermic ion-molecule reactions,

$$X^+ + Y_2 \longrightarrow XY^+ + Y, \tag{8-8}$$

are dominated by the orientation-independent part of the long-range attractive force between the positive ion X^+ and the neutral molecule Y_2, an interaction that leads, as we shall show, to a collision frequency independent of temperature.

By transforming to the center-of-mass system, the scattering of particles of masses M_1 and M_2 by an interaction which depends only on the distance R between the particles can be reduced to that of the scattering of a single particle of reduced mass μ by a center of force with potential $V(R)$ where $\mu^{-1} = M_1^{-1} + M_2^{-1}$. Defining the impact parameter p as the perpendicular distance from the center of force to the incident particle trajectory, we may write the effective potential as a sum of the polarization potential $V(R) = -\alpha e^2/2R^4$, α being the average electric polarizability of Y_2, and the centrifugal repulsion $\mu p^2 v^2/2R^2$ where v is the velocity of relative motion. The effective potential has a repulsive barrier with a maximum at $R^* = (2\alpha e^2/\mu p^2 v^2)^{1/2}$ and its magnitude is $(\mu p^2 v^2)^2/\alpha e^2$. For a given velocity, only in collisions with impact parameters less than the critical value $p^* = (4\alpha e^2/\mu v^2)^{1/4}$ will the barrier be surmounted. Thus the cross section for close collisions is given by $\sigma = \pi p^{*2} = (2\pi/v)(\alpha e^2/\mu)^{1/2}$ and the collision frequency $\sigma v = 2\pi(\alpha e^2/\mu)^{1/2}$, independent of velocity and therefore of temperature. If it is assumed that every close collision leads to reaction, $2\pi(\alpha e^2/\mu)^{1/2}$ is also the rate coefficient. For heavy ions colliding with H_2, which has a polarizability $\alpha = 7.8 \times 10^{-25}$ cm^3, the value of the rate coefficient is about 1.5×10^{-9} cm^3/sec. Laboratory measurements have shown that rate coefficients of this order are common for exothermic ion-molecule reactions though specific exceptions have been found where the assumption that every close collision leads to reaction is invalid.

For neutral reactions (8-7), the long-range interaction is the relatively weak Van der Waals interaction which decreases as R^{-6} at large R. Short-range forces are important and the cross section for close collisions is comparable to the geometric cross section of the order of 10^{-15} cm^2. Thus at interstellar temperatures the neutral-particle collision frequency is of the order of 10^{-10} cm^3/sec. This is an upper limit to the rate coefficient because potential barriers occur in most neutral particle reactions, giving rise to activation energies ΔE below which reaction does not occur. The rate coefficients then decrease rapidly as $\exp(-\Delta E/kT)$ as T decreases, where k is Boltzmann's constant.

Thus the gas phase chemistry of the interstellar medium is largely controlled by ion-molecule reactions in which the positive ions undergo a series of exothermic chemical exchange (reaction 8-8) and charge transfer processes

$$X^+ + YZ \longrightarrow X + YZ^+, \tag{8-9}$$

which terminate in the creation of positive ions of low ionization potentials with insufficient energy to react further with the ambient gas atoms or molecules.

The molecular ions can be removed by the process of dissociative recombination,

$$XY^+ + e \longrightarrow X + Y. \tag{8-10}$$

Partly because of the long-range Coulomb attraction, the rate coefficients for dissociative recombination are large for most ionic species with values of the order of 10^{-6} to 10^{-7} cm^3/sec at cloud temperatures.

Table 8-2 lists the cosmic abundances of several of the more common elements. The analysis of visible and ultraviolet interstellar absorption lines has suggested that the elements are depleted in interstellar clouds and the appropriate values to use for the gas phase chemistry may be smaller by up to a factor of five (cf. Spitzer and Jenkins, 1975). In a diffuse cloud, the ionization is maintained by the interstellar ultraviolet radiation field and all those elements with ionization potentials less than that of atomic hydrogen are ionized. Thus carbon exists as C^+, silicon as Si^+, iron as Fe^+ and the fractional ionization $n_e/n_H \sim 5 \times 10^{-4} \delta$ where δ is the depletion factor.

Because of the large relative abundance of H or H_2 in the clouds, any constituent that can react with hydrogen will do so rapidly. A sequence of hydrogen abstraction reactions

$$XH_n^+ + H_2 \longrightarrow XH_{n+1}^+ + H \tag{8-11}$$

often occurs until a complex ion is created for which the abstraction reaction is endothermic. Free electrons are as abundant as the minor constituents and reaction (8-10) usually has a large rate coefficient so that reaction (8-11) is

Table 8-2 Cosmic abundances of the elements relative to hydrogen.

H	1	He	0.1
O	6.8×10^{-4}	C	3.7×10^{-4}
N	1.2×10^{-4}	Si	3.5×10^{-5}
Fe	2.5×10^{-5}	S	1.6×10^{-5}

followed by the production of the neutral species XH_n. The neutral species may then undergo reactions with positive ions or with other neutral atoms and molecules but in most cases at a much slower rate.

All the atomic and molecular systems are subject to photoionization and photodissociation and a complex array of processes occurs.

The H^+ Chemistry

We can now discuss the chemistry of H^+ ions. H^+ does not react with H or H_2 and cannot undergo dissociative recombination. It can be removed by radiative recombination

$$H^+ + e \longrightarrow H' + h\nu, \tag{8-12}$$

the prime indicating that the hydrogen may be produced in excited states. Radiative recombination is the process that dominates in H II regions but because it involves a photon it is rather slow with a rate coefficient of about 10^{-11} cm³/sec (cf. Chapter 7). It happens that atomic hydrogen and atomic oxygen have nearly identical ionization potentials and the H^+ ions can charge transfer with O to form O^+ in the reaction

$$H^+ + O \longrightarrow H + O^+. \tag{8-13}$$

In the absence of any molecules the O^+ ions undergo radiative recombination,

$$O^+ + e \longrightarrow O + h\nu, \tag{8-14}$$

and the reverse process to reaction (8-13),

$$O^+ + H \longrightarrow O + H^+. \tag{8-15}$$

Reaction (8-13) is endothermic by 0.0196 eV or 232 K so that its rate coefficient k_{13} and that, k_{15}, of reaction (8-15) are related by

$$\frac{k_{13}}{k_{15}} = \frac{8}{9} \exp{(-232/T)},$$

the factor of 8/9 coming from the statistical weights of the available reaction paths. The value of k_{15} is uncertain at very low temperatures but a combination of atmospheric, laboratory, theoretical, and astrophysical studies suggests that in the warmer diffuse clouds a value of about 10^{-9} cm³/sec is appropriate. The sequence of reactions (8-13) and (8-15) significantly affects the concentration of O^+ but because the rate coefficients of reactions (8-12) and (8-14) are similar it does not greatly change the concentrations of the H^+ ions or of the electrons.

In the presence of only small amounts of molecular hydrogen, the situation is drastically altered. The O^+ ions react rapidly with H_2 in the hydrogen abstraction reaction

$$O^+ + H_2 \longrightarrow OH^+ + H, \tag{8-16}$$

which is followed by the abstraction sequence

$$OH^+ + H_2 \longrightarrow H_2O^+ + H, \tag{8-17}$$

$$H_2O^+ + H_2 \longrightarrow H_3O^+ + H. \tag{8-18}$$

The rate coefficients of reactions (8-16), (8-17), and (8-18) have been measured at room temperature in the laboratory. The reactions are rapid.

The polyatomic ion H_3O^+ is isoelectronic with ammonia, NH_3. It is chemically saturated and does not react further with H_2. It is removed by dissociative recombination

$$H_3O^+ + e \longrightarrow H_2O + H, \tag{8-19}$$

$$H_3O^+ + e \longrightarrow OH + H_2, \tag{8-20}$$

at a rate of about $10^{-6} n(e)$ sec^{-1} where $n(e)$ is the electron density.

A summary diagram of the chemical sequence is illustrated in Figure 8-7. Reaction (8-20), for example, is represented by the line joining H_3O^+ to H_2O, labeled by the constituent with which H_3O^+ reacts to form H_2O. The sequence can be cast in the form of a simultaneous set of algebraic equations in which the rate of production of any species is equated to the rate of destruction. The solutions are the equilibrium concentrations of the participating species. The ionizing rate ζn_H cm^{-3} sec^{-1} enters as a parameter that can be varied until the solution for $n(H^+)$ is equal to that inferred by the observations of HD. For the cloud in front of ζ Ophiuchi, the derived proton density is $n(H^+) = 0.01$ cm^{-3} and the corresponding ionizing flux ζ is 1.5×10^{-17} sec^{-1}, a value which is not much larger than that provided by the high-energy cosmic rays measured in the solar neighborhood (Spitzer, 1968).

The Chemistry of OH, CO, and HCO^+

The chemical scheme for the removal of H^+ leads to the formation of water vapor, H_2O, and the hydroxyl radical, OH. The branching ratio for the

paths (8-19) and (8-20) is unknown. In a diffuse cloud, H_2O is soon dissoci-
ated in a time of about 30 years in the obscured interstellar radiation field by
ultraviolet photons,

$$H_2O + h\nu \longrightarrow OH + H, \tag{8-21}$$

to produce OH so that the uncertainty in the branching ratio is unimportant
in the calculation of the abundance of OH. Thus the abundance of OH can
also be used as a measure of the ionizing flux since each H^+ ion leads to an
OH molecule.

However, to derive the ionizing flux from observations of OH, we must
understand the OH removal mechanisms. The OH radicals may be

Figure 8-7.   The H^+, OH, HCO^+, and CO chemistry

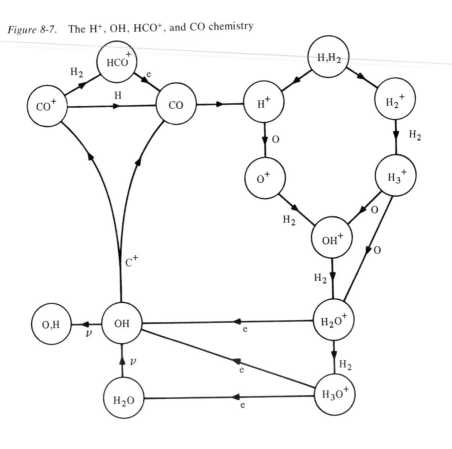

destroyed by direct photodissociation

$$OH + h\nu \longrightarrow O + H, \tag{8-22}$$

and by the chemical reactions

$$OH + C^+ \longrightarrow CO^+ + H \tag{8-23}$$

$$OH + C^+ \longrightarrow CO + H^+. \tag{8-24}$$

The mean time for photodissociation is about 300 years in the unshielded radiation field but it increases exponentially with depth into the cloud as the grains absorb the dissociating photons. The rate coefficient for the sum of the chemical reactions (8-23) and (8-24) is 1.6×10^{-9} cm^3/sec so that a cloud of density 100 cm^{-3} with a C$^+$ abundance of 10^{-2} cm^{-3}, say, the chemical time is comparable at the edge of the cloud but shorter in the interior. Thus chemical reactions determine the destruction of OH throughout most of the cloud.

Reaction (8-23) is often followed by the rapid reaction

$$CO^+ + H_2 \longrightarrow HCO^+ + H \tag{8-25}$$

but in clouds where $n(H)$ is comparable to $n(H_2)$,

$$CO^+ + H \longrightarrow CO + H^+ \tag{8-26}$$

also occurs.

Both reactions (8-24) and (8-26) produce H$^+$, which acts as a catalyst for the formation of OH and H$_2$O. Care must be taken in deriving the ionizing fluxes to recognize that the chemical scheme of Figure 8-7 can be repeated and a single H$^+$ ion can produce more than one HD or OH molecule. The recent detection of OH by ultraviolet absorption towards ζ Ophiuchi and o Persei is consistent with the scheme of Figure 8-7 and the ionizing fluxes derived from the measurements of HD.

The unknown branching ratio between reactions (8-19) and (8-20) directly affects the production of H$_2$O, and a reliable prediction of the abundance of H$_2$O cannot be given.

The chemical sequence leads to the formation of carbon monoxide, which is the second most abundant molecule in the interstellar medium. Carbon monoxide has also been detected in interstellar clouds in other galaxies. It is a very stable molecule with a large energy of dissociation and its destruction by photodissociation or by chemical reactions is slow. Thus a small source of CO can lead to a large equilibrium abundance.

The interstellar molecular ion HCO^+ is also produced in the scheme initiated by the H^+-O charge transfer both by reaction (8-25) and by

$$C^+ + H_2O \longrightarrow HCO^+ + H. \tag{8-27}$$

It is isoelectronic with the interstellar molecule hydrogen cyanide, HCN. It is chemically stable and does not react with molecular hydrogen. However, in diffuse clouds it is readily removed by dissociative recombination,

$$HCO^+ + e \longrightarrow H + CO, \tag{8-28}$$

to produce carbon monoxide and it has been detected only in denser regions.

The Chemistry of CH^+ and CH

Unlike O^+, C^+ does not react with H_2 and a different scheme is necessary for the formation of CH^+ and CH. In the terrestrial environment, molecules are built up from smaller systems by three-body reactions;

$$X + Y + M \longrightarrow XY + M. \tag{8-29}$$

Rate coefficients for three-body reactions rarely exceed 10^{-28} cm^6/sec so that at interstellar densities of 10^3 cm^{-3}, say, one molecule XY is formed in a mean time of not less than 10^9 years. Thus three-body association reactions can be ignored. A two-body association of X and Y requires the production of another particle if momentum and energy are to be simultaneously conserved. In practice, the particle is a photon and XY is formed by radiative association:

$$X + Y \longrightarrow (XY)^* \longrightarrow XY + h\nu. \tag{8-30}$$

Thus

$$C^+ + H \longrightarrow CH^+ + h\nu \tag{8-31}$$

and

$$C^+ + H_2 \longrightarrow CH_2^+ + h\nu \tag{8-32}$$

may occur. They will be followed by a complex of reactions summarized in Figure 8-8.

There is little difficulty in explaining the measured abundances of CH but

CH$^+$ remains a problem because of its rapid removal by

$$CH^+ + e \longrightarrow C + H \qquad (8\text{-}33)$$

and by

$$CH^+ + H_2 \longrightarrow CH_2^+ + H. \qquad (8\text{-}34)$$

The last reaction has been measured to be fast at room temperature. If it remains fast at the colder cloud temperatures, it seems that because H$_2$ is required to form CH and yet it destroys CH$^+$, CH and CH$^+$ must occupy different regions of space. Calculations show also that it is necessary to postulate that reaction (8-33) (which has not been measured) is anomalously slow. Alternatively, equilibrium considerations may not be appropriate and the CH$^+$ molecules may be a result of the interactions of the stars with the surrounding interstellar material.

Figure 8-8. The CH$^+$ and CH chemistry.

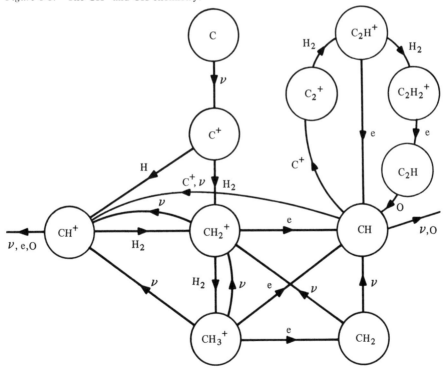

The reaction scheme presented in Figure 8-8 has other chemical consequences. In particular, the recently detected interstellar molecule C_2H is produced by abstraction reactions of C_2^+ with H_2 and the third interstellar molecule detected by visual absorption, CN, can be readily formed by exchange reactions of CH^+ and CH with atomic nitrogen.

The CH sequence leads also to CO and HCO^+ through the chemi-ionization process

$$O + CH \longrightarrow HCO^+ + e, \tag{8-35}$$

ion-molecule reactions such as

$$CH_3^+ + O \longrightarrow HCO^+ + H_2, \tag{8-36}$$

and chemical exchange processes such as

$$CH^+ + O \longrightarrow CO + H^+, \tag{8-37}$$

$$C_2^+ + O \longrightarrow CO + C^+. \tag{8-38}$$

Tentative models have been constructed of diffuse cloud chemistry based upon the schemes outlined here and they are reasonably successful in reproducing the available data on the abundances of simple molecules. Their success may be due in large measure to the flexibility allowed by the uncertainties that attend the rates of many of the processes that are invoked. Much further study is needed before the molecular abundances can be used as a reliable indication of the physical conditions prevailing in the clouds. Indeed a major problem exists of identifying a chemical sequence that leads to the efficient production of formaldehyde, H_2CO. Because the H_2 bond must be broken, formaldehyde cannot be produced from H_2 and CO in an exothermic series of reactions and it is easily photodissociated. Yet H_2CO is a ubiquitous interstellar molecule which apparently exists in relatively unobscured regions as well as in dense clouds.

The Chemistry of Dense Clouds

The molecules H_2 and CH^+ have been detected only in diffuse clouds. They do not have emission or absorption in the radio region of the spectrum but they may be detected eventually in the infrared. All the other molecules listed in Table 8-1 have been detected by radio observations of dense clouds through which visible and ultraviolet light does not penetrate because of absorption by the grains.

Dense molecular clouds are the most massive objects in the Galaxy and a substantial fraction of the mass of the Galaxy is in the molecular clouds. Molecules have been detected in numerous clouds associated with H II regions. The molecular cloud Sagittarius B2 associated with the H II region Sgr B, lying near the galactic center, is probably the most massive cloud in the Galaxy, with a mass exceeding 10^6 solar masses. It is a particularly rich source of interstellar molecules and all the radio molecules of Table 8-1 have been detected in the Sgr B2 cloud. Star formation has undoubtedly occurred inside the cloud.

Dark clouds which are observed photographically are also sources of interstellar molecules. They are dense clouds not associated with known H II regions. Mapping of the distribution of the intensity of the emission from CO in dark clouds has revealed intensity peaks that may indicate regions of incipient star formation and the dark clouds probably contain protostellar objects and young stars in the earliest stages of stellar evolution.

The determination of the physical characteristics of the dense clouds from the measured molecular line intensities is not straightforward. It involves detailed assessments of the collision processes that contribute to the population and depopulation of the emitting or absorbing level and of the photon scattering mechanisms that control the transfer of radiation through the cloud. Of special significance is the ratio of the mean collision time τ_c to the spontaneous radiative time τ_r. The uncertainties in the interpretation are much reduced if measurements are available for several transitions of several molecules with different values of the ratio τ_c/τ_r.

The observations and their analysis have been reviewed by Solomon (1973) and by Zuckerman and Palmer (1974). It appears that dense molecular clouds have densities in the range 10^3 cm^{-3} to 10^6 cm^{-3} with temperatures of about 10 K prevailing in the dark clouds and temperatures between 30 and 100 K prevailing in the molecular clouds associated with H II regions.

There are differences between the chemistries of diffuse and dense clouds. In diffuse clouds, a molecule may undergo many chemical reactions but the sequence is finally terminated and the molecule destroyed by photodissociation or by dissociative recombination of a positive molecular ion whose origin is the photoionization of a neutral system. In a dense cloud photodissociation and photoionization are rare events. The constituents are essentially neutral and the fractional ionization n_e/n_H is low. The lifetime of a molecular species is determined by chemical reactions, and it can be so long that a small formation rate leads to a large equilibrium abundance. A detailed numerical study of the chemistry of a dense cloud is described by Herbst and Klemperer (1973). Subsequent developments have been reviewed by Watson (1975).

The observation of the positive ions HCO^+ and N_2H^+ demonstrates that there is some source of ionization in the dense clouds. The ion HCO^+ can be formed by the chemi-ionization process (8-35), and CH could conceivably be formed in the absence of C^+, but N_2H^+ cannot be produced by an analogous reaction. The origin of the ionization could be internal, a newly-born star yet to dissipate the surrounding obscuring cloud material, but the measured high-energy cosmic rays provide an alternative source.

High-energy cosmic rays will ionize the molecular hydrogen to form some H^+ but mostly H_2^+. The reaction

$$H_2^+ + H_2 \longrightarrow H_3^+ + H \tag{8-39}$$

is very fast. The triatomic ion H_3^+ can react with most neutral systems according to

$$H_3^+ + H \longrightarrow H_2X^+ + H \tag{8-40}$$

and

$$H_3^+ + X \longrightarrow HX^+ + H_2. \tag{8-41}$$

Thus if X is atomic oxygen, these reactions provide entries into the sequence of Figure 8-7 that produces OH, H_2O, CO, and HCO^+. If X is carbon monoxide, reaction (8-41) leads to $HCO,^+$ and if X is atomic carbon, reactions (8-40) and (8-41) produce CH^+ and CH_2^+, which lead through the sequence of Figure 8-8 to CH.

Ionization of He by cosmic rays is also important because in cold clouds He^+ is removed mostly by the fast reaction

$$He^+ + CO \longrightarrow He + C^+ + O, \tag{8-42}$$

which destroys CO and produces C^+. The C^+ then participates in the scheme of Figure 8-8.

If X is atomic nitrogen, the sequence summarized in Figure 8-9 can occur, leading to HCN, CN, and NH_3. However, the last step in the formation of NH_4^+,

$$NH_3^+ + H_2 \longrightarrow NH_4^+ + H,$$

is now known to be slow, so that unless the fractional ionization is very low, the abstraction sequence terminates in NH_3^+ and produces NH_2 and NH. The observation of NH_3 in interstellar clouds then argues in support of a grain formation mechanism, though the gas phase charge transfer of NH_3^+ with a heavy neutral element such as Na may suffice in clouds of low fractional ionization.

Figure 8-9.   The NH_3 chemistry.

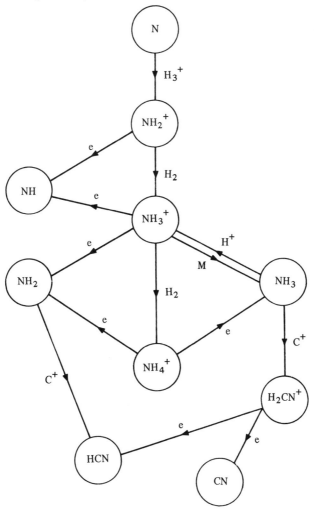

The role of grain chemistry is difficult to explore quantitatively. Atoms collide with grains at a rate of $10^{-17} n_H (T/A)^{1/2}$, where A is the atomic weight of the atom and T is the temperature. If the grains are negatively charged the rate will be increased by a factor of about three for positive ions but will be greatly reduced if the grains are positively charged. The history of the atom that collides with the grain depends upon the largely unknown nature of the grain surface. To form a molecule the atom must attach to the surface long enough to encounter a second atom. If the grain is cold, thermal evaporation will be slow and molecular formation is probable. The release of the molecule into the interstellar gas is easily achieved for H_2 because of its low adsorption energy, and some reactive molecules may be released directly from the surface in the formation process. It is not clear how saturated molecules such as NH_3, H_2O, and CH_4 which are surely formed by condensation on grains can be returned to the gas. The possible mechanisms have been reviewed by Watson (1975). Unsaturated molecules may be formed and released immediately. Formaldehyde might result from a recombination of C^+ with water and an electron on a negatively charged grain, but the data are not available that would permit a reliable quantitative assessment.

Grains are surely insufficient as a source of the interstellar HD molecules, and it is difficult to envisage the interstellar molecular ions HCO^+ and N_2H^+ emerging from a grain surface. The measured relative abundance of the chemically analogous molecules CS and CO in interstellar clouds is about 10^{-5}, very much less than the relative cosmic abundance of sulfur and oxygen. A small ratio is readily explained if gas phase mechanisms are responsible for the production of the molecules. In the sulfur analogue of Figure 8-7, the reaction of S^+ with H_2 is endothermic, and CS is formed with low efficiency.

It appears that gas phase mechanisms can be constructed for the production of most of the simpler molecules that have been observed. No specific schemes have been advanced to explain the production of the more complex species like methanol CH_3OH and dimethyl ether $(CH_3)_2O$, though a large variety of possibilities exists. It may be that the more complex molecules are produced by ultraviolet processing of molecules absorbed on the grains so that their presence is an indication of the formation of internal energy sources in the dense clouds.

Energy Balance of Interstellar Clouds

The sources of heating of interstellar clouds are many. In addition to macroscopic sources such as hydromagnetic waves and the dissipation of turbulent motion, molecular and grain processes working in response to the absorption of ultraviolet radiation and galactic cosmic rays are potentially large sources (Dalgarno and McCray, 1972). In particular, after absorption of an interstellar ultraviolet photon by a grain there is a finite probability that a photoelectron will be ejected into the gas with an energy of the order of an electron volt. The photoelectron will lose its energy in elastic collisions with the ambient thermal electrons and the heated thermal electrons share their energy with the interstellar gas by elastic collisions with hydrogen atoms and molecules. The net effect is a heating of the interstellar gas. The efficiency depends upon grain composition and in particular on the existence of an ice mantle. If the grains are not composed of H_2O ice or metals, they have large interstellar photoemission yields for photons between 1200 A and 912 A and the estimated heating rate is of the order $10^{-26} n_H$ erg cm^{-3} sec^{-1}.

Cosmic rays and X-rays produce energetic electrons by ionization. The energetic electrons lose energy in exciting and ionizing the atomic and molecular hydrogen, but eventually the energy is reduced to the point where elastic collisions with the ambient electrons are the major energy loss processes. The last several electron volts of the initial electron energy are converted into heat and the heat input arising from the ionizing rate 1.5×10^{-17} sec^{-1} indicated by our earlier discussion of the HD/H$_2$ ratio is of the order $10^{-28} n_H$ erg cm^{-3} sec^{-1}. The ionizing heat source penetrates to greater depths than the grain photoelectric source.

Chemical sequences such as that of Figure 8-8 also contribute to the heating but the contribution is not large. The chemistry is more important in its control of the cooling processes in an interstellar cloud.

Interstellar clouds lose energy by the conversion of kinetic energy into internal excitation energy of the atomic and molecular constituents of the cloud. The threshold for rotational excitation of molecular hydrogen is equivalent in energy to 450 K, and rotational excitation of H$_2$ is important only in the warmer diffuse clouds. It is negligible in the typical diffuse clouds which have temperatures near 80 K and in the dense clouds. In diffuse clouds, only C^+ has a level structure with a threshold energy near 80 K and the major energy loss is produced by the fine-structure excitation of C^+. Ionized carbon has a ^2P ground electronic state, and the interaction between the spin and orbit angular momentum splits the level into $^2P_{1/2}$ and $^2P_{3/2}$ levels of different total angular momenta. The $^2P_{1/2}$ level lies 64.0 cm^{-1} below the $^2P_{3/2}$ level and the processes

$$e + C^+(^2P_{1/2}) \longrightarrow e + C^+(^2P_{3/2}) \tag{8-43}$$

and

$$H + C^+(^2P_{1/2}) \longrightarrow H + C^+(^2P_{3/2}) \tag{8-44}$$

extract thermal energy from the cloud. The excited $C^+(^2P_{3/2})$ ion can radiate a photon at $156\mu m$ in the spontaneous radiative transition

$$C^+(^2P_{3/2}) \longrightarrow C^+(^2P_{1/2}) + h\nu. \tag{8-45}$$

The probability of absorption elsewhere in the cloud is small and the 156 μm photon escapes from the cloud.

The threshold energy of processes (8-43) and (8-44) is equivalent to a temperature of 92 K so that a diffuse cloud requires a substantial energy source to maintain a temperature above 92 K but a smaller source is required for temperatures much below 92 K. The temperatures near 80 K derived for many diffuse clouds are consistent with the threshold energies of processes (8-43) and (8-44).

Because of molecular formation and photon trapping, the energy balance of a dense cloud is a much more complicated problem. The formation of molecular hydrogen as well as its destruction are heating mechanisms which lead to nonthermal energetic hydrogen atoms. The original energy source is the ultraviolet radiation that is absorbed. The formation of H_2 actually reduces the cooling efficiency because the excitation of C^+ in collision with H_2,

$$C^+(^2P_{1/2}) + H_2 \longrightarrow C^+(^2P_{3/2}) + H_2, \tag{8-46}$$

occurs more slowly than the similar process with atomic hydrogen.

Greater consequences stem from the formation of heavier molecules such as carbon monoxide. Molecules other than hydrides have closely spaced rotational levels, and they can be excited by collisions at low temperatures. The energy threshold for the excitation of the $J = 1$ rotational level of carbon monoxide,

$$CO(J = 0) + H_2 \longrightarrow CO(J = 1) + H_2, \tag{8-47}$$

is equivalent to about 5 K. The $J = 1$ level, populated by reaction (8-47), decays spontaneously with the emission of a photon at a wavelength of 2.6 mm; it is the observation of this emission that is used to map the distribution of carbon monoxide in the Galaxy.

In dense clouds, the cooling efficiency is drastically reduced by photon trapping; the emitted photon can be reabsorbed by another CO molecule in the $J = 0$ level elsewhere in the cloud. The emissions and absorptions tend to populate the lower levels, and collisions cause the distributions to approach a thermal equilibrium characterized by the kinetic temperature. The energy loss is similar to that from a black body at the kinetic temperature of the cloud.

The efficiency of photon trapping depends upon the velocity field of the cloud. If the Doppler shift moves the frequency of the photon outside the line profile of the absorbing molecule, the photon will escape reabsorption. The velocity fields are unknown, and there is a considerable uncertainty which affects not only the description of the energy balance but also the interpretation of the radio observations of those photons that do escape from the cloud.

Calculations have been carried out assuming various velocity fields, and they show that for a collapsing cloud the most efficient energy loss processes in CO are not the $1 \rightarrow 0$ and $2 \rightarrow 1$ transitions detected by radio observations but transitions such as $5 \rightarrow 4$ which emit in the unobserved submillimeter region of the spectrum. Thus collisional excitation of molecules and the radiative transfer of molecular emissions crucially affect the thermal balance of an interstellar cloud and consequently its dynamical evolution. Different molecules have different formation time-scales and different photon-trapping lengths. The molecular control of the energy loss and ionization processes will lead to thermal and pressure gradients and possibly to thermal and dynamic instabilities that cause fragmentation of the clouds and lead to star formation. A more detailed understanding of the chemistry of the interstellar medium must be achieved to establish a complete picture of the evolution of molecular clouds and star formation; a beginning has been made.

References

Dalgarno, A., and McCray, R. A.   1972. Heating and ionization of H I regions, *Annu. Rev. Astron. Astrophys. 10,* 375–426.

Foley, H. M.   1972. Introduction to molecular spectra, in T. R. Carson and M. J. Roberts, eds., *Atoms and Molecules in Astrophysics,* pp. 156–199, Academic Press, London and New York.

Herbst, E., and Klemperer, B. W.   1973. The formation and depletion of molecules in dense interstellar clouds, *Astrophys. J. 185,* 505–533.

Hobbs, L. M.   1973. Interstellar Na I, K I, Ca II, and CH^+ line profiles toward Zeta Ophiuchi, *Astrophys. J. (Letters) 180,* L79–82.

Solomon, P. M.   1973. Interstellar molecules, *Phys. Today 26,* 32–40.

Spitzer, L.   1968. *Diffuse Matter in Space,* Interscience, New York.

Spitzer, L., and Cochran, W. D.   1973. Rotational excitation of interstellar H_2, *Astrophys. J. (Letters) 186,* L23–28.

Spitzer, L., and Jenkins, E. B.   1975. Ultraviolet studies of the interstellar gas, *Annu. Rev. Astron. Astrophys. 13,* 133–164.

Townes, C. H., and Schawlow, A.   1955. *Microwave Spectroscopy,* McGraw-Hill, New York.

Watson, W. D.   1975. Physical processes for the formation and destruction of interstellar molecules, in R. Balian, P. Encrenaz, and J. Lequeux, eds., *Atomic and Molecular Physics and the Interstellar Matter,* pp. 117–324, North-Holland, Amsterdam.

Zuckerman, B., and Palmer, P.   1974. Radio radiation from interstellar molecules, *Annu. Rev. Astron. Astrophys. 12,* 279–313.

9

Radio Observations of
Galactic Masers

J. M. Moran

More than thirty molecules in interstellar clouds have been detected by their radio emission during the past twelve years. Most of these molecular species are in approximate thermodynamic equilibrium and their emission spectra can be readily understood and quantitatively interpreted. However, the radiation from several species is so peculiar that it can be due only to maser amplification. The word "maser" is an acronym for *m*icrowave *a*mplification by *s*timulation *e*mission of *r*adiation. The essential requirement for maser amplification is population inversion. That is, there must be a larger number of molecules in the upper level of the transition than in the lower. Under such conditions stimulated emission will exceed absorption, and signals propagating through a cloud of molecules in such a state may undergo exponential amplification, instead of decay, with distance. On close scrutiny there may be many cases of population inversion in astronomical situations —such as the radio-frequency recombination lines of hydrogen. However, in only a few cases such as in clouds of OH (hydroxyl) and H_2O (water vapor) is there sufficient optical depth for the exponential gain to take hold and produce very intense radiation. This requires not only sufficient molecules but also a source of energy capable of overpopulating the upper level of the masering transition. This energy source is called the pump. The spectacular masers of OH and H_2O, along with two less spectacular ones, SiO (silicon monoxide) and CH_3OH (methyl alcohol), are the subject of this chapter. Little is known about the SiO masers since they were discovered only a year ago, and there is only one known CH_3OH maser. They are included here for completeness.

The most outstanding maser in the sky is the water vapor maser associated with the H II region called W49. Its spectrum at two different times is shown in Figure 9-1. The power output, which occurs over a range of about 30 MHz in a large number of discrete spectral components typically having widths of 50 kHz, has at times exceeded 10^{33} erg/sec or 1 $L_{\odot}$. Hence its luminosity in a small spectral window is comparable to that of the sun over the

Figure 9-1. Part of the spectrum of the water vapor maser in the H II region W49 at two different times. Such large changes are typical of this type of maser. The velocity axis is determined with respect to the local standard of rest and assumes a rest transition frequency of 22235.080 MHz. (Adapted from Sullivan, 1973.)

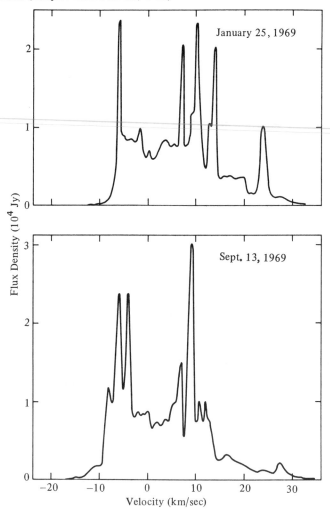

entire electromagnetic spectrum. The spectral components have diameters of only about 10^{-3} arcsec so that this power output is equivalent to that of a black body at a temperature of 10^{15} K. Since the distance to W49 is 15,000 pc, the spots have linear diameters of about 2×10^{14} cm or 10 astronomical units. The various spectral components or "spots" are distributed over an area having a diameter of about 1 arcsec, as shown in Figure 9-2. In addition, the various features come and go over a time scale of about a year so that the

Figure 9-2. A map of the spectral components, labeled by velocity, of the water vapor maser in W49. The error bars show the relative measurement accuracy with respect to the -1.8 km/sec component. The position of the map is uncertain by about 1 arcsec. The sizes of the components vary between 0.0003 and 0.003 arcsec. This map was made from an analysis of the relative rates of change of fringe phase (equation 9-24) in a VLBI experiment with an antenna spacing of 845 km (Moran et al., 1973).

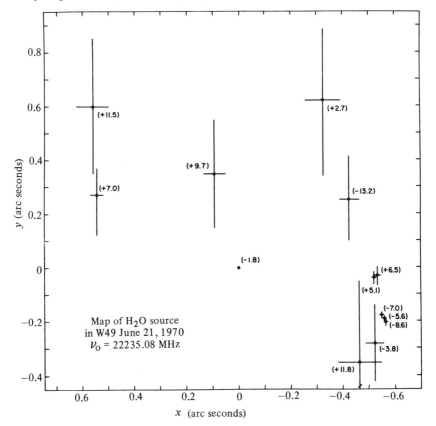

appearance of the spectrum changes character dramatically from year to year.

What we see therefore is a cluster of tiny spots of tremendous brightness over a narrow spectral range, which turn on and then turn off within a year or so. This maser with its incredible properties is so localized in space and wavelength that it was overlooked for a long time. However, in retrospect it seems as though it could have been detected with a diode and an ear phone. The water vapor maser in W49, like most other masers of its class is closely associated with a source of infrared radiation and a small region of ionized hydrogen known as a compact H II region. These three phenomena are now considered characteristic features of the birth of an O or B star. In the mid-1960s, the study of masers became isolated from molecular astronomy as more and more of their bizarre features became known, and because quantitative results such as the estimation of the density of molecules could not be directly inferred by ordinary methods. But now this field is beginning to tie in much more closely with studies of molecular clouds and star formation. The role of the astronomical maser has changed from that of a physical curiosity to that of a new-star indicator. Surprisingly, there is another class of masers associated with evolved stars, which will also be described here.

In this chapter we consider only the molecular sources of strong maser emission. The nature of nonmasering molecular sources is discussed in excellent review articles by Zuckerman and Palmer (1974), Turner (1974), and Rank et al. (1971). Masers can be distinguished from nonmasering molecular clouds by six observable characteristics: line width, number of distinct spectral features, polarization, temporal variability, size, and brightness temperature. The first four can be considered indicators of possible maser action, while size and brightness temperature are the definitive characteristics. The range of these quantities is listed in Table 9-1. As an example, the spectra of the four masers and one nonmaser, HDO, in Orion are shown in Figure 9-3. There are examples of masers with only one feature, stable in time, unpolarized but still with high brightness temperature. On the other hand, SiO has multiple narrow features which vary with time and is almost certainly a strong maser even though such a source has not been observed with interferometers to determine its size and brightness temperature. It should be emphasized that the relationship between masers of different molecular species is unclear. They occur near one another but do not coexist in the same volumes of space. Different types of masers seem to appear at different stages of stellar evolution.

Table 9-1 Characteristics of maser and nonmaser molecular clouds.

Characteristic	Maser	Nonmaser
Line width (km/sec)	0.1–3	2–100
Number of distinct spectral features	1–100	few
Polarization (%)	0–100	none known
Size (arcsec)	10^{-4}–1	10^2–10^4
T_B (K)	10^9–10^{15}	<100
Time scale of variability (sec)	10^5–10^8	>10^8

Before discussing the energy levels of the molecules, the observational characteristics, and the models, we give a brief historical account of the rapid development of this field over the last decade.

History

The amount of information collected on the subject of cosmic masers over the past twelve years is very large. The brief chronological account of some of the significant discoveries in the field presented here serves to show how the subject grew. Some of the highlights are listed in Table 9-2.

During the 1950s, a period of concentrated effort in spectral-line radio astronomy on the 21-cm radiation from neutral hydrogen, at least one attempt was made to detect radiation from OH. It failed because the transition frequency was not known accurately enough. With great effort, Dousmanis, Sanders, and Townes (1955) and Ehrenstein, Townes, and Stevenson (1959) were able to make enough ground-state OH in the laboratory to measure the radio transition frequencies. With accurate frequencies, Weinreb et al. (1963) detected OH in absorption against Cassiopeia A in October 1963 with a novel spectrometer Weinreb built at the Massachusetts Institute of Technology for his thesis. An Australian group measured the absorption in all of the four hyperfine transitions in the ground state against the galactic center and recognized the nonthermal character of the radiation. In 1965 groups at Berkeley and Harvard, using H II regions as background sources for absorption surveys, accidentally found OH in emission. The character of this emission was peculiar and the Berkeley group attributed their spectra to OH and an unidentified molecule which they dubbed "mysterium." The mysterium idea was quickly dismissed. The emission was found to be both

Figure 9-3. The spectra of the four masers in the Orion Nebula—H_2O (Sullivan, 1973), OH (Manchester, Robinson, and Goss, 1970; Menon, 1967), SiO (Snyder and Buhl, 1974), and CH_3OH (Barrett, Ho, and Martin, 1975)—and one nonmaser, HDO (Turner et al., 1975). The maser spectra are essentially free of measurement noise. The ripples in the HDO spectrum are due to receiver noise, and probably only one spectral component is present. The top four spectra show a principal characteristic of maser emission—a multiplicity of narrow spectral features.

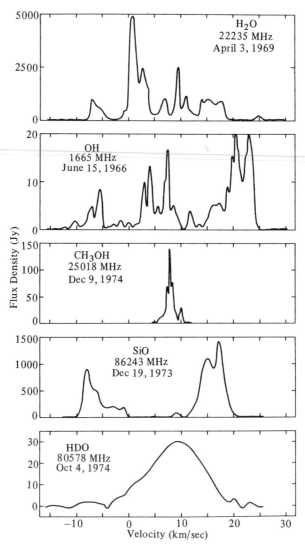

Table 9-2   Some highlights in the study of cosmic masers.

Year	Event	Source
1917	Prediction of stimulated emission	Einstein (1917)
1951	Laboratory maser built	Townes (1965)
1955	Laboratory measurement of OH frequencies	Dousmanis, Sanders, and Townes (1955) Ehrenstein, Townes, and Stevenson (1959)
1963	OH absorption discovered	Weinreb et al. (1963)
1964	OH absorption anomalies discovered	Gardner et al. (1964)
1965	OH emission discovered	Weaver et al. (1965) Gundermann (1965)
1965	Linear and circular polarization measured	Weinreb et al. (1965) Barrett and Rogers (1966) Davies, de Jager, and Verschuur (1966)
1966	First interferometry of OH masers	Rogers et al. (1966) Cudaback, Read, and Rougoor (1966)
1966	Cosmic maser theory developed	Litvak et al. (1966)
1967	OH maser associated with IR source	Raimond and Eliasson (1967)
1967	Compact H II regions associated with OH masers	Mezger et al. (1967)
1968	First map of maser emission	Moran et al. (1968)
1968	Normal OH emission discovered	Heiles (1968)
1968	OH emission in IR stars discovered	Wilson and Barrett (1968)
1968	Excited-state OH maser detected	Zuckerman et al. (1968)
1969	Discovery of cosmic H_2O maser	Cheung et al. (1969)
1970	OH classification scheme developed	Turner (1970)
1972	Size of H_2O spots determined	Burke et al. (1972)
1972	Theory of radiative transfer in spherical masers developed	Goldreich and Keeley (1972)
1974	OH and H_2O time variations correlated with IR variations	P. M. Harvey et al. (1974) Schwartz, Harvey, and Barrett (1974)
1974	SiO masers discovered	Snyder and Buhl (1974)
1975	CH_3OH maser in Orion identified	Hills, Pankonin, and Landecker (1975)
1975	Separation between OH and H_2O masers measured	Mader et al. (1975)

linearly and circularly polarized. The line widths were characteristic of gas temperatures as low as 4 K but typically 30 K. The equivalent black-body temperature, on the other hand, was greater than 10^3 K. Only a lower limit could be placed on the temperature because the sources were smaller than the beamwidths of the largest parabolic antennas, about 12 arcmin. Maser emission was proposed as a radiation mechanism to explain the temperature discrepancy and the polarization, and the first viable model was calculated by Litvak et al. (1966). The initial interferometric investigations in 1966 showed that the masering regions were smaller than 5 arcsec and did not coincide with any stars or other optical sources on the Palomar sky survey plates. Further interferometry, culminating in 1968 with an experiment using very-long-baseline interferometry (VLBI) carried out between the United States and Sweden, demonstrated that the emission from the masers came from a cluster of spots, the sizes of the individual spots being as small as 0.004 arcsec.

Meanwhile, Raimond and Eliasson (1967) found that the OH maser in Orion was coincident to within a few arcsec with the newly discovered Becklin-Neugebauer infrared (IR) object in Orion, tentatively identified as a protostar. This led I. S. Shklovsky to suggest that IR photons from this object were powering the maser. Mezger and his associates (1967) discovered that the maser emission occurred near compact H II regions and suggested protostellar associations. Wilson and Barrett (1968), following the Orion connection, discovered OH emission from a number of IR stars in 1968. That same year Heiles (1968) detected nonmasering OH emission in dark clouds of interstellar dust.

In 1969, H_2O was discovered by Cheung et al. (1969) in the H II region W49 and was shortly thereafter recognized as a masering molecule. This discovery was somewhat surprising since the H_2O transition was far above the ground state. Two years later, intercontinental interferometry between the United States and Russia established the size of the smallest spot of the H_2O maser in W49 as 0.0003 arcsec (Burke et al., 1972). In 1972 Goldreich and Keeley (1972) solved the radiative transfer equation for spherical masers with various degrees of saturation and showed how the observed size could be much smaller than the true size of the maser. During this period there was much additional theoretical work by many people on pumping and polarization mechanisms (Litvak, 1974). The time variations in the OH and H_2O emission from infrared stars was convincingly demonstrated to be correlated with the infrared variations, by P. M. Harvey et al. (1974) and Schwartz, Harvey, and Barrett (1974).

Maser emission from a molecule tentatively identified as SiO in the first vibrationally excited state was discovered in Orion by Snyder and Buhl (1974). Their identification was confirmed by subsequent observation of other rotational transitions. A series of rotational lines of CH_3OH were detected in Orion and just recently these lines have been convincingly demonstrated to be due to maser emission with the 30-arcsec beam of the new 100-m antenna at Bonn, Germany. Hills, Pankonin, and Landecker (1975) found that the various features were spatially separated from each other, similarly to the H_2O emission. Finally, simultaneous observations of the OH and H_2O masers in several sources by Mader et al. (1975) confirmed the earlier work of Hills et al. (1972) that the two masers were not spatially coincident in several cases.

The initial work mentioned here has been followed up by more complete studies, many of which are referred to in the next sections. As a footnote it is interesting to mention how the career of C. H. Townes winds through the study of cosmic masers. Townes and others developed the first laboratory maser. The ground state OH frequencies were measured in his laboratory at Columbia, which made it possible for his student, A. H. Barrett, and his coworkers to detect interstellar OH. He subsequently moved to Berkeley and participated in the discovery of the first H_2O maser and worked on several pumping schemes for masers.

Radio Astronomical Measurements

Before proceeding to describe in detail the observations of masers it is useful to digress and discuss how these measurements are made. Of particular importance to the study of masers are the new methods for measuring angular diameters as small as 10^{-4} arcsec.

The radiation field from a source at frequency ν can be characterized by its intensity I_ν, the energy per unit time, solid angle, area, and frequency. It can also be characterized by a brightness temperature T_B which is the temperature of an equivalent black body giving the same intensity. Hence

$$I_\nu = \frac{2h\nu^3}{c^2} \frac{1}{e^{h\nu/kT_B} - 1}, \tag{9-1}$$

where h is Planck's constant, k is Boltzman's constant, and c is the velocity of light. Since $h\nu/kT_B \ll 1$ for most radio astronomical applications, the exponential term in equation (9-1) can be expanded, giving the so-called Rayleigh-Jeans approximation,

$$T_B = \frac{c^2}{2k\nu^2} I_\nu. \qquad (9\text{-}2)$$

T_B and I_ν are intrinsic properties of the source and do not depend on the distance between the source and the observer. The flux density at the earth will be the intensity integrated over the solid angle subtended by the source,

$$F = \int_{\text{source}} I \, d\Omega. \qquad (9\text{-}3)$$

A source of uniform intensity and of angular diameter θ_S will have a solid angle Ω_S equal to $\pi\theta_S^2/4$ so that the flux density is

$$F = I\Omega_S = I\,\frac{\pi}{4}\,\theta_S^2 = \frac{\pi k T_B \nu^2 \theta_S^2}{2c^2}. \qquad (9\text{-}4)$$

The common unit of flux density is the Jansky, which is equal to 10^{-26} watt m^{-2} Hz^{-1}. An antenna such as a parabolic reflector of diameter D will have a pencil beam whose angular width is governed by the usual diffraction effects and given by

$$\theta_A \sim \lambda/D, \qquad (9\text{-}5)$$

where λ is the wavelength which is equal to c/ν. The collecting area of the antenna will be its geometric cross section times an efficiency factor η,

$$A = \eta\pi D^2/4. \qquad (9\text{-}6)$$

The received power W from a source smaller than the beam width can be expressed in terms of an antenna temperature T_A by the relation

$$W = kT_A = FA/2. \qquad (9\text{-}7)$$

The factor of 2 accounts for the fact that an antenna with a single feed can extract only half the power from an unpolarized radiation field. The solid beam angle of the antenna Ω_A is defined as λ^2/A which is, from equation (9-5), approximately equal to $(4/\pi\eta)\theta_A^2$. It follows immediately from equations (9-3) and (9-7) that for the small sources

$$T_A = T_B\,\frac{\Omega_S}{\Omega_A}. \qquad (\Omega_S \ll \Omega_A) \qquad (9\text{-}8)$$

For large sources,

$$T_A \cong T_B. \qquad (\Omega_S \gg \Omega_A) \qquad (9\text{-}9)$$

These situations are depicted in Figure 9-4. Hence T_A, a measurable quantity, is the brightness temperature times a dilution factor, i.e., the brightness temperature averaged over the antenna beam. T_B cannot be inferred without knowledge of the source size. It is frequently assumed that the shape of the antenna beam, i.e., the collecting area versus angle from the antenna axis, and the brightness temperature distribution source are described by gaussian functions with widths θ_A and θ_S respectively. The observed antenna temperature profile, obtained by scanning the antenna across the source, will also be gaussian with width θ_M given by

$$\theta_M = \sqrt{\theta_S^2 + \theta_A^2}. \qquad (9\text{-}10)$$

Hence from measurements of θ_M and T_A, θ_S and T_B can be estimated by the relations

$$\theta_S = \sqrt{\theta_M^2 - \theta_A^2}, \qquad (9\text{-}11)$$

Figure 9-4. The antenna temperature for the two cases $\Omega_S > \Omega_A$ (top) and $\Omega_S < \Omega_A$ (bottom). The bottom case is applicable to cosmic masers which are much smaller than any antenna beam so that it is possible to estimate only a lower limit on the brightness temperature and an upper limit on the angular size.

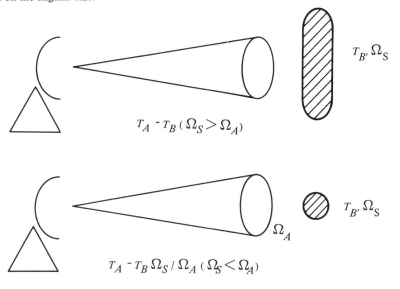

$$T_B = T_A \left[\frac{\theta_A{}^2}{\theta_M{}^2 - \theta_A{}^2} \right]. \qquad (9\text{-}12)$$

These estimates of T_B and θ_S will be imprecise because noise in the receiver limits the accuracy of any antenna temperature measurement to

$$\Delta T_A = T_r / \sqrt{\Delta \nu \bar{t}}, \qquad (9\text{-}13)$$

where T_r is the receiver temperature, $\Delta \nu$ is the receiver bandwidth and $\bar{t}$ is the integration time. Hence in practice θ_S cannot be determined reliably if it is smaller than roughly $\theta_A/4$. The detailed structure of the source on scales smaller than θ_A is obliterated by the smoothing properties of the antenna. Since cosmic masers are unresolvable with presently conceivable parabolic antennas, only upper limits on θ_S and lower limits on T_B, based on equations (9-11) and (9-12), can be given. The new 100-m telescope of the Max-Planck-Institute for Astrophysics has the smallest beamwidth of any filled-aperture telescope, 30 arcsec at $\lambda = 1.2$ cm. The cost of building such a telescope is approximately proportional to D^3 and it is not likely that very much larger ones will be constructed. However, other methods can be used to study the small-scale source structure.

One technique for making high-resolution radio maps is called aperture synthesis. Sir Martin Ryle, who won the Nobel Prize for his development of this technique, provided a very informative description of it in his Nobel lecture (Ryle, 1975). We briefly describe the principles here. Consider the aperture plane of a parabolic antenna as being divided into a grid of N squares. The electric field at the focal point of the parabola will be the sum of the N electric fields on the aperture grid, E_i. The received power is proportional to the square of the field at the focal point, or

$$P \propto \left\langle \left[\sum_i E_i \right]^2 \right\rangle = \sum_i \langle E_i^2 \rangle + \sum_{i \neq j} \langle E_i E_j \rangle, \qquad (9\text{-}14)$$

where $\langle \rangle$ denotes a time average. It therefore seems reasonable that the response of a large parabola of diameter D can be synthesized by moving a pair of small antennas over an area having diameter D, and patiently measuring each $E_i E_j$ term separately.

Aperture synthesis can be understood more quantitatively by studying the operation of two antennas whose received voltages, $v_1(t)$ and $v_2(t)$, which are proportional to the electric fields E_i and E_j in equation (9-14), are multiplied as shown in Figure 9-5. This setup is called a two-element interferometer.

Consider a point source of monochromatic radiation at frequency ν_0 having intensity b, i.e., a sine-wave generator at infinity. The incident plane wave reaches the second antenna a time $\tau = D \cos \phi/c$ after reaching the first antenna where c, D, and ϕ are defined in Figure 9-5. The integration time $\bar{\tau}$ is short with respect to changes in τ produced by the rotation of the earth. If the received voltages are represented in complex form as

$$v_1 = \sqrt{b} \exp (i2\pi\nu_0 t),$$
$$v_2 = \sqrt{b} \exp [i2\pi\nu_0(t - \tau)], \tag{9-15}$$

where i is $\sqrt{-1}$, then the response of the interferometer, which is a correla-

Figure 9-5.   A schematic diagram of a two-element interferometer which measures the Fourier components of the radio image.

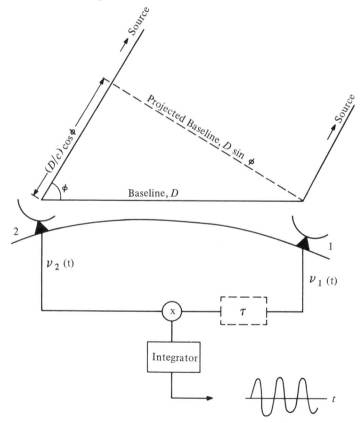

tion function, is

$$R(\tau) = \frac{1}{\bar{t}} \int_0^{\bar{t}} v_1(t)v_2^*(t - \tau) = b \exp (i2\pi v_0\tau), \qquad (9\text{-}16)$$

where * denotes the complex conjugate. In terms of ϕ,

$$R(\phi) = b \exp \left[i \frac{2\pi D}{\lambda} \cos \phi(t) \right], \qquad (9\text{-}17)$$

where $\lambda = c/v_0$. The slow oscillation described by equation (9-17) is just the beat between the two signals which have slightly different apparent frequencies because of the differential Doppler shift caused by the rotation of the earth. $R(\phi)$ is called a fringe pattern. The fringe phase is

$$\Phi(t) = \frac{2\pi D}{\lambda} \cos \phi(t). \qquad (9\text{-}18)$$

The maxima of $R(\phi)$ occur when the fringe phase is a multiple of 2π, that is, when

$$\frac{2\pi D}{\lambda} \cos \phi(t) = 2\pi n, \qquad (9\text{-}19)$$

where $n = 0, \pm 1, \pm 2, \cdots$. This defines a set of small circles on the sky as shown in Figure 9-6, spaced at intervals $\Delta\Phi = \lambda/D \sin \phi$ where $\Delta\Phi$ is called the fringe spacing and $D \sin \phi$ is the component of the baseline projected in the direction of the source. Hence the fringe phase gives information about the source location. If the source is allowed to have finite diameter θ_S, it is clear that if $\theta_S > \lambda/(D \sin \phi)$ no fringe pattern will be observed because, while some parts of the source are interfering constructively (i.e., $\Phi = 0$), others are interfering destructively (i.e., $\Phi = \pi$). The net effect is the cancellation of the fringe pattern. The ratio of amplitude of the fringes to the amplitude expected from a point source of the same strength is called the fringe amplitude.

The restrictions on the source size and bandwidth can be removed in a quantitative way. The received signal will have a bandwidth Δv, due either to the source's own spectral width or the bandpass filter width of the receiver. The interferometer response, corrected for the increased power by dividing by Δv, is obtained by integrating equation (9-17), over frequency so that

$$R(\phi) = \frac{b}{\Delta v} \int_{v_0 - \Delta v/2}^{v_0 + \Delta v/2} \exp \left(i 2\pi \frac{Dv}{c} \cos \phi \right) dv$$

or

$$R(\phi) = b \left[\frac{\sin \left(\pi \dfrac{D\Delta\nu}{c} \cos \phi \right)}{\pi \dfrac{D\Delta\nu}{c} \cos \phi} \right] \exp \left(i2\pi \frac{D}{c} \nu_0 \cos \phi \right). \qquad (9\text{-}20)$$

The fringe phase is unaffected. However, the amplitude of the fringes will be reduced unless the term in brackets in equation (9-20) is approximately unity, which, recalling that $D\nu_0 \cos \phi/c = \tau$, requires that

$$\Delta\nu \ll 1/\tau. \qquad (9\text{-}21)$$

This is the well-known concept that the correlation time of a signal is equal to the reciprocal of its bandwidth. However, from equation (9-13), $\Delta\nu$ must be as large as possible to ensure adequate sensitivity for the detection of weak sources. Therefore, in order to ensure that the fringes are not weakened, a delay line is inserted between antenna 1 and the multiplier, which is set to exactly compensate for the propagation delay corresponding to a certain reference direction. This reference direction is some convenient point on

Figure 9-6.   Small circles showing loci of maximum constructive interference as defined by equation (9-19). An extended source (lower right) will produce only weak fringes because of phase cancellation among its various parts. A double source (lower left) can produce no fringes or strong fringes, depending on how the components are aligned with respect to the fringe pattern.

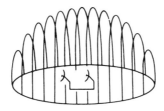

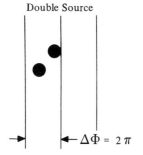
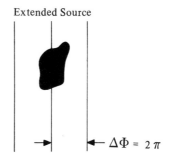

Double Source   Extended Source

$\leftarrow \Delta\Phi = 2\pi$ $\leftarrow \Delta\Phi = 2\pi$

the sky near the source under study. The fringe phase of a point source at the reference position always will be zero, by definition. The fringe phase will then differ from zero for radiation of intensity b from a point which is offset from the reference point by the angle x in the right ascension direction and the angle y in the declination direction. From the Taylor expansion of Φ in equation (9-18), the fringe phase will be

$$\Delta\Phi \simeq \frac{\partial\Phi}{\partial x}\, x + \frac{\partial\Phi}{\partial y}\, y. \qquad (9\text{-}22)$$

If we define

$$u = -\frac{D \sin \phi}{\lambda}\frac{\partial\phi}{\partial x},$$

$$v = -\frac{D \sin \phi}{\lambda}\frac{\partial\phi}{\partial y}, \qquad (9\text{-}23)$$

we have

$$\Delta\Phi = 2\pi(ux + vy) \qquad (9\text{-}24)$$

and the response is

$$R(u,v) = b \, \exp\,[i2\pi(ux + vy)]; \qquad (9\text{-}25)$$

u and v are just the projected lengths of the baseline in units of wavelengths in the x and y directions respectively. Being a linear system, the response to an extended incoherent source whose intensity distribution is $b(x,y)$ will be

$$R(u,v) = \iint b(x,y) \, \exp\,[i2\pi(ux + vy)]\, dxdy. \qquad (9\text{-}26)$$

Hence, the correlation function $R(u,v)$ measured by the interferometer is the Fourier transform of the brightness distribution; $b(x,y)$ can be found by the inverse transform,

$$b(x,y) = \iint R(u,v) \, \exp\,[-i2\pi(ux + vy)]\, dudv. \qquad (9\text{-}27)$$

An interferometer measures Fourier components of the source brightness distribution. Modern antenna arrays at Cambridge, England, and Westerbork, the Netherlands, use many antennas to simultaneously measure many values of R so that, with the help of the rotation of the earth, enough coefficients can be measured in 12 hours to produce a complete radio image. The very large array (VLA) of telescopes under construction near Socorro, N.M., by the National Radio Astronomy Observatory (NRAO) will have 27 antennas in a Y configuration of length 35 km. The resolution at 1.3 cm wavelength will be 0.1 arcsec.

By the mid-1960s, interferometers with antennas separated by more than 100 km and connected by microwave links were unable to resolve all the interesting structure in quasars and masers. Much higher resolution was achieved by the technique of very-long-baseline interferometry (VLBI) whereby the signals were recorded on magnetic tape and later brought together and correlated. With this system the separation between antennas could be arbitrarily large since no real time communication was required.

There are several stringent requirements which must be met for such a system to work. Tape recorders cannot directly record the received microwave signal. The signal band from ν_0 to $\nu_0 + \Delta\nu$ must first be translated to a video band from zero frequency to $\Delta\nu$. This is done by mixing (i.e., multiplying and filtering) the microwave signal with a local oscillator waveform which is a sine wave of frequency ν_0. Any deviations from a perfect sine wave in the local oscillator become impressed on the signal and cause deviations in the fringe phase. If the fringe phase is to deviate less than 2π over an integration period t_p the deviation in the frequency of the local oscillator $\delta\nu$ must satisfy the relation

$$2\pi\delta\nu t_p < 2\pi. \tag{9-28}$$

Hence,

$$\frac{\delta\nu}{\nu_0} < \frac{1}{\nu_0 t_p}. \tag{9-29}$$

If $\nu_0 = 10^{10}$ Hz and $t_p = 10^2$ sec, $\delta\nu/\nu_0$, a parameter used to characterize the stability of oscillators, must be less than 10^{-12}. Stability of this order is achievable only with atomic frequency standards. Hydrogen-maser frequency standards provide $\delta\nu/\nu_0 \sim 10^{-14}$ for commonly used averaging periods. There are only about a dozen in existence and they are very expensive. A cheaper substitute is a rubidium vapor standard which provides $\delta\nu/\nu_0 \sim 10^{-12}$. A hydrogen maser oscillator connected to a counter becomes a clock whose accuracy over a year is better than 1 μsec. Hence if the bandwidth is less than 10^6 Hz, the proper alignment of the data on playback as required by equation (9-21) is no problem, provided the clocks are set properly. Most radio telescopes as well as national time-keeping facilities are synchronized to within 10^{-5} sec by atomic clocks which are carried among the stations every few months. The highest resolution obtained by the VLBI technique was achieved with the Haystack Observatory antenna in West-ford, Mass., and with the Crimean Astrophysical Observatory's antenna in Semeiz, Crimea, USSR, operating at 1.35 cm. The separation was 7400 km or 5.5×10^8 wavelengths, giving a fringe spacing of 0.0004 arcsec.

The interpretation of VLBI results, beyond the estimate of the approxi-mate source diameter, is difficult because the correlation function is very poorly sampled; i.e., the brightness must be deduced from the measurement of only a few Fourier coefficients. In the case shown in Figure 9-7 for the maser spot at -43.7 km/s in W3(OH), a simple model of a uniformly bright disk source, which gives values of the correlation function predicted by equation (9-26) and normalized to unity at zero baseline length, fits the data well. Most sources are more complex. The current procedure is to fit the data to a source model with a few (<20) parameters rather than attempt a Fourier transformation. Plans to connect a large number of antennas over the continent into a VLBI network are underway (Swenson and Kellermann, 1975); they will allow Fourier transform analysis of data.

Figure 9-7. The fringe amplitude of the smallest component in W3(OH) at -43.7 km/sec. The declining amplitude with increasing baseline length is modeled as a uniformly bright disk source of diameter 0.004 or 0.005 arcsec.

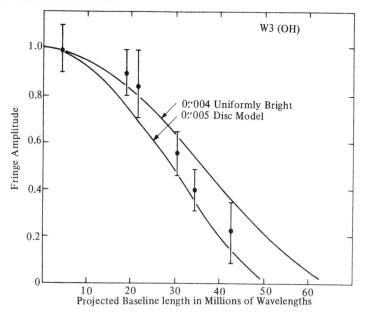

It is reasonable to ask why resolutions of 0.001 arcsec can be achieved at radio frequencies while optical resolution is limited to 1 arcsec by atmospheric fluctuations. A wavefront at high radio frequencies or optical frequencies impinging on the atmosphere is distorted, because of variations in the index of refraction caused by inhomogeneities in the distribution of the neutral atmospheric gas. The optical observations are affected by small-scale inhomogeneities or turbulence having scale sizes on the order of a few feet or less, which change rapidly and make telescope images "dance" on time scales of a few milliseconds. Optical images are thereby blurred to about an arc second on long photographic exposures. The radio interferometer is influenced by large-scale inhomogeneities associated with large-scale weather patterns. The resulting differences in the propagation through the atmosphere to the two widely spaced telescopes can amount to many centimeters. This causes the apparent angle of arrival of the wavefront, as measured by the differential delay τ, to change slowly on a time scale of hours. Adjustments can be made during the data processing for this effect and the magnitude of the correlation function and hence the source size can be measured to the intrinsic accuracy of the interferometer, λ/D. However, the determination of the source position is limited currently to about 0.1 arcsec, because of limitations in the determination of atmospheric propagation delay and the stability of the frequency standards. The interstellar ionized plasma may cause seeing problems along the galactic plane (Harris et al., 1970).

Molecular Energy Levels

The internal energy of a molecule can be approximately described as the sum of the rotational, vibrational, and electronic energies. That is,

$$E = E_r + E_v + E_e. \tag{9-30}$$

The rotational energy is associated with the end-over-end rotation of the molecule. Transitions between rotational levels are generally in the radio or far infrared part of the spectrum. Vibrational energy associated with nuclear oscillation gives rise to transitions in the near infrared and the visible. Transitions between electronic energy levels are in the visible and the ultraviolet. In addition there are important interactions between electronic and rotational motions which give rise to radio frequency transitions.

The rotational spectrum of a diatomic molecule like SiO is easily described in semiclassical terms. The moment of inertia of the molecule I is μr^2, where μ is the reduced mass and r is the separation between the nuclei. The angular momentum is quantized,

$$I\omega = \frac{h}{2\pi} \sqrt{J(J + 1)}, \tag{9-31}$$

where ω is the rotation frequency, h is Planck's constant, and J is the rotational quantum number. The rotational energy, $I\omega^2/2$, can be written as

$$E = hBJ(J + 1), \tag{9-32}$$

where B is the rotational constant given by

$$B = \frac{h}{8\pi^2 I}. \tag{9-33}$$

The values of B for OH and SiO are 5.5×10^{11} Hz and 2.2×10^{10} Hz respectively. Selection rules allow electric dipole transition between levels for which $\Delta J = \pm 1$. The spectrum consists of a uniformly spaced set of lines at frequencies

$$\nu = \frac{E_{J+1} - E_J}{h} = 2B(J + 1). \tag{9-34}$$

In diatomic molecules with identical atoms such as O_2 or H_2, the centers of positive and negative charge coincide and there is no electric dipole moment and consequently no electric dipole rotational spectrum.

The vibration energy spectrum is given by

$$E = h\nu_{osc} \left(v + \frac{1}{2} \right), \tag{9-35}$$

where v is the vibrational quantum number ($= 0, 1, 2, \cdots$) and ν_{osc} is the oscillation frequency. ν_{osc} for OH and SiO is 1.1×10^{14} Hz and 3.7×10^{13} Hz respectively. Real diatomic molecules are not rigid rotators so that B changes slightly with both J and v. These effects are described by Herzberg (1950). The energy levels of SiO are accurately described by equations (9-32) and (9-35) since the electronic angular momentum is zero and causes no perturbations. They are shown in Figure 9-8 along with the observed maser transitions between the rotational levels.

The description of the electronic energy levels is more complex for molecules than for atoms but the nomenclature is similar. Whereas the field of an atom has spherical symmetry, the field of a diatomic molecule has axial symmetry about the internuclear axis. Hence the significant quantum number will be the projection of the electronic angular momentum L on the internu-

Figure 9-8. Some of the low rotational and vibrational energy levels of SiO showing the maser transitions which have been detected. There may be one case of maser emission in the $V = 0$, $J = 2 \rightarrow 1$ transition (Buhl et al., 1975).

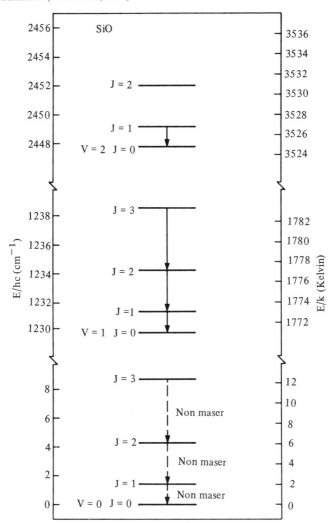

clear axis. This is denoted Λ and takes on integer values from $-L$ to L. For nonrotating molecules, states with positive or negative Λ are degenerate, having the same energy. States with $|\Lambda| = 0, 1, 2, \cdot \cdot \cdot$ are designated $\Sigma, \Pi, \Delta, \cdot \cdot \cdot \cdot$. The molecule may possess a net spin angular momentum s if all

Figure 9-9. Part of the rotational spectrum of OH. The rotational ladder has two branches due to spin splitting. The Λ doubling and hyperfine splitting which split each rotational level into four sublevels are not shown to scale. The number on the right side of each energy level is the total angular-momentum quantum number F. The known maser transitions are indicated.

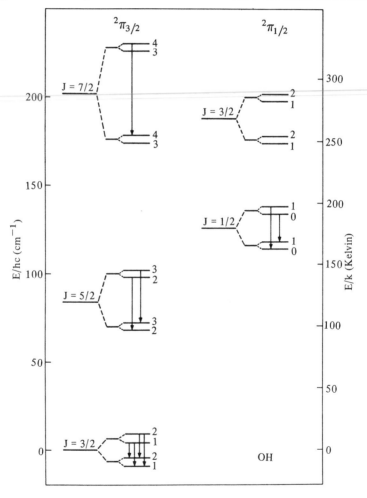

electrons are not paired off. The levels are designated by $2s + 1$ as a super-script; e.g., $^2\Sigma$ for $|\Lambda| = 0$ and $s = 1/2$.

In the ground state of most molecules, including SiO, CH_3OH, and H_2O, the electronic angular momentum is zero. However, OH has a $^2\Pi$ ground state, i.e., $\Lambda = 1$, $s = 1/2$. This leads to significant interactions among the spin and orbital angular momenta of the electrons and the rotational angular momentum, which give rise to the observed OH maser transitions. The total electronic angular-momentum quantum number will be either $\Lambda + s$ or $\Lambda - s$, i.e., 3/2 or 1/2. The 3/2 state of OH has a lower energy than the 1/2 state. This is called spin splitting. The total angular momentum quantum number J, including rotation, will be 3/2, 5/2, 7/2, $\cdot$ $\cdot$ $\cdot$ for the first case and 1/2, 3/2, 5/2, $\cdot$ $\cdot$ $\cdot$ for the second. The interaction of the electron angular mo-mentum and the rotational angular momentum splits each level into two, re-moving the two-fold Λ degeneracy. This phenomenon, called Λ doubling, was first investigated by Van Vleck (1929). Each of these levels is again split by a hyperfine spin interaction between the hydrogen nuclear angular mo-mentum I and the electronic spin angular momentum. The grand total angu-lar momentum is $\mathbf{F} = \mathbf{J} + \mathbf{I}$. The rotational energy levels along with the Λ doubling and hyperfine splitting are shown in Figure 9-9. Most extensive OH observations have been made of the four transitions in the lowest rotation level, designated $^2\Pi_{3/2}$, $J = 3/2$. The transition frequencies are 1612.231, 1665.401, 1667.358, and 1720.528 MHz.

The rotational spectra of H_2O and CH_3OH are more complex than those of diatomic molecules because the moments of inertia are different about the possible rotation axes. CH_3OH has the further complication that the OH rad-ical rotates internally with respect to the CH_3 radical. However, in the ground state there is no coupling of spin and orbital angular momenta. Their energy levels and the observed astronomical transitions are shown in Figures 9-10 and 9-11.

All the known maser emission lines are listed in Table 9-3 along with the frequency, wavelength, energy above ground state in units of both cm^{-1} and K, and the spontaneous transition probability, A, for each transition. The energy in temperature units is just E/k, which is the temperature required to significantly populate the level. There are many other transitions in these molecules which have not been detected in astronomical sources. For ex-ample, H_2O has a transition at 183 GHz but the atmospheric water vapor ab-sorbs too strongly for it to be observed from the ground. The Λ doublet transitions of OH in the first excited vibrational state have been sought but not found.

Observations and Interpretations

There are two main classes of masers—those associated with cool, large stars radiating primarily in the infrared, and those associated with H II regions. We will discuss them separately because they have few common characteristics, as can be seen from Tables 9-4 and 9-5. Indeed, the two types are probably associated with quite different phenomena, the IR masers being

Figure 9-10. Part of the rotational energy levels of H_2O, an asymmetrical rotator. The microwave maser transition is due to a chance proximity of the 6_{16} and 5_{23} levels. Data taken from de Jong (1973).

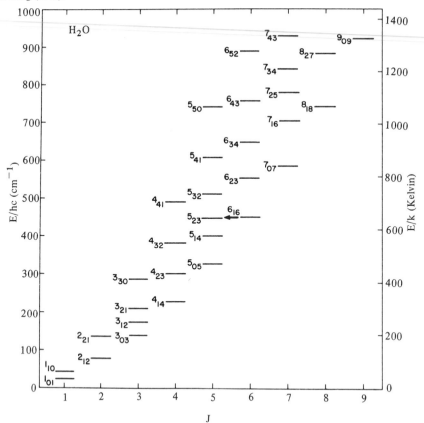

related to the later stages of stellar evolution and the H II masers being related to the early stages. In addition, a few masers have been found in association with other objects such as early-type stars, Herbig-Haro objects, and nonthermal radio sources.

Figure 9-11. Part of the rotational energy spectrum of CH_3OH showing only the "E stack." The "A stack," the levels of the molecule without internal rotation, are not shown. Adapted from Lees (1973).

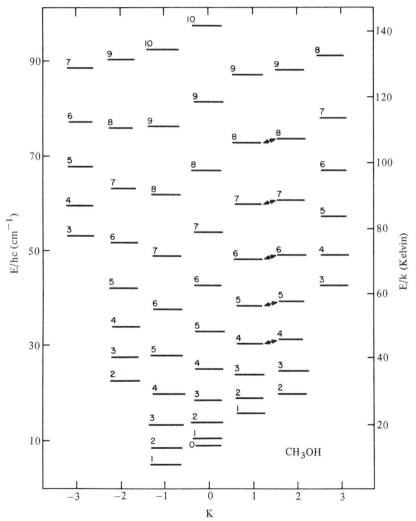

Table 9-3 Parameters of observed maser transitions.

Molecule	Transition	ν MHz	λ cm	E/hc cm^{-1}	E/k K	A sec^{-1}
OH	$^2\Pi_{3/2}\, J = \tfrac{3}{2}\, F = 1 \longrightarrow 2$	1612.231	18.6	0	0	1.29×10^{-11}
	$^2\Pi_{3/2}\, J = \tfrac{3}{2}\, F = 1 \longrightarrow 1$	1665.402	18.0	0	0	7.11×10^{-11}
	$^2\Pi_{3/2}\, J = \tfrac{3}{2}\, F = 2 \longrightarrow 2$	1667.359	18.0	0	0	7.11×10^{-11}
	$^2\Pi_{3/2}\, J = \tfrac{3}{2}\, F = 2 \longrightarrow 1$	1720.530	17.4	0	0	9.42×10^{-12}
	$^2\Pi_{3/2}\, J = \tfrac{5}{2}\, F = 2 \longrightarrow 2$	6030.747	5.0	84	120	1.53×10^{-9}
	$^2\Pi_{3/2}\, J = \tfrac{5}{2}\, F = 3 \longrightarrow 3$	6035.092	5.0	84	120	1.57×10^{-9}
	$^2\Pi_{1/2}\, J = \tfrac{1}{2}\, F = 0 \longrightarrow 1$	4660.242	6.4	126	181	1.08×10^{-9}
	$^2\Pi_{1/2}\, J = \tfrac{1}{2}\, F = 1 \longrightarrow 0$	4765.562	6.3	126	181	3.86×10^{-10}
	$^2\Pi_{3/2}\, J = \tfrac{7}{2}\, F = 4 \longrightarrow 4$	13441.371	2.2	184	265	9.26×10^{-9}
H$_2$O	$6_{16} \longrightarrow 5_{23}$	22235.080	1.35	447	644	1.91×10^{-9}
SiO	$^1\Sigma\; v = 1,\, J = 1 \longrightarrow 0$	43122.03	0.70	1230	1770	3.00×10^{-6}
	$v = 1,\, J = 2 \longrightarrow 1$	86243.27	0.35	1232	1774	2.87×10^{-5}
	$v = 1,\, J = 3 \longrightarrow 2$	129363.12	0.23	1236	1780	1.04×10^{-4}
	$v = 2,\, J = 1 \longrightarrow 0$	42820.48	0.70	2448	3525	2.93×10^{-6}
CH$_3$OH	$J = 4\; k = 2 \longrightarrow 1$	24933.468	1.20	31	45	8.40×10^{-8}
	$5\; k = 2 \longrightarrow 1$	25959.080	1.20	39	56	8.74×10^{-8}
	$6\; k = 2 \longrightarrow 1$	25018.123	1.20	49	71	8.98×10^{-8}
	$7\; k = 2 \longrightarrow 1$	25124.873	1.20	61	87	9.21×10^{-8}
	$8\; k = 2 \longrightarrow 1$	25294.411	1.20	74	106	9.48×10^{-8}

Table 9-4 Characteristics of masers associated with IR stars.

Quantity	H$_2$O	OH	SiO	CH$_3$OH
Transitions observed	1	3	4	0
Number known	~50	~50	16	0
Line width (km/sec)	1–2	1–2	0.5–2	—
T_K (K)[a]	400–1600	400–1600	250–3500	—
Number of spectral features	1–10	2–10[b]	1–10	—
Velocity range (km/sec)	5–50	5–80	2–15	—
Polarization (%)	none	small	none (?)	—
Lifetime of feature (sec)	$>10^7$	$>10^7$	?	—
Spot size	10^{14}	10^{15}	$<10^{16}$	—
T_B (K)	10^{11}–10^{12}	10^9–10^{11}	$>10^3$	—
Cluster size (cm)	10^{15}	2×10^{15}	?	—
Power[c] (ergs/sec)	10^{24}–10^{28}	10^{24}–10^{28}	10^{26}–10^{27}	—

[a] Assuming no line narrowing or mass motions.

[b] More for supergiants.

[c] Assuming isotropic radiation.

Table 9-5 Characteristics of masers associated with H II regions.

Quantity	H_2O	OH	SiO	CH_3OH
Transitions observed	1	9	4	4
Number known	~50	~50	1[a]	1[a]
Line width (km/sec)	0.5–2	0.1–1	2	0.5
K_K (K)[b]	100–1500	4–400	3500	150
Number of spectral features	1–100	1–50	5	3
Velocity range (km/sec)	1–300	1–30	25	4
Polarization (%)	linear (0–10)	linear (0–100) circular (0–100)	circular (0–20)	none
Lifetime of feature[c] (sec)	10^6–10^7	10^7–10^8	?	?
Spot size (cm)	10^{13}–10^{14}	10^{14}–10^{15}	$<10^{16}$	$<10^{16}$
T_B (K)	10^{13}–10^{15}	10^{12}–10^{13}	$>10^3$	$>10^3$
Cluster size (cm)	10^{16}–10^{17}	10^{16}–10^{17}	?	10^{17}
Power[d] (ergs/sec)	10^{27}–10^{33}	10^{27}–10^{30}	10^{29}	10^{27}

[a] The one source known is Orion A.
[b] Assuming no line narrowing or mass motions.
[c] There are some cases of shorter time scales.
[d] Assuming isotropic radiation.

IR Star Masers

The IR stars which have masers associated with them are either Mira variables or irregular supergiant variables. These are oxygen-rich stars [O/C > 1] with surface temperatures of 1800–2800 K. They are thought to be evolved stars, undergoing mass loss and surrounded by dust shells having temperatures of 600–800 K. The dust absorbs the stellar radiation and emits strongly between 3 and 20 μm, giving the objects their highly reddened appearance. Their luminosities are about $10^4 L_\odot$. Whereas many of these stars have been known for hundreds of years, many new ones were discovered in the infrared survey at 2 μm by Neugebauer and Leighton (1969).

Some measurements of the structure of IR stars have been obtained by means of lunar occultation. For example, the structure of IRC 10011 was found to contain a stellar core of diameter 7 astronomical units (AU) at 2000 K, surrounded by a dense shell of diameter 50 AU and in turn surrounded by a tenuous shell extending outwards to a diameter of at least 500 AU (Zappala et al., 1974). Sensitive but usually fruitless searches for continuum radio emission from IR stars have placed stringent upper limits on the ionized plasma density.

Wilson and Barrett (1972) in a comprehensive survey of 456 IR stars discovered 25 OH masers. Hence it seems that about 6% of IR stars exhibit maser emission. A striking characteristic of these OH masers is that the emission always occurs in two distinct velocity ranges separated typically by 5 to 80 km/sec. There seems to be no preference for the strongest component to be at the higher or the lower velocity. The ground state emission is almost always strongest at 1612 MHz and emission at 1720 MHz never occurs. Excited state emission has not been detected except in the star NML Cygnus. The radiation is usually unpolarized. The spectrum of IRC 10011, which is a classic example of an IR/OH maser, is shown in Figure 9-12. The H_2O and SiO emission usually consists of one or a few components located between the two OH velocity ranges.

The most extensive interferometry has been done on the red-giant stars NML Cygnus and VY Canis Majoris. The OH spots are typically 0.1 arcsec in size (100 AU or 10^{15} cm) and are spread over an area of about 2 arcsec (2000 AU or 3×10^{16} cm). The spatial pattern of the velocity features gives some suggestion that the OH is in an expanding shell (Herbig, 1974). The H_2O spots are about 0.01 arcsec (10 AU) in size and spread over a smaller area of about 0.1 arcsec (100 AU). Current models place the H_2O maser close to the star and the OH maser in an outer dust shell. The observations of the OH masers in several Mira variables show the spot sizes to be on the order of 0.1–1.0 arcsec.

Since the VLBI results are incomplete and at best give only the two-dimensional projected image of the source, the geometry of the masers is still unclear. Models which have been considered involve expansion, con-

Figure 9-12. The spectrum of OH in the $^2\Pi_{3/2}$ $J = \frac{3}{2}$, $F = 1 \rightarrow 2$ transition (ν_0 = 1612 MHz) towards the infrared star IRC + 10011. The spectrum, with two widely separated velocity components, is typical of this type of maser (Wilson and Barrett, 1972).

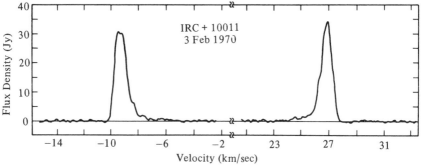

traction, or rotation. Some include a stationary shock front surrounding the star to explain the two OH emission groups. Contraction and rotation are not likely to be dominant because the gravitational energy needed would require the star to be much too massive. It would be very helpful in understanding the maser emission if the velocity of the star were known. The optical spectra of these stars is rich in absorption and emission features whose velocities are well determined. It has generally been assumed that the velocity of the absorption features, which usually coincides with the velocity of the red-shifted OH emission groups, is the same as the velocity of the star. In this view, then, one OH group is at rest with respect to the star and the other is moving towards the earth. This viewpoint makes the OH emission rather difficult to explain. However, Reid (1975) has given statistical arguments to show that the stellar velocity may lie between the OH groups. If this is correct, the OH emission could come from a simple expanding shell. The maser spots would be seen principally along the line of slight directly to the star where the velocity gradients are minimum. The two velocity components would arise from the OH in front of and in back of the star.

The IR flux density from Mira variables oscillates in roughly sinusoidal fashion with periods in the range from 100 to 1000 days. The flux density of the OH and H_2O emission varies along with the infrared, with no observable phase lag. This correlation has been well documented for OH by P. M. Harvey et al. (1974) and for H_2O by Schwartz et al. (1974). Samples of their data are shown in Figures 9-13 and 9-14. Such a correlation probably exists for the SiO masers. The infrared, OH, and H_2O maxima lag the optical maximum by typically 0.1 to 0.2 periods.

The correlation between the infrared and maser variations apparently is due to a radiative rather than a mechanical coupling mechanism. VLBI measurements, supported by the observation that the velocities of the radio features do not change during the period while the velocities of the optical lines do, indicate that the masers are at least 10^{14} cm from the star and mechanically decoupled from it. Since the velocity of mass motions is less than 50 km/sec, the minimum time required for matter or pressure waves to travel from the star to the maser is much larger than the observed phase lag.

The most likely and obvious radiative connection between the infrared and the masers is through the mechanism of radiative pumping of the maser. The number of photons in the range 3 to 35 μm is about equal to the observed number of microwave photons, so there is enough energy to account

Figure 9-13. The infrared flux density from the star IRC + 10011 at 2.2 μm (top), and the flux density integrated over frequency of the OH maser at 18 cm (bottom), versus time toward the star R Aquilae. The infrared and OH vary together with no observable phase lag (P. M. Harvey et al., 1974).

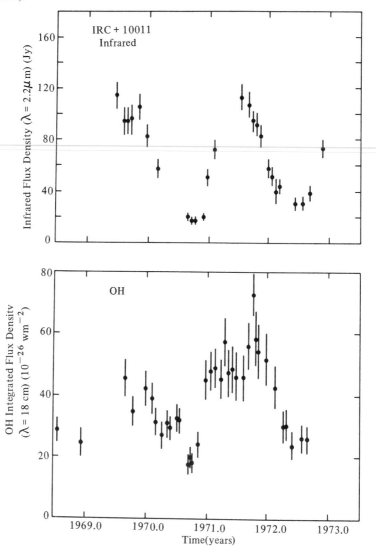

for the microwave emission. The detailed relation between the infrared flux and the microwave flux can be complex. A simple model for a saturated maser, as discussed in the next section, suggests that the microwave flux density is proportional to the pump rate, which in turn is proportional to the infrared flux density.

Figure 9-14.   The flux density of the H_2O maser (top), and the integrated flux density of the OH maser (bottom), versus the continuum flux density at 2.2 μm (P. M. Harvey et al., 1974; Schwartz et al., 1974).

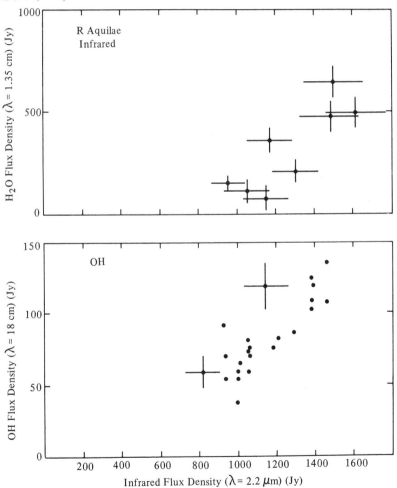

H II Region Masers

A large number of H II regions have been searched for maser emission. About 50 OH masers have been identified. The probability of occurrence is about 25%. H_2O masers generally occur along with OH masers. No CH_3OH or SiO masers, except the ones in Orion, have yet been found near H II regions.

The OH masers near H II regions radiate most strongly at 1665 or 1667 MHz in the ground state. The emission from excited states of OH is generally much weaker than in the ground state and is seen in only a few sources. The spectra generally show many features having widths of 0.1–1 km/sec, spread over a velocity range of 2–20 km/sec. There is no correspondence among features at various transitions. The radiation is usually strongly polarized; indeed, 100% left- or right-circularly polarized features often occur. The various features come from spots typically having diameters of 10^{14} cm distributed over an area having a diameter of 10^{16} cm.

The formation of the spectral components in OH masers may be influenced by the Zeeman effect. In the presence of a magnetic field, each energy level is split into $2F + 1$ sublevels designated by the quantum number m_F which ranges in value from F to $-F$. Transitions for which $\Delta m_F = 0, \pm 1$ are allowed. The Zeeman pattern for the 1612-MHz ($F = 1 \rightarrow 2$) and the 1720-MHz ($F = 2 \rightarrow 1$) transitions has a total of 9 components (3 for $\Delta m_F = 0$, 3 for $\Delta m_F = 1$, and 3 for $\Delta m_F = -1$). The radiation involving transitions where there is a change in angular momentum ($\Delta m_F = \pm 1$) are circularly polarized when viewed parallel to the magnetic field, and linearly polarized when viewed perpendicular to the magnetic field lines. Radiation for transitions where $\Delta m_F = 0$ is not seen when viewed parallel to the magnetic field and is linearly polarized otherwise. The typical spacing between Zeeman components is 1 kHz or 0.2 km/sec per milligauss of magnetic field. No complete Zeeman patterns have been identified in the complex spectra observed. Attempts to identify selected Zeeman components have led to magnetic fields of about 5 milligauss, which is not surprising since the spacing between observed features is about one km/sec (Zuckerman et al., 1972; Chaisson and Beichman, 1975). In the case of several 1720-MHz sources with simple spectra, left- and right-handed pairs seem to be coincident, thus supporting the Zeeman identifications (Lo et al., 1975). Fields of this magnitude are not unreasonable. The general magnetic field of the Galaxy is 10^{-6} gauss where the hydrogen density N_H is about 1 cm^{-3}. If the gas is compressed to $N_H = 10^6$ cm^{-3}, a rough but somewhat low value for a maser region, the magnetic field would be 10^{-2} gauss since the field is proportional to $N_H{}^{2/3}$, which follows from the conservation of flux if the magnetic field is frozen to the medium. However, propagation in the maser is probably very nonlinear and any statements about magnetic field are uncertain.

The H_2O masers, like the OH masers, exhibit complex spectra with many features. The spot sizes are generally a factor of at least 10 smaller than the OH spot sizes. There is no detailed correspondence between the OH and H_2O components in velocity or position. In fact, there is generally no overlap between the OH and H_2O clusters for the sources whose absolute positions are known. A few H_2O masers have features over a range of 300 km/sec, which is difficult to explain in terms of mass motion. The amplitudes of the features in both the OH and H_2O masers vary erratically. Small changes are noticeable in some cases in days or weeks. The lifetime of OH features is generally 10^7 to 10^8 sec, whereas for H_2O features it is only 10^6 to 10^7 sec (Sullivan, 1973).

The H II regions associated with the masers have typical sizes of 1 to 10 pc, electron densities of 10^3 cm, and temperatures of 10^4 K. These regions often contain several "compact" H II regions, sometimes called bright knots, which have diameters of 10^{-2} to 10^{-1} pc, electron densities of 10^5 to 10^6 cm^{-3}, and electron temperatures of 10^4 K. Almost every maser is closely associated with a compact H II region. In addition, in at least half the cases, there is an unresolved infrared source close to the masers. The H II region W3 provides a good example of maser emission of this type. There are two maser regions, as shown on the Palomar plate in Figure 9-15. Both are removed from the main nebulosity. The spectra of all the ground-state OH transitions for the southern maser W3(OH) are shown in Figure 9-16. Figure 9-17 presents a graphic representation of the OH and H_2O masers along with the IR source and H II condensation. The measured positions of the objects are not quite good enough to define precisely the spatial relationship among these objects. Goss, Lockhart, and Fomalont (1975) have listed all the measured positions. The H II condensation, the IR source, and the OH maser are very close together. Indeed the OH maser spots may occur in a shell around the H II condensation. The main OH spots have not moved significantly over the past 8 years, as can be seen by comparing the measurements from Moran et al. (1968), P. J. Harvey et al. (1974), and the most recent ob-

servations. The H_2O maser is definitely removed slightly from this association, a situation which may be typical. The association of the masers, the
compact H II sources, and the infrared objects is certainly exciting but the
physical connection among them is not clear.

A plausible picture has been developed by Habing et al. (1974), de Jong
(1973), and extended by Lo, Burke, and Haschick (1975) to include H_2O
masers. The compact H II region is the ionized remnant of the prestellar
cloud surrounding a new O or B star just reaching the main sequence. The
OH maser occurs in the dust envelope, which hides the star from view, just
outside the Strömgren sphere of ionized gas. The dust shell shields the OH
from photodissociation by the ultraviolet radiation of the star. In a time of

Figure 9-15.   The region around IC 1795 or W3 from the National Geographic Society-Palomar
Sky Survey. The crosses indicate the positions of the two masers in this region. The northern
one is called W3(C) and the southern one is called W3(OH).

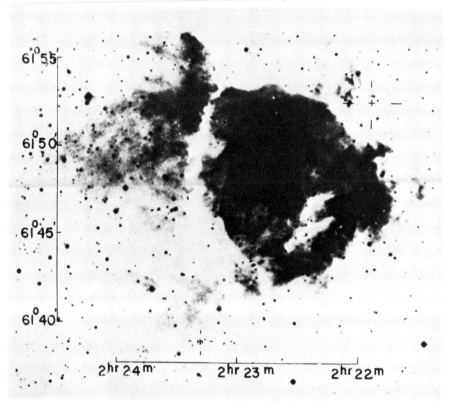

Figure 9-16. The spectrum of the $1 \to 2$, $1 \to 1$, $2 \to 2$, and $2 \to 1$ transitions in the ground state ($^2\Pi_{3/2}$, $J = \frac{3}{2}$) at 18 cm for the maser W3(OH). The solid line is right circular polarization and the broken line is left circular polarization. This is typical of the H II/OH masers—many narrow highly polarized components with no repetition of components among the different transitions (Barrett and Rogers, 1966).

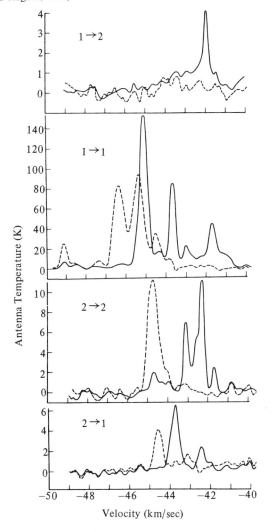

about 10^4 years the ionization front expands through the maser, dissociating the OH molecules, so that the maser disappears. This idea is supported by the fact that no OH masers coincide with the "compact" H II regions larger than 0.07 pc (2×10^{17} cm).

In the gas being compressed ahead of the ionization front, conditions are

Figure 9-17.   The region near W3(OH) showing the infrared source IRS9 (Wynn Williams, Becklin, and Neugebaurer, 1972), the compact H II region (Baldwin et al., 1973), the H_2O maser (Hills et al., 1972), and OH maser (Mader et al., 1975). The errors on the position of the H II condensation are about 0.1 arcsec. The uncertainties in position of the infrared, OH, and H_2O sources are about 2 arcsec. The origin of the coordinates is located at right ascension (1950) $2^h23^m16^s$ and declination (1950) $61°39'0''$.

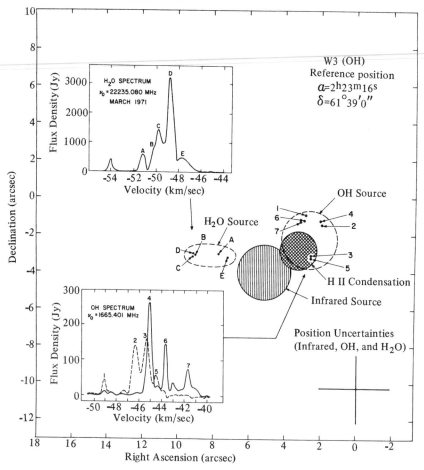

favorable for secondary star formation. The H_2O maser might exist in the infalling envelope of such a protostellar cloud where the temperature and densities are high enough to populate the masering transition. In this view the H_2O masering occurs at an earlier stage of stellar evolution than OH masering. In any event, the dynamic lifetime of both masers, i.e., the size of the masering region divided by the velocity spread, is only about 10^3–10^4 years. The masers therefore appear to be fleeting phenomena associated with the birth of one or more O or B stars.

A fundamental unanswered question is whether each maser spot corresponds to a star (local hypothesis) or whether all the spots are filaments surrounding a single star as in the case of evolved IR stars (global hypothesis). Kwan and Thuan (1974) argue in favor of the local hypothesis. However, as time goes on and more and more spots are identified, the number of stars may be too large for the local hypothesis. A combination of the two views seems reasonable.

Maser Models

In this section attention is focused on a simple maser cloud model which can describe the observed emission from a single feature or spot. The model will explain several characteristics of a maser amplifier: exponential gain, saturation, and line narrowing. The appearance of spherical cosmic masers and the sources of their energy are also discussed.

Consider a cloud of molecules having two energy levels, denoted 1 and 2. The transition frequency between them is ν_0. The statistical weights of the levels are taken to be the same, to simplify the presentation. This is in fact often the case; however, in exact calculations it will be necessary to consider the difference in the statistical weights. If N_1 and N_2 are the total population densities of the two levels, then the number of molecules in a given velocity interval dv along the line of sight can be defined as $N_1 f(v) dv$ and $N_2 f(v) dv$. For the case where thermal motion dominates, the distribution function $f(v)$ will be a gaussian function,

$$f(v) = \frac{1}{\sqrt{2\pi}} \frac{1}{u} \exp\left[-\frac{v^2}{2u^2}\right], \tag{9-36}$$

where $u = (kT_k/M)^{1/2}$, M is the mass of the molecule, T_k is the kinetic temperature. The distribution can be written as a function of frequency through the Doppler effect,

$$\frac{\nu - \nu_0}{\nu_0} = \frac{v}{c}. \tag{9-37}$$

Hence

$$f(\nu) = \frac{1}{\sqrt{2\pi}} \frac{1}{w} \exp\left[- \frac{(\nu - \nu_0)^2}{2w^2} \right] \tag{9-38}$$

where

$$w = (\nu_0/c)(kT_k/M)^{1/2}. \tag{9-39}$$

If $\Delta\nu_D$ is the full width at half maximum, then to a good approximation

$$f(\nu) = \frac{1}{\Delta\nu_D} \exp\left[- \frac{4 \ln 2 (\nu - \nu_0)^2}{\Delta\nu_D^2} \right]. \tag{9-40}$$

The intensity of the radiation propagating along a given ray path will be decreased by absorption and increased by emission. The equation describing the radiation transfer can be written as

$$\frac{dI_\nu}{ds} = -k_\nu I_\nu + \eta_\nu, \tag{9-41}$$

where

$$k_\nu = (N_1 - N_2) B \frac{h\nu}{4\pi} f(\nu), \tag{9-42}$$

$$\eta_\nu = N_2 A \frac{h\nu}{4\pi} f(\nu); \tag{9-43}$$

k_ν and η_ν are the volume absorption and emission coefficients respectively. A and B are the Einstein coefficients, which are related by the equation

$$A = \frac{2h\nu^3}{c^2} B. \tag{9-44}$$

There is only one Einstein B coefficient since the statistical weights are assumed to be equal. If k_ν and η_ν are constant through the cloud, equation (9-41) can be integrated to give

$$I_\nu = I_\nu(0) e^{-\tau_\nu} + S_\nu(1 - e^{-\tau_\nu}), \tag{9-45}$$

where $I_\nu(0)$ is the intensity of a background source. τ_ν is the optical depth defined as

$$\tau_\nu = k_\nu L, \qquad (9\text{-}46)$$

where L is the cloud thickness, and S_ν is the source function

$$S_\nu = \frac{\eta_\nu}{k_\nu} = \frac{N_2 A}{(N_1 - N_2)B}. \qquad (9\text{-}47)$$

The excitation temperature describing the ratio of the level populations is defined by the relation

$$\frac{N_2}{N_1} = e^{-h\nu_0/kT_x}. \qquad (9\text{-}48)$$

If the gas is in thermal equilibrium at the temperature T_k, equation (9-48) is the Boltzmann equation with T_x equal to T_k. Notice that T_x can be negative or positive. For small positive values of T_x most of the molecules are in the lower state, and for small negative values of T_x most of the molecules are in the upper state. Substituting equations (9-44) and (9-48) into (9-47) shows that S_ν is the Planck function, equation (9-1), for temperature T_x, so that with the Rayleigh-Jeans approximation, equation (9-2), it follows that equation (9-45) can be written as

$$T_B = T_c e^{-\tau_\nu} + T_x (1 - e^{-\tau_\nu}), \qquad (9\text{-}49)$$

where T_c is the brightness temperature of the background source.

Maser emission occurs when $N_2 > N_1$ so that the absorption coefficient and hence the opacity are negative, as is the excitation temperature. Equation (9-49) will describe a masering cloud as long as the intensity, and therefore the brightness temperature, which grows exponentially, is small enough not to disturb the population distribution. Eventually, however, the growth will be limited as the upper level becomes depleted, thus changing T_x, and the maser is said to be saturated.

The effect of the radiation on the level populations will now be considered. The radiative transfer equation (9-41) for the line center frequency ν_0, with equations (9-40), (9-42), and (9-43), will be

$$\frac{dI}{ds} = \frac{h\nu_0}{4\pi\Delta\nu_D} [(N_2 - N_1) BI + N_2 A]. \qquad (9\text{-}50)$$

The population inversion is

$$\Delta N = N_2 - N_1. \tag{9-51}$$

To determine how ΔN depends on I we assume that the two energy levels are in statistical equilibrium; that is, the number of transitions per unit of time from levels 2 to 1 is equal to the number of transitions per unit of time from levels 1 to 2. This equilibrium equation can be written:

$$N_2(C_{21} + M_{21} + P_{21} + A) = N_1(C_{12} + M_{12} + P_{12}) \tag{9-52}$$

C_{21} and C_{12} are the collision rates from levels 2 to 1 and from levels 1 to 2, respectively. If collisions dominated all other terms, then

$$\frac{N_2}{N_1} \approx \frac{C_{12}}{C_{21}} = e^{-h\nu_0 / kT_k}, \tag{9-53}$$

so that the excitation temperature would equal the kinetic temperature. In any event, the collision rates are nearly equal for microwave transitions since $h\nu_0/kT_k$ is small. M_{21} and M_{12} are the microwave transition rates due to the stimulated emission and absorption, which are equal to B times the intensity averaged over all directions, $\bar{I}$. Since a stimulated photon has the same direction as the stimulating photon, the radiation usually becomes beamed in a maser so that we can write for both M_{12} and M_{21}:

$$M = B\bar{I} = BI \frac{\Omega_m}{4\pi}, \tag{9-54}$$

where Ω_m is the beam solid angle. P_{12} and P_{21} are the "up" pump and "down" pump rates respectively, which describe how some unspecified agent transfers population between the levels using other energy levels in the molecule as intermediate steps, as shown in Figure 9-18. If the pump rates P_{12} and P_{21} were zero, and the radiation field were specified by a radiation temperature T_R, so that

$$I = 2kT_R\nu_0^2/c^2, \tag{9-55}$$

then the solution to equation (9-52), using equation (9-53) and (9-54) with $\Omega_m = 4\pi$, would yield an excitation temperature T_x given by

$$T_x = T_k \frac{T_0 + T_R}{T_0 + T_k} \tag{9-56}$$

where

$$T_0 = \frac{h\nu_0}{k} \frac{C_{12}}{A}.$$
(9-57)

T_x is bounded between T_R and T_k and cannot be negative. Hence a two-level cosmic maser is not possible. The influence of other levels is essential in establishing the population inversion. Two-level laboratory masers can be built in which the molecules in the lower level are spatially separated from those in the upper level by magnetic or electric fields. In order to simplify the solution of the equilibrium equation we assume to good approximation that A is negligible and that $C_{12} \approx C_{21} \equiv C$, so that

$$\frac{N_2}{N_1} = \frac{P_{12} + M + C}{P_{21} + M + C}.$$
(9-58)

The population inversion which the pump could establish by itself if $M = C = 0$ is, from equations (9-51) and (9-58),

$$\Delta N_0 = N \frac{P_{12} - P_{21}}{P_{12} + P_{21}},$$
(9-59)

where $N = N_1 + N_2$, the total population. The so-called pump efficiency η_p,

$$\eta_p = \Delta N_0 / N,$$
(9-60)

Figure 9-18. A maser system with two energy levels showing the means of population transfer due to spontaneous emission, A; induced microwave transitions, M_{12} and M_{21}; collisions, C_{12} and C_{21}; and pumping, P_{12} and P_{21}. The pumping cycle may actually include more than one additional level and can invert the population in the maser transition if $P_{12} > P_{21}$.

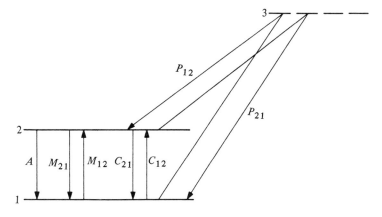

will be small in cosmic masers (~ 0.01). Letting $P = P_{12} + P_{21}$ and using equations (9-51), (9-58), and (9-59) leads to the useful relation for the population inversion,

$$\Delta N = \frac{\Delta N_0}{1 + \dfrac{2(C + M)}{P}}. \tag{9-61}$$

As expected, collisions and microwave transitions reduce the inversion. Substituting equations (9-61) and (9-51) into (9-50) gives

$$\frac{dI}{ds} = \frac{\alpha_0 I}{1 + \left(\dfrac{I}{I_s}\right)} + \epsilon, \tag{9-62}$$

where

$$\alpha_0 = \frac{h\nu_0 B}{4\pi\Delta\nu_D} \frac{\Delta N_0}{1 + 2C/P}, \tag{9-63}$$

$$I_s = \frac{1 + 2C/P}{1\dfrac{B\Omega m}{2\pi P}}, \tag{9-64}$$

$$\epsilon = \frac{h\nu_0}{4\pi\Delta\nu_D} N_2 A, \tag{9-65}$$

and ϵ is assumed to be a constant, a good assumption in the unsaturated region. This term is usually dropped altogether since the stimulated emission generally greatly exceeds the spontaneous emission. However, if no background source is available as an input to the maser, this term, representing the maser's own spontaneous emission, becomes the input signal to the amplifier. I_s is the intensity at which the maser becomes saturated. For $I \ll I_s$ the solution is

$$I = I_0\, e^{\alpha_0 l} + \frac{\epsilon}{\alpha}(e^{\alpha_0 l} - 1); \tag{9-66}$$

or, converting to temperature,

$$T_B = T_c\, e^{\alpha_0 l} + |T_x|(e^{\alpha_0 l} - 1), \tag{9-67}$$

which will be recognized as being the same as equation (9-49), where $\tau = -\alpha l$; αl is usually called the unsaturated gain. $|T_x|$ and T_c represent respectively the maser input signals due to its own spontaneous emission and to a background source. If $I \gg I_s$, the right-hand side of equation (9-62) is a constant. The maser is saturated and the intensity is

$$I = I_0 + (\alpha_0 I_s + \epsilon)l. \tag{9-68}$$

The intensity grows linearly with distance in a saturated maser. Reasonable parameters for an H_2O/H II maser might be (Sullivan, 1973) $\Omega_m = 10^{-4}$, $C = 10^{-2}$ sec^{-1}, $P = 10^{-1}$ sec^{-1}, $T = 500$ K, $\Delta\nu_D = 85$ kHz, and $\Delta N_0 = 5$ cm^{-3}. Such a maser would become saturated after a distance of 10^{15} cm with $\alpha_0 l = 25$.

In a cosmic maser that is unsaturated, small changes in the gain produce large changes in the output intensity. If $\alpha_0 l$ is equal to 25 and then changes by 1%, the output intensity will change by 25%. Changes in the gain, for example, could occur if the length of the maser changed because of turbulent gas motions. Notice that the intensity is independent of pump rate as long as $P \gg C$.

The situation is different for a saturated maser. The intensity will vary linearly with the gain. Since the saturated intensity, I_s, is proportional to the pump rate, changes in the pump rate will cause linear changes in the intensity. The growth of intensity versus distance in a maser is shown in Figure 9-19.

During unsaturated growth, the width of the spectral profile will decrease. Consider a case in which the input signal to the maser is a background continuum source so that the brightness temperature, with frequency dependence is, from equation (9-67),

$$T_B(\nu) = T_c e^{\alpha(\nu)l}, \tag{9-69}$$

$\alpha(\nu)$ is given by

$$\alpha(\nu) = \alpha_0 e^{-4\ln 2(\nu - \nu_0)^2/\Delta\nu_D^2} \tag{9-70}$$

due to the frequency dependence of the population inversion in equation (9-40); that is,

$$\Delta N_0(\nu) = \Delta N_0 f(\nu). \tag{9-71}$$

The profile of T_B is approximately gaussian and has a width, $\Delta\nu$, of

$$\Delta\nu \cong \frac{\Delta\nu_D}{\sqrt{\alpha_0 l}} \tag{9-72}$$

for $\alpha_0 l \gg 1$. Hence the line width decreases until the center of the line begins to saturate. The line width then broadens as the wings of the line continue to experience exponential growth until it again equals $\Delta\nu_D$, as shown in Figure 9-19. Goldreich and Kwan (1974a) describe a mechanism which prevents $\Delta\nu$ from increasing after it reaches its minimum value.

The maser gains are probably on the order of 20–30 so that the input temperature is increased by a factor of 10^9–10^{13} to give the observed brightness temperatures, which range from 10^9 for OH/IR masers to 10^{15} for the brightest H_2O/H II maser. Unsaturated masers will exhibit line widths which are about one-fifth of the Doppler widths. Hence, for example, where the OH/H II masers exhibit line widths of about 2 kHz, corresponding to an apparent kinetic temperature of 50 K, the true kinetic temperature, if the maser were unsaturated, would be $\sim 10^3$ K.

The literature is filled with discussion as to whether particular types of masers are saturated. The general consensus is that most masers are at least partially saturated. There is a general but persuasive argument which supports the saturated maser hypothesis. The spectra of most masers contain several components having nearly the same amplitude. Furthermore the ratio of the strongest feature to the weakest is rarely greater than 10 or 100. This would be hard to explain in an unsaturated maser where small varia-

Figure 9-19. The gain and line width of a spectral component versus distance in a one-dimensional maser (Sullivan, 1973).

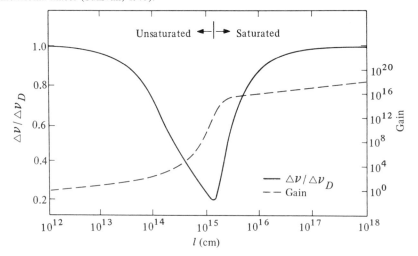

tions in the path length at different velocities would cause much larger differences in the observed amplitude. Saturation would tend to limit the growth of the features to values depending on the pump power from a common pump source.

The foregoing discussion cannot account for nonlinear effects due to quantum coherence phenomena which are observed in laboratory masers such as "spiking." Rosen (1974) suggests that such effects might be important. It is interesting to note, however, that the amplitude of the received electric field from a maser is describable as a gaussian random process to high accuracy (Evans et al., 1972).

It is difficult to achieve the required gain in models where the gain path l is as small as the observed dimension of cosmic masers. For example, with $N_H = 10$, $N_{OH}/N_H = 10^{-4}$, $\eta_p = 0.01$, and all the OH molecules in the ground state, a maser of observed diameter 10^{14} cm would have a gain of only 1. Two different geometric models of cosmic masers alleviate this difficulty. If, for example, the masers are long tubes or filaments of length L and diameter L^*, where $L \gg L^*$, the gain can be increased by L/L^* or up to about 100. The maser radiation would be beamed into the solid angle $\Omega_m = (L^*/L)^2$ and only those filaments pointed towards the earth would be observed. The time variability could be explained by rotation of the filaments or by turbulent motions which change the path length over which the velocity is constant.

The second model is a spherical maser whose appearance is not immediately obvious. Consider a spherical maser of diameter D which is unsaturated. Rays passing through the sphere and reaching the observer will be nearly parallel since the angular size of the maser is very small, as shown in Figure 9-20. A ray passing through the center has a gain path D, while a ray at projected distance X has a gain path of only $(D^2 - 4X^2)^{1/2}$. Hence the brightness distribution as a function of X would be

$$T_B(X) = T_c \exp\left[\alpha_0 D \left(1 - \frac{4X^2}{D^2}\right)^{1/2}\right]. \tag{9-73}$$

The observed width at half maximum of the source, D^*, will be

$$\frac{D^*}{D} \approx \frac{1}{\sqrt{\alpha_0 D}} \tag{9-74}$$

The molecules at the surface will see primarily radiation coming from the diametric direction and the beam angle appropriate for equation (9-64) will be

$$\Omega_m \approx \left(\frac{D^*}{D}\right)^2. \tag{9-75}$$

The maser will of course appear isotropic to the observer.

Now assume that the pump source weakens and the pump rate decreases uniformly throughout the sphere. Also, assume that spontaneous emission drives the maser so that there is no preferred direction. The maser will first saturate at the outer edge where the intensity is greatest because of rays which have been amplified over the total diameter D.

Figure 9-20. Top: The rays, passing through an unsaturated, uniformly pumped spherical maser. Off-center rays will have less gain and the sphere will appear to be smaller than it is by the amount given in equation (9-74). Bottom: The rays passing through a partially saturated spherical maser. The rays that pass through the unsaturated core will be very much stronger than those that do not.

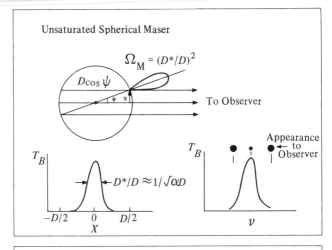

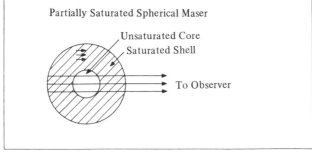

The maser will then have an unsaturated core of diameter D' surrounded by a saturated shell. The observer will see only those rays which pass through the unsaturated core and are exponentially amplified. Radiation originating in the saturated shell and not passing through the unsaturated one will grow only linearly and will be very weak at the edge of the maser. Hence, as the pump rate continually decreased, the maser would appear to get smaller down to some critical size after which it would appear to increase in size as the entire maser became saturated. In this model, most of the observed power comes from the unseen saturated shell. Goldreich and Keeley (1972) worked out the radiative transfer equation in detail for a spherical maser. The model parameters they selected are listed in Table 9-6.

Little has been said about the nature of the pump mechanism. The physics involved is very complex and the consensus on the models is far from unanimous. Litvak (1974) reviewed all the known pump mechanisms and gave references to original work. Goldreich and Kwan (1974b) summarized the operation of various pump models for H_2O. Here, we will only mention briefly various types of pumping mechanisms.

The fundamental requirement for population inversion is the presence of heat reservoirs at different temperatures which provide non-thermodynamic equilibrium conditions in the maser gas cloud. A simple example is that of a low-density gas cloud near a star. The radiation temperature that the gas molecules see is high in the direction of the star on one side and low in the direction of cold space on the other. Another situation is one in which the gas coexists with dust. If the dust is hotter or colder than the gas, then popu-

Table 9-6 Typical parameters for spherical masers related to H II regions.[a]

Quantity	H_2O maser	OH maser
N_H (cm^{-3})[b]	10^9	10^8
N_{OH} or N_{H_2O} (cm^{-3})	3×10^4	10^2
C (sec^{-1})	10^0	10^{-1}
P (sec^{-1})	10^{-1}	10^{-2}
n_p (cm^{-3})	10^{-2}	10^{-2}
T_k (K)	10^3	10^2
T_B (K)	6×10^{14}	6×10^{11}
D (cm)	2×10^{15}	2×10^{15}
D^* (cm)	4×10^{13}	1.2×10^{14}

[a] Data taken from Goldreich and Keeley (1972).
[b] Hydrogen density which governs the collision rate, C.

lation inversion is possible. The heat source responsible for the excitation of the gas is the pump which supplies the necessary power to the maser.

To determine whether population inversion actually occurs between a specific pair of energy levels, it is necessary to solve the equations of statistical equilibrium for all the energy levels in the molecule which affect the population of the levels in question. Selection rules and the magnitude of the collision rates and the transition probabilities play important roles in the solution to these equations. After the pumping cycle has been determined, it can be characterized simply by the two rates P_{12} and P_{21} as in equation (9-52). Pumping mechanisms in which radiation from some source induces molecules into excited states, whereupon they cascade down and invert the masing transition, are called radiative pumps. Most cosmic maser pumps are of this type. Radiatively pumped masers require at least one pump photon for every microwave photon. Hence pumping schemes using ultraviolet radiation are inefficient and require very luminous sources. For example, the UV pumping model for the OH source in W3(OH) requires a luminosity of $10^2 L_\odot$ since the microwave rate is 4×10^{46} photons/sec and the ratio of pump frequency to the microwave frequency is 10^6. Infrared pumping seems more attractive, especially given the proximity of nearly every maser to an infrared source. Nonradiative pumping is possible. Collisional pumping may be significant in H_2O masers where the gas density is high. At first, collisional pumping seems implausible since collisional processes suggest thermal equilibrium. Inversion occurs, for example, because there are more spontaneous decays from a high-lying state, populated by collisions, to the upper level of the maser transition than to the lower level. The inversion will be maintained if the photons emitted during the spontaneous decay are removed from the gas. For example, they can escape from the gas if the maser is thin in one dimension, or they can be absorbed on cold grains (de Jong, 1973). Another type of pumping is chemical pumping, whereby molecules are formed in selected excited states and cascade down to the maser transition.

Conclusions

The growth of information on the subject of cosmic masers, a phenomenon unsuspected twelve years ago, has been very great. The very basic questions are still open to discussion: What do the masers amplify? Are they saturated or unsaturated? How are they pumped? What is their geometry? Is the pump inside or outside the maser? Does each maser spot correspond to a star, or do all the spots correspond to a single star? Many of these questions could be answered by statistical studies of how the spots change in appearance and spectral character with time. The precise connection among the various molecular masers, the infrared sources, and the H II condensations is uncertain. Improved positions for the sources will help clarify this picture. Much exciting work remains to be done.

I was assisted in the preparation of the chapter by discussions with C. Gottlieb, C. Lada, M. Litvak, K. Y. Lo, H. E. Radford, M. J. Reid, P. R. Schwartz, and B. Zuckerman.

References

Baldwin, J. E., Harris, C. S., and Ryle, M. 1973. 5 GHz observations of the infrared star MWC 349, and the H II condensation W3(OH), *Nature 241*, 38–39.

Barrett, A. H., and Rogers, A. E. E. 1966. Observations of circularly polarized OH emission and narrow spectral features, *Nature 210*, 188–190.

Barrett, A. H., Ho, P., and Martin, R. N. 1975. Time variations and spectral structure of the methanol maser in Orion A, *Astrophys. J. (Letters) 198*, L119–122.

Buhl, D., Snyder, L. E., Lovas, F. J., and Johnson, D. R. 1975. Is there a maser in the silicon monoxide ground state! *Astrophys. J. (Letters) 201*, L29–31.

Burke, B. F., Johnston, K. J., Efanov, V. A., Clark, B. G., Kogan, L. R., Kostenko, V. I., Lo, K. Y., Matveenko, L. I., Moiseev, I. G., Moran, J. M., Knowles, S. H., Papa, D. C., Papadopoulos, G. D., Rogers, A. E. E., and Schwartz, P. R. 1972. Observations of maser radio sources with an angular resolution of 0".0002, *Sov. Astron. 16*, 379–382.

Chaisson, E. J., and Beichman, C. A. 1975. Further evidence for magnetism in the Orion region, *Astrophys. J. (Letters) 199*, L39–42.

Cheung, A. C., Rank, D. M., Townes, C. H., Thornton, D. D., and Welch, W. J. 1969. Detection of water in interstellar regions by its microwave radiation, *Nature 221*, 626–628.

Cudaback, D. D., Read, R. B., and Rougoor, G. W. 1966. Diameters and positions of three sources of 18-cm OH emission, *Phys. Rev. Letters 17*, 452–455.

Davies, R. D., de Jager, G., and Verschuur, G. L. 1966. Detection of circular and linear polarization in the OH emission sources near W3 and W49, *Nature 209*, 974–977.

de Jong, T. 1973. Water masers in a protostellar gas cloud, *Astron. Astrophys. 26*, 297–313.

Dousmanis, G. C., Sanders, T. M., and Townes, C. H. 1955. Microwave spectra of the free radicals of OH and OD, *Phys. Rev. 100*, 1735–1754.

Ehrenstein, G., Townes, C. H., and Stevenson, M. J. 1959. Ground state Λ doubling transitions of the OH radical, *Phys. Rev. Letters 3*, 40–41.

Einstein, A. 1917. Zur quantentheorie der strahlung, *Phys. Zeitschr. 18*, 121–128.

Evans, N. J., Hills, R. E., Rydbeck, O. E. H., and Kollberg, E. 1972. Statistics of the radiation from astronomical masers, *Phys. Rev. A 6*, 1643–1647.

Gardner, F. F., Robinson, B. J., Bolton, J. G., and van Damme, K. J. 1964. Detection of the interstellar OH lines at 1612 and 1720 Mc/sec, *Phys. Rev. Letters 13*, 3–5.

Goldreich, P., and Keeley, D. A. 1972. Astrophysical masers. I. Source size and saturation, *Astrophys. J. 174*, 517–525.

Goldreich, P., and Kwan, J. 1974a. Astrophysical masers. IV. Line widths, *Astrophys. J. 190*, 27–34.

Goldreich, P., and Kwan, J. 1974b. Astrophysical masers. V. Pump mechanism for H_2O masers, *Astrophys. J. 191*, 93–100.

Goss, W. M., Lockhart, I. A., and Fomalont, E. B. 1975. An improved 1665 MHz position for W3(OH), *Astron. Astrophys. 40*, 439–440.

Gundermann, E. J. 1965. Observation of the interstellar hydroxyl radical, Ph.D. thesis, Harvard University.

Habing, H. J., Goss, W. M., Matthews, H. E., and Winnberg, A. 1974. Identification of Type I OH masers with very small H II regions, *Astron. Astrophys. 35*, 1–5.

Harris, D. E., Zeissig, G. A., and Lovelace, R. V. 1970. The minimum observable diameter of radio sources, *Astron. Astrophys. 8*, 98–104.

Harvey, P. J., Booth, R. S., Davies, R. D., Whittet, D. C. B. and McLaughlin, W. 1974. Interferometric observations of the structure of main line OH sources, *Monthly Not. Roy. Astron. Soc. 169*, 545–576.

Harvey, P. M., Bechis, K. P., Wilson, W. J., and Ball, J. A. 1974. Time variations in the OH microwave and infrared emission from late type stars, *Astrophys. J. Suppl. 27*, 331–357.

Heiles, C. E. 1968. Normal OH emission and interstellar dust clouds, *Astrophys. J. 151*, 919–934.

Herbig, G. H. 1974. Structure of the OH/infrared object in NML Cygnus. II. Analysis of the OH interferometry, *Astrophys. J. 189*, 75–79.

Herzberg, G. 1950. Molecular spectra and molecular structure. I. Spectra of diatomic molecules, D. van Nostrand Co., Princeton, N.J.

Hills, R., Janssen, M. A., Thornton, D. D., and Welch, W. J. 1972. Interferometric positions of the water vapor emission sources in H II regions, *Astrophys. J. (Letters) 175*, L59-64.

Hills, R., Pankonin, V., and Landecker, T. L. 1975. Evidence for maser action in the 1.2 cm transitions of methanol in Orion, *Astron. Astrophys. 39*, 149–153.

Kwan, J., and Thuan, T. X. 1974. On the interpretation of the interferometric maps of H_2O masers near H II regions, *Astrophys. J. 194*, 293–300.

Lees, R. M. 1973. On the $E_1 - E_2$ labeling of energy levels and the anomalous excitation of interstellar methanol, *Astrophys. J. 184*, 763–771.

Litvak, M. M. 1974. Coherent molecular radiation, *Annu. Rev. Astron. Astrophys. 12*, 97–112.

Litvak, M. M., McWhorter, A. L., Meeks, M. L., and Zeiger, H. J. 1966. Maser model for interstellar OH microwave emission, *Phys. Rev. Letters 17*, 821–826.

Lo, K. Y., Burke, B. F., Haschick, A. D. 1975. H_2O sources in regions of star formation, *Astrophys. J. 202*, 81–91.

Lo, K. Y., Walker, R. C., Burke, B. F., Moran, J. M., Johnston, K. J., and Ewing, M. S. 1975. Evidence for Zeeman splitting in 1720 MHz OH line *Astrophys. J. 202*, 650–654.

Mader, G. L., Johnston, K. J., Moran, J. M., Knowles, S. H., Mango, S. A., Schwartz, P. R., and Waltman, W. B. 1975. The relative positions of the OH and H_2O masers in W49N and W3(OH), *Astrophys. J. (Letters) 200*, L111–114.

Manchester, R. N., Robinson, B. J., Goss, W. M. 1970. 18 cm observations of galactic OH from longitudes 128° to 300°, *Aust. J. Phys. 23*, 751–775.

Menon, T. K. 1967. The nature of OH emission sources in the galaxy, *Astrophys. J. (Letters) 150*, L167–170.

Mezger, P. G., Altenhoff, W. J., Schraml, J., Burke, B. F., Reifenstein, E. C., Wilson, T. L. 1967. A new class of compact H II regions associated with OH emission sources, *Astrophys. J. (Letters) 150*, L157–166.

Moran, J. M., Burke, B. F., Barrett, A. H., Rogers, A. E. E., Ball, J. A., Carter, J. C., and Cudaback, D. D. 1968. The structure of the OH source in W3, *Astrophys. J. (Letters) 152*, L97–101.

Moran, J. M., Papadopoulos, G. D., Burke, B. F., Lo, K. Y., Schwartz, P. R., Thacker, D. L., Johnston, K. J., Knowles, S. H., Reisz, A. C., and Shapiro, I. I. 1973. Very long baseline interferometric observations of the H_2O sources in W49N, W3(OH), Orion A, and VY Canis Majoris, *Astrophys. J. 185*, 535–567.

Neugebauer, G., and Leighton, R. B. 1969. Two-micron sky survey, a preliminary catalog, N69-37993, National Aeronautics and Space Administration, Washington, D.C.

Raimond, E., and Eliasson, B. 1967. Possible relation between an OH source and an infrared object in the Orion nebula, *Astrophys. J. (Letters) 150*, L171–172.

Rank, D. M., Townes, C. H., and Welch, W. J. 1971. Interstellar molecules and dense clouds, *Science 174*, 1083–1101.

Reid, M. J. 1975. The structure of hydroxyl masers and circumstellar envelopes of long period variable stars, Ph.D. thesis, California Institute of Technology.

Rogers, A. E. E., Moran, J. M., Crowther, P. P., Burke, B. F., Meeks, M. L., Ball, J. A., Hyde, G. M. 1966. Interferometric study of cosmic line emission at OH frequencies, *Phys. Rev. Letters, 17*, 450–452.

Rosen, R. A., 1974. A nonlinear model for the intensity, line width and coherence of astrophysical masers, *Astrophys. J. (Letters) 190*, L73–76.

Ryle, M. 1975. Radio telescopes of large resolving power, *Science 188*, 1071–1083.

Schwartz, P. R., Harvey, P. M., and Barrett, A. H. 1974. Time variation of the H_2O maser and infrared continuum in late type stars, *Astrophys. J. 187*, 491–496.

Snyder, L. E., and Buhl, D. 1974. Detection of possible maser emission near 3.48 millimeters from an unidentified molecular species in Orion, *Astrophys. J. (Letters) 189*, L31–33.

Sullivan, W. T. 1973. Microwave water vapor emission from galactic sources, *Astrophys. J. Suppl. 25*, 393–432.

Swenson, G. W., and Kellermann, K. I. 1975. An intercontinental array. A next-generation radio telescope, *Science 188*, 1263–1268.

Townes, C. H. 1965. Production of coherent radiation by atoms and molecules, *Science 149*, 831–841.

Turner, B. E. 1970. Anomalous emission from interstellar hydroxyl and water, *J. Roy. Astron. Soc. Canada*, 221–237, 282–304.

Turner, B. E. 1974. Interstellar molecules, in G. L. Verschuur and K. I. Kellermann, eds., *Galactic and Extragalactic Radio Astronomy,* pp. 199–255, Springer-Verlag, New York.

Van Vleck, J. H. 1929, On σ-type doubling and electron spin in the spectra of diatomic molecules, *Phys. Rev.* 33, 467–506.

Weaver, H., Williams, R. W., Dieter, N. H., and Lum, W. T. 1965. Observations of a strong unidentified microwave line and of emission from the OH molecule, *Nature 208,* 29–31.

Weinreb, S., Barrett, A. H., Meeks, M. L., and Henry, J. C. 1963. Radio observations of OH in the interstellar medium, *Nature 200,* 829–831.

Weinreb, S., Meeks, M. L., Carter, J. C., Barrett, A. H., and Rogers, A. E. E. 1965, Observations of polarized OH emission, *Nature 208,* 440–441.

Wilson, W. J., and Barrett, A. H. 1968. Discovery of hydroxyl radio emission from infrared stars, *Science 161,* 778–779.

Wilson, W. J., and Barrett, A. H. 1972. Characteristics of OH emission from infrared stars, *Astron. Astrophys. 17,* 385–402.

Wynn-Williams, C. G., Becklin, E. E., and Neugebauer, G. 1972. Infrared sources in the H II region W3, *Monthly Not. Roy. Astron. Soc. 160,* 1–14.

Zappala, R. R., Becklin, E. E., Matthews, K., and Neugebauer, G. 1974. Angular diameter of IRC + 10011 at 2.2, 10, and 20 microns, *Astrophys. J. 192,* 109–112.

Zuckerman, B., and Palmer, P. 1974. Radio radiation from interstellar molecules, *Annu. Rev. Astron. Astrophys. 12,* 279–313.

Zuckerman, B., Palmer, P., Penfield, H., and Lilley, A. E. 1968. Detection of microwave radiation from the $^2\Pi_{1/2}$, $J = \frac{1}{2}$ state of OH, *Astrophys. J. (Letters) 153,* L69–76.

Zuckerman, B., Yen, J. L., Gottlieb, C. A., and Palmer, P. 1972. Observations of the $^2\Pi_{3/2}$, $J = \frac{5}{2}$ state of interstellar OH, *Astrophys. J. 177,* 59–78.

10

Active Galaxies

K. Brecher

We know a great deal about active galaxies; we understand very little about them. Quasars, radio galaxies, Seyfert galaxies, the nuclei of otherwise "normal" galaxies all show surprising signs of activity. Explosions, ejected clouds of relativistic particles, visible jets and wisps are all phenomena which in the past two or three decades have been observed to be associated with galaxies of one kind or another. Discovery of these phenomena parallels, and indeed was given impetus, by the development of several nontraditional branches of astronomy: radio, infrared, ultraviolet, and X-ray. Hints of activity in galaxies, occurring on time scales short compared with the Hubble time or even with their dynamic times, appeared in 1943 with the discovery by Carl Seyfert of a number of spiral galaxies with anomalously broad emission lines. Nevertheless, the great interest in galactic "violence" has been spurred primarily by nontraditional means of observation, and even more exotic theoretical speculations and ideas, such as the suggestion by Ambartsumian (1958) that the centers of galaxies might be the sites of their birthplaces.

To begin this brief discussion of "active galaxies," we should define the term. This is not an easy task, for there is no general agreement on what constitutes activity—activity with respect to what? For the present purposes, we will consider all galaxies which are not "normal" to be "active." By a normal galaxy we simply mean a collection of about 10^9 to 10^{11} stars, which evolve primarily by thermal equilibrium processes in a gravitationally stable configuration. By contrast, we consider a system to be active if it gives rise to nonthermal radiative phenomena and/or shows evidence of gravitational disequilibrium. Since normal galaxies radiate predominantly at optical wavelengths (the radiation arising from *atomic* opacities in stellar photospheres), the first key to activity is the detection of radiative fluxes from a galaxy in excess of that produced by, say, a 5,000–10,000 K black-body spectrum. Radio, infrared, ultraviolet and X-ray luminosities comparable to or greater than optical luminosities must, in general, have a nonthermal, i.e., nonblackbody, and nonstellar origin. (Scattering of optical radiation by dust grains to produce infrared radiation, and "thermal" bremsstrahlung emission by hot, optically *thin* gas clouds are two important exceptions.)

Gravitational disequilibrium can display itself in a variety of guises. During the first half of this century, astronomers considered galaxies to be systems evolving on a time scale of $\sim 10^{10}$ years (apart from the more rapid changes due to single occasional novae or supernovae). Typical stellar velocities of revolution about the centers of spiral galaxies or typical mean dispersion velocities in elliptical galaxies were found to be a few hundred kilometers per second. However, objects showing variability on a time scale short compared with 10^{10} years (or even with the 10^8-year galactic rotation times), or displaying expansion or ejection or dynamical velocity of thousands of kilometers per second, have been discovered over the past decade, all indicating the presence of gravitationally unstable systems.

Several separate developments occurring simultaneously over the past 10 to 20 years have led to a revised view of the nature of galaxies. This chapter surveys the properties of active galaxies; considers individual galaxy types and tries to explain what can be deduced directly from the observations with a minimum number of astronomical assumptions; discusses the sources of the activity in those systems; and finally, explores the relation of such objects to normal galaxies. Before proceeding to a more detailed look at these objects, let us briefly review the history of the subject. A good survey of the early work on active galaxies appears in the paper by Burbidge, Burbidge, and Sandage (1963).

Although most of the interest in active galaxies dates to the late 1950s, Seyfert (1943) wrote a most remarkable and unusual paper. In it, he discussed the properties of the nuclei of several spiral galaxies, which appear stellar on short enough exposure (cf. Figure 10-3, shown later). Further, their spectra showed broad emission lines (up to 8000 km/sec wide), unlike those seen in nebulae contained in other spiral galaxies. As noted in 1959 by Burbidge, Burbidge, and Prendergast in their study of the rotation curve of NGC 1068, no mass could exist large enough to explain these large dispersion velocities as an equilibrium phenomenon; they concluded that these objects were truly explosive. Since the idea had developed from the late 1920s onward that the spiral nebulae were equilibrium configurations, formed billions of years ago, the idea that they were at present undergoing some form of activity was thought to be most remarkable.

In 1944, Grote Reber, an "amateur" astronomer, discovered the radio source Cygnus A with his $2,000 homemade radio telescope. By 1951, a good enough position of this "radio star" had been determined for Walter Baade, using the powerful 200-inch Palomar telescope, to determine that the radio emission arose from the region of a very peculiar extragalactic object. He initially proposed a model consisting of two colliding galaxies for the system, an event which has a probability of only one in a hundred million. The discovery of literally millions of other radio sources makes this explanation unlikely, in retrospect. However, it created great excitement at the time. Following the original suggestion of Alfvén and Herlofson (1950), the Russian astrophysicists V.L. Ginzburg and I. S. Shklovsky proposed that the radio sources arise from synchrotron radiation by relativistic electrons moving in magnetic field regions extended over millions of light years, a theory that pointed toward the importance of nonthermal phenomena. Shortly thereafter, G. Burbidge calculated that Cygnus A would require a minimum of about 10^{60} erg of fast electrons and magnetic fields to explain the observed radio flux (see Appendix for details). This is the equivalent of $10^6 \, M_\odot$ converted entirely into relativistic particles. Even an entire galaxy of $10^{10} \, M_\odot$ would need to be spectacularly efficient compared with the sun, say, to produce such events. As we shall see, such events probably repeat more than once per Hubble time, making the energetics still more difficult to understand.

The discovery of quasi-stellar sources emitting optical fluxes several times

Table 10-1 Properties of active galaxies.

---

1. High luminosity
2. Nonthermal emission, with excess ultraviolet, infrared, radio and X-ray flux (compared with normal galaxies)
3. Rapid variability and/or small size
4. Peculiar photographic appearance: high contrast in brightness of nucleus and large-scale structure
5. Explosive appearance or jet-like protuberances
6. Gravitational disequilibrium, evidenced by high internal velocities
7. Broad emission lines (sometimes) and nonstellar spectrum

---

greater than that of the Milky Way, infrared fluxes 10 times larger still, and, even more surprising, huge fluxes of X-rays, has led to one mystery after another during the past decade. We shall proceed, then, to consider some of the observational facts concerning active galaxies—in particular, their general features which can be derived directly from the observations. Table 10-1 lists some of the features exhibited by active galaxies. Then we consider possible theoretical interpretations. The vastness of the subject precludes a complete treatment in this brief review.

Observed Properties of Active Galaxies

Seyfert Galaxies

Spiral galaxies are considered to be gravitationally bound configurations of stars which evolve on a time scale of billions of years. Properties of the nuclei of 12 unusual spiral galaxies, now known as Seyfert galaxies, were described by Seyfert (1943). Their most interesting features are the following:

1. Their nuclei are very small and extremely bright for their size, relative to the rest of the spiral galaxy surrounding it.

2. The optical spectra of their nuclei contain emission lines not usually seen in the spectra of normal galaxies, indicating extremely high excitation states, and a nonstellar origin.

3. The emission features, especially of the hydrogen lines, are very wide; considered as Doppler motions, these widths correspond to velocities of 500 to 4000 km/sec. Interpreting the line widths as indicators of large-scale motions, one immediately concludes that matter is leaving the central regions of these galaxies in a time less than 10^8 years. This interpretation by Burbidge et al. (1959) was one of the first suggesting violence at the centers of otherwise normal galaxies.

How common is such activity? Of the 1500 galaxies in the Shapley-Ames catalogue, about 2–5% of them appear to be of the Seyfert type. Whether this implies that *all* spirals spend 10^8 years in an active phase, or whether only one in a hundred spends the majority of its lifetime in such a phase, is unclear.

The presence of broad emission lines in Seyfert galaxies is accompanied by strong infrared fluxes superimposed on a typical stellar continuous spec-

Figure 10-1. Wavelength distribution of energy emitted by several active galaxies, compared with that of a normal galaxy. The relative placement of the curves indicates only the decreasing order of luminosity, from top to bottom. Adapted from Weyman (1969).

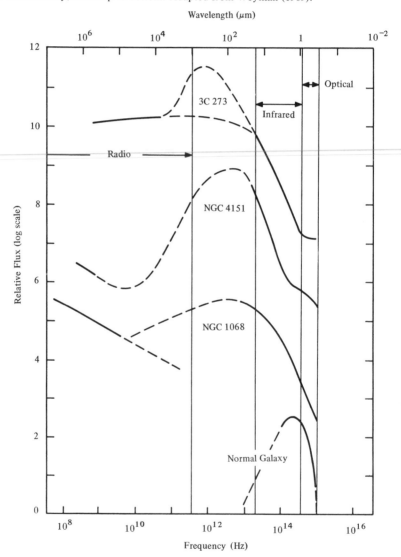

trum (Figure 10-1). There is also a strong ultraviolet excess which may be due either to an excess of bright young stars or to some nonthermal mechanism. Similarly, the infrared excess may be due to scattering of optical radiation by dust grains, or to nonthermal processes. The division of Seyfert galaxies into two classes by Weedman (1972) may imply that both types of descriptions are valid. Type 1 Seyferts show very broad emission lines (several thousand kilometers per second), intense Hβ emission, and extremely small nuclear size. The source of their activity is unclear. Type 2 Seyferts, with emission lines of 500–1000 km/sec, relatively stronger forbidden lines (compared with hydrogen lines) than for the type 1, and somewhat larger nucleus size, could well be understood as a collection of very hot, luminous stars, reddened by an excess of dust. Specific properties of several prominent Seyfert galaxies follow.

NGC 1275. One of the most interesting objects in the sky, this Seyfert is also a strong radio source lying in a rich cluster of galaxies (the Perseus cluster). It is not obviously a spiral, and in fact appears explosive and filamentary in Hα photographs. A comparison with the Crab Nebula (Figure 10-2) makes such an interpretation all the more appealing. Outside its nucleus, at distances up to 10 kpc, Doppler line widths indicating ejection velocities of 3000 km/sec are apparent. NGC 1275 shows a continuous nonthermal optical spectrum which is polarized as much as 5%; it also has ultraviolet and infrared excesses compared with a stellar black-body spectrum. It contains a complex radio source having at least three components. The longest component extends over a region of order 80 kpc, and contains a minimum of perhaps 10^{60} erg of relativistic electrons and magnetic fields. The second component, which becomes optically thick at about 800 MHz, has a size of perhaps 10–20 pc. The third component, becoming optically thick to synchrotron self-absorption at about 1-cm wavelength must be smaller than about 0.25 pc. The sources are variable on time scales of days to years.

NGC 4151. This is one of the best examples of type 1 Seyfert galaxies. Figure 10-3 shows three increasingly exposed photographs of the object. On the shortest exposure it could almost be confused with a star. The intermediate exposure shows the faint outer spiral regions. Balmer lines as broad as 7500 km/sec are found in its nucleus. The source is variable in its continuum emission but not in its emission line spectrum, indicating a continuum-emission core and a line-emission halo structure for the object.

NGC 1068. This is one of the brightest Seyferts. Burbidge and colleagues

Figure 10-2.   Hα photograph of the Seyfert galaxy NGC 1275 and a red-light photograph of the Crab Nebula, showing similar wispy structure. Although the two objects differ in mass by a factor of 10^{11}, their similar appearance suggests an explosive nature for both. Photographs courtesy of Kitt Peak National Observatory.

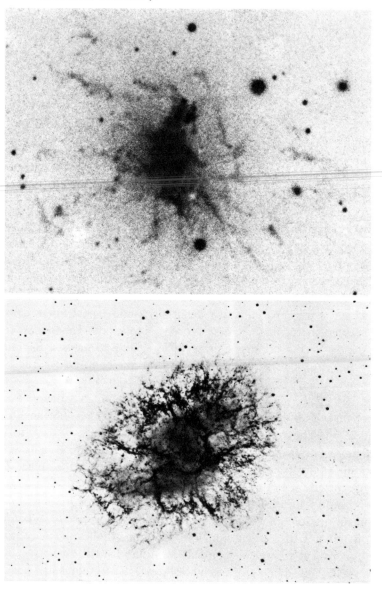

Figure 10-3. The stellar-like nucleus of the Seyfert galaxy NGC 4151 appears only in the short-exposure photograph at the top. The two successively longer exposures (middle and bottom) emphasize the outer spiral structure. From Morgan (1968). Photograph courtesy of Hale Observatories.

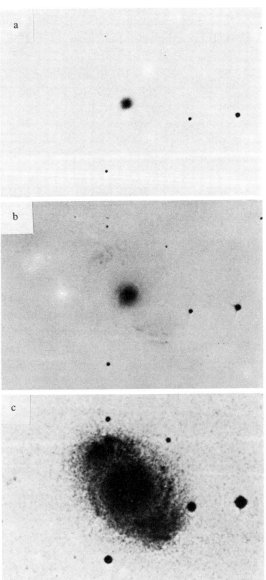

(1959) analyzed its rotation curve, finding a mass of $\sim 3 \times 10^{10} M_\odot$ within 2000 kpc of its center. The nucleus itself (with a radius of about 200 kpc) contains only about 10% of the mass, which is not enough to gravitationally confine the apparently high-velocity gas clouds present. This is the prototype of a type 2 Seyfert. Less concentrated in the center and less violent in activity than NGC 4151, nonetheless it indicates gravitational disequilibrium in its center.

3C 120. This strong radio galaxy is the brightest known Seyfert. Its infrared luminosity is at least a hundred times greater than that of the Milky Way. With emission lines of ~ 3000 km/sec in width, huge excess radio, infrared, and ultraviolet fluxes, this type 1 Seyfert may represent a kind of energetic bridge between the other Seyferts and the even more powerful quasars.

Some of the luminosities of Seyfert and other active galaxies are listed in Tables 10-2 and 10-3. A much more extensive discussion may be found in an article on the nuclei of galaxies by Burbidge (1970); see also Pacholczyk and Weymann (1968). We shall return to some conclusions about these objects after discussing several other types of active galaxies.

N, Compact, and Other Active Galaxies

Many other types of galaxies show signs of activity. Matthews, Morgan, and Schmidt (1964) observed the properties of several types of radio galaxies "having brilliant stable nuclei containing most of the luminosity of the system. A faint nebulous envelope of small visible extent is observed." This definition is similar to that applied to the Seyfert galaxies. The main difference is that the majority of N galaxies are identified with strong radio sources.

Table 10-2 Luminosities of active galaxies.

Galaxy type	Radio	Luminosity		
		Infrared	Optical	X-ray
Normal spiral	5×10^{38}	3×10^{42}	4×10^{43}	3×10^{39}
Radio galaxy	$10^{42}-10^{45}$	2×10^{42}	10^{44}	$\lesssim 3 \times 10^{41}$
N-type or Seyfert	$10^{40}-10^{45}$	3×10^{46}	5×10^{43}	$10^{42}-10^{44}$ (?)
Quasar	$10^{44}-10^{46}$	$4 \times 10^{47[a]}$	$10^{46}-10^{47}$	$10^{46[a]}$

[a] 3C 273

Table 10-3 Infrared luminosities of individual active galaxies.

Object	Type	Infrared luminosity
Sag A	Galactic nucleus	9×10^{38}
M 82	Exploding (?) spiral	8×10^{42}
Cyg A	Radio galaxy	2×10^{44}
NGC 1068	Seyfert type 2	4×10^{43}
NGC 4151	Seyfert type 1	3×10^{42}
3C 120	Seyfert type 1	7×10^{43}
3C 273	Quasar	4×10^{47}

"Compact" galaxies are a class defined by Zwicky (1964) as objects of a high surface brightness that can just be distinguished from stars on the Palomar 48-inch Schmidt telescope plates. These objects do not form a well-defined class because of their varying distances and the selection effect thereby introduced. "Intergalactic H II regions," therefore, have found their way into the lists of compact galaxies compiled by Zwicky.

The Russian astronomer Markarian (1972) has uncovered a class of galaxies which exhibit strong ultraviolet continuum emission. Two types of galaxies appear in his lists: those that are diffuse and frequently have an ultraviolet continuum as well as emission-line radiation distributed throughout the entire object, and those in which the ultraviolet continuum is emitted by the bright nucleus of a galaxy (often a spiral). Again, Seyfert galaxies often appear in the classification. The properties of those and other classes of active galaxies (e.g., Haro galaxies) are reviewed in detail in an article by van den Bergh (1975).

One interesting example of a supposedly active galaxy should be mentioned in passing. It is perhaps particularly important now because of the impetus it gave to this subject. In 1963 Lynds and Sandage suggested that the definitely peculiar-looking galaxy M 82 (Figure 10-4) was exploding. This suggestion was based primarily on two pieces of evidence. First it *looked* as if it were exploding. Spectral lines in the wispy outer regions show widths of only a few hundred kilometers per second. Lynds and Sandage suggested that the wisps were in fact moving almost perpendicular to our line of sight, and concluded that true gas ejection velocities of thousands of kilometers per second were involved. Second, the continuum radiation from the wisps was polarized. The polarization could most easily be understood, by analogy with the Crab Nebula, as arising from synchrotron radiation produced by fast electrons moving in strong magnetic fields. It appears now that this interpretation of the data is unwarranted. Morrison, Solinger, and Markert (1976) have shown that the polarization and Doppler shifts may arise simply by the scattering of optical light from the central galaxy on slowly drifting extragalactic dust. The observed line widths of a few hundred kilometers per second are then taken to be representative of the relatively normal underlying galaxy.

Quasars

The most active of all active galaxies are the quasi-stellar radio sources, or quasars. Before examining some of their properties, we should note that

Figure 10-4. Hα photograph of the galaxy M 82. While its appearance is peculiar, it may not have the explosive characteristics suggested by Lynds and Sandage (1963). Photograph courtesy of Hale Observatories.

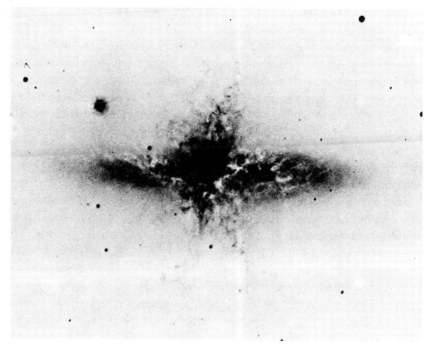

calling them "galaxies" at all is a dangerous procedure, for it prejudges one of the most exciting and controversial issues surrounding them. Schmidt (1964) proposed the following criteria for classifying an object as a quasar: 1, the object must be starlike, and preferably be identified with a radio source; 2, it must have variable optical light; 3, must have a large ultraviolet flux of radiation; 4, must have broad emission lines in the spectra, often together with narrow absorption lines; and 5, the spectral lines must exhibit a large redshift.

It is this last criterion which makes the quasars so spectacular and is itself so controversial. Interpreting the redshifts as due to the expansion of the universe puts the quasars at such great distances that their luminosities are in a class by themselves: as large as ten thousand times the brightness of the Milky Way, or 10^{48} erg/sec. Yet quasars appear starlike and are variable on time scales of the order of a year (see Figure 10-5 for a picture of a quasar, 3C 273, emitting a jet). If quasars vary coherently (as one object), then they must be smaller than a few light years across, and this means that they emit from a volume a billion times smaller than our Galaxy! From here on we will side-step the issue of the origin of quasar redshifts. Arp, G. Burbidge, and others have raised enough doubt concerning the question of redshift that it should be examined on its own. Most of the controversy centers on the statistical significance of associations between nearby galaxies and quasars, and on the reality of "bridges" connecting high-redshift quasars with low-redshift galaxies. The interested reader is referred to Field, Arp, and Bahcall (1973) for the proceedings of a debate on the question. Despite this controversy, we will proceed under the *assumption* that quasar redshifts are indeed indicators of their distance. We shall return to the point only in connection with the BL Lacerta type objects.

Many objects have been loosely called quasars even when they show no radio emission. In fact, there are now thought to be more than 10^6 "radio-quiet" QSOs (quasi-stellar objects, but not quasi-stellar radio sources, the term from which "quasar" was originally derived) down to the 19th-mag plate limit of the Palomar sky survey plates. Of these, about 200 have now been identified and have had redshifts determined for them. The ratio of radio-quiet to radio-active quasi-stellar objects is probably about 100:1. Nevertheless, the first quasar to be optically identified was the strong radio source 3C 48. In 1963, Matthews and Sandage of the Hale Observatories noticed a peculiar stellar object in the radio-position error box for the Cambridge radio source 3C 48. Another object, 3C 273, also appeared starlike, with the addition of a jetlike protuberance (see Figure 10-5), and had a peculiar spectrum. Schmidt (1963) found the spectrum of 3C 273 to be well described by a hydrogen spectrum—assuming it had a redshift $z \simeq 0.13$, corresponding to a velocity of about 36,000 km/sec. Such a high redshift implied

Figure 10-5. This long exposure of the quasar 3C 273 shows a jet of radiating material. The central object is unresolvable on shorter exposures. Photograph courtesy of Kitt Peak National Observatory.

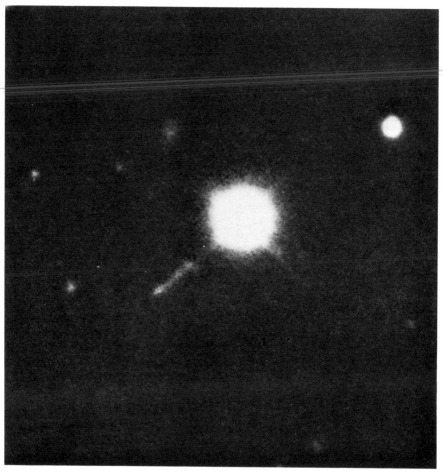

that the source lay outside our Galaxy. Within the past 10 years about 250 more such objects have been identified with redshifts as high as 3.53, which indicates a velocity of recession from us of about 0.9 times the velocity of light.

The emission line spectra of quasars imply that the line spectrum arises from a region of density $n \sim 10^5$, and temperature $T \sim 2 \times 10^4$ K. The continuum spectrum arises from some other mechanism than that producing the emission lines. This is further indicated by large (factor of 2) variations on time scales of days to years in the continuum, both at optical and radio wavelengths, but not in the lines. Furthermore, the continuum often appears polarized, implying a synchrotron origin for the continuum emission. The rapid variability of 3C 273 (and 3C 120) at radio wavelengths is shown in Figure 10-6.

In addition to the emission line spectrum, one often finds one (or more) absorption line systems. Typically, the absorption lines have narrow widths and have redshifts less than (or equal to) that of the emission lines. These observed characteristics suggest that separate regions produce the emission lines, the absorption lines, and the continuum.

At least one quasar, 3C 273, is a powerful X-ray source as well, emitting 10^{46} erg/sec at kilovolt X-ray energies. The infrared emission from 3C 273 is two orders of magnitude larger than this, making it one of the most luminous objects in the universe.

An excellent and extensive discussion of the properties of quasars appears in the book by Burbidge and Burbidge (1967). Early discussions of the properties of quasars appear in conference proceedings edited by Robinson, Schild, and Schucking (1965). More recent observational and theoretical work is reported by Evans (1972). It suffices to conclude this brief review by emphasizing the main features of these objects. Aside from frequent extended radio lobes (see section on radio galaxies, below), optical quasars are small compared with galaxies. The continuum must arise from a region that is small compared with a few light-years, yet it can produce up to 10^{48} erg/sec in luminosity.

BL Lac Type Objects

Several objects, extreme N-type galaxies, have been found in recent years which have all the characteristics of quasars, with the exception of the presence of broad emission lines. In fact, these spectra at first glance appear to be devoid of any features except a continuum with infrared and ultraviolet excesses. However, the prototype of this class, BL Lacertae, has recently been found to have a fuzzy halo about the bright central nuclear source. Oke and Gunn (1974) have managed to get a spectrum of this fuzzy region by occulting the central source. They find the region to have stellar-like emission not unlike that from a normal elliptical galaxy. Although this result has not been confirmed by other workers (Baldwin et al., 1975), if it turns out to be correct it will provide an extra argument for the theory that quasar redshifts are correct indicators of their distance. Another object of the same class, AP

Figure 10-6.   Radio variability of the Seyfert galaxy 3C 120 and the quasar 3C 273. Since 3C 273 is about 20 times farther from us than 3C 120, it is about 1000 times brighter at radio wavelengths. Adapted from Weymann (1969).

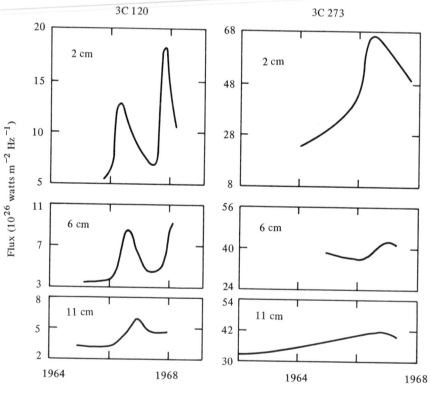

Librae, has also been found to have weak circumnuclear stellar emission lines, indicating a cosmological redshift (Disney, Peterson, and Rodgers, 1974). As possible links between quasars and elliptical galaxies, they may not only help to resolve the issues of the nature of quasar redshifts, but may also help to indicate the morphological or evolutionary connection between normal elliptical galaxies and active galaxies. In Table 10-4 we summarize some of the observed properties of active galaxies.

Radio Galaxies

Finally, we mention the classical radio galaxies. Historically, these were among the first "active" galaxies to be considered. However, unlike the above examples, the activity detected is not mainly in the nucleus of the galaxies, but outside of them. Typically, the radio emission arises from two giant radio lobes situated on opposite ends of the central galaxy. These lobes

Table 10-4 Properties of some variable compact objects.

Object	Redshift	Optical luminosity (erg/sec)	Optical variability[a]	Radio variability[b]
Seyfert galaxies				
3C 120	0.0323	2×10^{46}	30 days	1.5 yr
NGC 1068	0.00364	4×10^{44}	4 hr	740 days
NGC 1275	0.0180	4×10^{45}	700 days	
			5 days	
NGC 4151	0.00330	3×10^{44}	77 days	
			1 day	30 days
Quasars				
3C 48	0.367	8×10^{45}	1 yr	1 yr
3C 273	0.158	4×10^{47}	10 days	13 yr
3C 345	0.594	2×10^{46}	19 days	80 days
			2 hr	
BL Lac objects				
BL Lac	0.07	4×10^{45}	74 min	5 min
			23 min	
AP Lib	0.0486	$\sim 10^{45}$	10 days	?

Data from Elliot and Shapiro (1974).

[a] Minimum observed time for $\Delta m \geq 0.3$.

[b] Minimum observed time between two flux maxima.

may range from 10 to 1000 kpc in extent, and be separated by as much as 10 Mpc. Cygnus A has already been cited as an example. Recently, astronomers using the Westerbock synthesis radio array in the Netherlands have detected the largest single object in the universe to date: the double radio source 3C 236 (see Figure 10-7). One sees two extended radio lobes, 18 million light years from end to end, as well as a central source of radio emission associated with an elliptical galaxy. A good survey of the properties of such objects can be found in the compendium on radio astronomy edited by Verschuur and Kellerman (1974).

Here, we mention only that the extended radio lobes offer the possibility

Figure 10-7. Contour map of the radio components associated with the galaxy 3C 236. The linear extent of the radio emission is about 5.7 Mpc or 19 million light years, making it the largest known object. From Willis, Strom, and Wilson (1974). Reproduced from *Nature,* vol. 250, p. 625, with the permission of Macmillan Journals, Ltd.

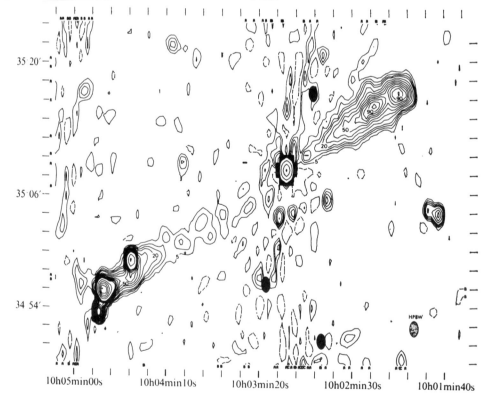

of estimating the energetics involved in violent activity of galaxies, without the difficulty associated with a full understanding of the central object. The extended radio lobes are believed to emit radio waves by incoherent synchrotron emission. Under the assumption of a uniformly filled region of constant magnetic field of strength B, electrons having the energy distribution $N(E) \propto E^{-\alpha}$, one can calculate the minimum energy required in nonthermal particles in a magnetic field to produce the radiation (see Appendix). One finds energies as large as 10^{60}–10^{61} erg. Since these giant double lobes are apparently ejected by the parent nuclei, one must consider central objects powerful enough to produce such huge fluxes of nonthermal matter. Of course, repetition of ejection more than once per Hubble time only increases the energy requirements. There is clear evidence in radio galaxies such as Centaurus A for repeated ejections on a time scale of perhaps 10^8 years or less, increasing the requisite energy output of some galactic nuclei during their lifetimes to as much as 10^{62} erg, or more.

Physical Properties of Active Galaxies

We now examine quantitatively what can be derived from the observed properties of active galaxies. This will lead to the physical characteristics of what may be termed a canonical active-galaxy nucleus, which, on the whole, will be representative of the phenomenon even though one or more features may be missing. Deducing the mass, size, energetics, composition, and gross structure of the object should lead to an understanding of the ultimate source of activity. We shall discuss the source of activity in a later section.

Mass

For the most part, the masses of quasars and other unresolvable active galaxies (e.g., N systems, BL Lac type objects) are difficult if not impossible to establish. The proximity of several quasars to clusters of galaxies could set a limit of, say, $10^{15}\ M_\odot$ on their masses, on the assumption that they do not tidally disrupt the clusters. Such a limit is not a very strong constraint on models of quasars.

More significant is the result first obtained by Burbidge and associates (1959) from the rotation curve of NGC 1068. They found an upper limit on the mass of the central 500-pc nucleus of $5 \times 10^9\ M_\odot$. As we shall see, this places severe constraints on models of Seyfert activity.

Lower mass estimates come from examination of total emission line strengths. If one sees a flux of Hβ emission $L(H\beta)$ from a region of radius R, the total luminosity is given by

$$L(H\beta) = N_e N_p f_{H\beta}(T) \frac{4\pi R^3}{3}$$

where $f_{H\beta}$ is the emission rate per particle, only weakly dependent on T. Observations of NGC 1068, for example, indicate an Hβ luminosity of $\simeq 10^{41}$ erg/sec. Inside a radius of $R \simeq 50$ pc, one finds $N_e \simeq 10^4$, or $M_{gas} \simeq 10^6\ M_\odot$. This is a minimum mass for a typical active galaxy nucleus.

Extensive discussions of mass estimates for active galaxies appear in O'Connell (1971). They center on two central ideas: dynamics and line radiation. Although results vary from object to object, one can in general say that the nuclei of galaxies have masses lying in the range 10^6–$10^{10}\ M_\odot$. Quasars may be more massive, although no obvious evidence exists one way or the other. In Table 10-5 we list estimates for the mass ejection rate from active galaxies.

Physical Scale Size

Four kinds of arguments are used to set limits on the scale size of galactic nuclei. First and most obvious is the angular resolution of the telescopes used, both optical and radio. Since Seyfert and other galactic nuclei do not have well-defined boundaries, one must be careful in assigning meaning to a size. With the few arcsec angular resolution of optical telescopes (atmosphere limited) for objects at a distance of 10 Mpc, one can resolve structures down to a distance of order 100 pc. Nuclei of Seyferts must contain active regions on scales that are smaller than this. For quasars at distances of 10^9 pc, one can conclude only that the objects are smaller than kpc sizes.

Emission and absorption line studies yield more significant details. In order for Seyfert nuclei to show both permitted and forbidden lines, more than one emitting region must be present. Typically in type 1 Seyferts, the forbidden lines can be described by emission from a region of radius 300 pc, density $n \sim 10^4$, velocity dispersion about 300 km/sec. However, as shown by Rees and Sargent (1972), the regions emitting the wings of the permitted lines (in NGC 4151, for example) must come from a region of scale size 0.1 pc, $n \simeq 10^8$, and dispersion velocity $\sim 10^8$ cm^{-3}. Ultimately, most of the continuum emission must come from a still smaller region.

The third measure of size of active galaxies comes from their variability. Table 10-4 lists the time scales of variability seen in several Seyfert galaxies, quasars, and BL Lac type objects. In both radio and optical radiation, variability on time scales from minutes to years by factors of 2 or more are observed. How are these variabilities to be interpreted? Variability on a time scale usually implies that if the object is acting *coherently* its size is smaller than R, where $R \lesssim c\tau$.

For a relativistically moving object, $R \lesssim \gamma c\tau$, where γ is the Lorentz

Table 10-5 Estimates of the masses ejected or radiated from galactic nuclei.

Source	Estimated mass loss rate ($M_\odot$/year)	Estimated time scale (years)
Galactic center		
Neutral hydrogen	1–10	10^7
Gravitational radiation	10–10^3 ?	?
Nonthermal radiation	10^{-5}	?
M 31 (outflow of gas: optical data)	1	?
Seyfert nuclei		
Nonthermal radiation	0.1	10^8
NGC 4151 (outflow of gas assumed steady)	10–1000	10^8
NGC 4151 (outflow of gas assumed sporadic)	0.6–0.006	10^8
NGC 1275 (outflow of gas)	1–10	10^8
M 82 (outflow of gas) if any	1	10^6
Strong radio galaxies		
Ejection of relativistic particles and magnetic energy, $\geq 10^{59}$–10^{61} erg	$> 10^{-1}$–10	$(10^6$–$10^9)$?
Quasars		
Nonthermal optical radiation	0.1–0.001	10^7 ?
Nonthermal infrared radiation	$(10$–$0.1)$?	10^7 ?
Ejected gas	?	?

Adapted from Burbidge (1971).

factor. Taking $\gamma \sim 1$ in general, variability of the whole light output on a time scale of minutes implies that the object is smaller than $\sim 10^{13}$ cm, which is about 1 AU. Unless this time scale represents only the time scale of flaring on a much larger object, one is forced to conclude that the ultimate objects underlying active galaxies reside in fantastically small volumes. Admittedly, we must be exceedingly careful in applying this simple argument. Variability may arise from phase velocities rather than from proper group velocities, so that the sizes of the underlying objects would not bear a one-to-one relation to the shortest time scale of variability.

Finally, the radio emission from compact objects gives a clue as to their size. If the emission is by incoherent synchrotron radiation of relativistic electrons, at a low enough frequency, the radiating electrons become optically thick to their own emission. To see this, suppose the source has a surface brightness T_b and is observed at a frequency ν to have a flux $F(\nu)$ and solid angle Ω. Then

$$F(\nu) \simeq kT_b(\nu^2/c^2)\Omega.$$

For sources radiating by the synchrotron mechanism the energy of the particles will be $E \simeq kT_b$, where

$$E = \beta^{-1/2}B^{-1/2}\nu^{1/2} = kT_b$$

(cf. equation 10-3 of the Appendix). Therefore, at low frequencies

$$F(\nu) \propto \beta^{-1/2}\nu^{5/2}.$$

One can then estimate the value of B for a compact (self-absorbing) source, to independently determine its properties. The fact that many radio sources in galactic nuclei and many quasars appear to be self-absorbed indicates that they are small.

Energetics

We have already discussed one approach to determine the energetics of active galaxies. Line emission requires extended regions of gas $\cong 10^4$ K and masses of $10^6 M_\odot$. The total thermal energy involved is perhaps $E_{th} \simeq NkT \simeq 10^{52}$ erg. Much larger amounts of energy are required to produce the motion of the gas, $E_v \simeq (\frac{1}{2})mv^2 \simeq 10^{54}$ erg. Such thermal or dynamic energies are not unusual for normal galaxies and stars. However, much larger nonthermal energy sources are required for active galaxies. In the Appendix we outline the argument that allows one to estimate the *minimum* total energy required in fast electrons and magnetic fields to explain the radio luminosity of a cloud emitting incoherent synchrotron radiation. Applying these conditions to radio galaxies like Cygnus A, G. Burbidge showed in the early 1950s that at least 10^{60} erg of fast electrons and magnetic fields are required to explain the emission from Cyg A. Other radio sources and quasars show similar extreme energetics; typical values for the energetics of individual objects are shown in Table 10-6; for a detailed discussion see Demoulin and Burbidge (1968).

Note that the same arguments, when applied to compact objects, *may* lead to logical difficulties. As first noted by Hoyle, Burbidge, and Sargent (1966), if a compact object emits synchrotron radiation from a very strong magnetic field in a compact region of space, the same electrons must scatter on the synchrotron photons which have been produced. This Compton scattering will then tend to make the electrons radiate even faster. Since the energy loss rate of electrons by Compton scattering is the same as that for synchrotron radiation, with the replacement of $B^2/8\pi$ by ρ_ν, where ρ_ν is the photon energy density, it follows that for an isotropic distribution of electrons

$$B^2/8\pi > \rho_\nu,$$

in order for the observed radiation to be predominantly synchrotron (and therefore polarized). Applying the criteria of size and energy density to quasars, combined with the above restriction, leads to a logical dilemma. Hoyle and co-workers (1966) concluded that quasars must be closer than their redshifts would indicate, thus lowering their calculated central energy density. Others have pointed out that a nonisotropic distribution of electrons reduces the above restriction. For particles with pitch angles to the field $\theta \ll 90°$, the above restriction is then

$$B^2 \sin^2\theta/8\pi > \rho_\nu.$$

This condition can be met for small enough values of θ, implying that electrons are streaming along the fields.

The conclusion is that galactic nuclei must be the sites of extremely efficient nonthermal energy production. Since regions of only $M \lesssim 10^9\,M_\odot$ must give rise to total nonthermal fluxes of $E \gtrsim 10^{60}$ erg or more, an efficiency of $\epsilon > E/Mc^2 \sim 0.1\text{–}1\%$ must be involved. It is worth noting that nuclear energy generation is just barely this efficient. Since the energy appears in a nonthermal form, we must conclude that energy production mechanisms other than those usually present in stars are involved in galactic nuclei.

Table 10-6 Typical energetic properties of optical synchrotron sources in the nuclei of galaxies and quasars.

Object	B_c (gauss)[a]	Total energy if $B = B_c$[b] (erg)	B_{eq} (gauss)[c]	Total energy if $B = B_{eq}$ (erg)	Typical electron energy (GeV)	Distance traveled by electron (cm)
NGC 1068	5.4 (−4)[d]	1.4 (54) 6.1 (55)	5.0 (−4)	5.0 (53) 1.9 (55) 4.6 (54)	270 ($B = B_{eq}$)	9 (19)
NGC 4151 (Seyfert)	2.6 (−1)	2.8 (49) 1.7 (50)	1.1 (−1)	1.6 (49) 5.9 (50) 1.4 (50)	12 ($B = B_c$)	9 (15)
3C 120 (Seyfert)	3.2 (−1)	1.1 (51) 2.7 (51)	8.8 (−2)	3.0 (50) 1.1 (52) 2.7 (51)	11 ($B = B_e$)	7 (15)
3C 390.3 (N)	1.6 (−1)	1.0 (52) 2.3 (52)	4.3 (−2)	2.5 (51) 9.3 (52) 2.2 (52)	15 ($B = B_c$)	2 (16)
3C 109 (N)	4.6 (−2)	5.2 (51) 2.6 (52)	1.7 (−2)	2.5 (51) 9.3 (52) 2.2 (52)	30 ($B = B_c$)	1 (17)
Quasar (cosmological)	1.6 (+1)	4.4 (52) 4.5 (52)	1.2 (0)	9.0 (50) 3.5 (52) 8.4 (51)	1.5 ($B = B_c$)	2 (13)

Adapted from Table 2, Demoulin and Burbidge (1968).

[a] Minimum magnetic field if Compton effect is unimportant.

[b] Upper line, $x = 1$; lower line, $x = 100$; when x_α = proton-electron cosmic ray ratio.

[c] Equipartition magnetic field.

[d] Numbers in parentheses are powers of 10 by which value is to be multiplied.

Composite Description of Active Galaxies

We are now in a position to give a simple over-all description of active galaxies. Active galaxies emerge as composite systems. At the very center is a machine which runs the show. Its nature is unclear, and will be discussed in the next section. In any case, it must be the ultimate source of the energy and the mass ejected from these systems in the form of cosmic ray electrons, magnetic fields, and hot gases.

Outside of the central region is the synchrotron-radiation emission region, having a size scale between 10^{13} and 10^{15} cm. This region is thought to be responsible for the radio emission, which has the typical power loss and polarization characteristics of such a source. The source of intense central infrared emission is less clear. It may also be nonthermal in origin, although the reprocessing of synchrotron radiation by dust cannot be ruled out. Using the energy minimization procedure, we find that these synchrotron emitting regions of scale size perhaps 10^{16} cm require 10^{50}–10^{52} erg to be present in the nuclei of Seyfert galaxies, and as much as $E \sim 10^{54}$ erg to be present in quasars, in the form of fast particles and magnetic fields. Such energies must be replenished on a time scale as short as $T \sim E/L$, where L is the source luminosity. For 3C 273, $L \simeq 10^{48}$ erg/sec, $E \simeq 10^{54}$ erg, so $T = 1$ month. Such a phenomenon must therefore be short-lived, and cannot go on in a quasar over a time scale of 10^{10} years.

Beyond this central source is a region of radio emitting, but diffuse, clouds. In some sources, such as 3C 120, outbursts each year are observed which give rise to clouds of electrons and magnetic fields of $\sim 10^{52}$ erg. These individual clouds themselves evolve according to the simple model of Van der Laan (1966).

Located farther out is the region of rapidly ejected gas. This region produces the broad emission lines observed in all but the BL Lac type systems. How this gas is ejected and heated is still unclear. Several mechanisms of excitation are possible—fast particles (protons?), ultraviolet radiation, plasma waves, or other means.

Still farther out (but still in the "nucleus") is the region which produces absorption lines. Perhaps as much as 10^{21}–10^{22} cm from the central exciting object, this region is the most remote part of the active galaxy directly affected by activity in the nucleus.

Finally, outside of the entire galaxy at distances of 0.1 to 10 Mpc, one often finds the great radio lobes. These radio emitting regions may have been ejected as a plasma cloud from the central object. They contain as much as 10^{60} erg of energy (more if allowance is made for the possible presence of cosmic ray protons). It is unclear whether these clouds are ejected as smaller blobs which then expand to fill large volumes, or are the product of ejection of systems similar to the underlying "demon" in the nucleus itself. (A third, and much less likely possibility, is that these sources are only illusory: that they are the result of interaction between the hypothetical intergalactic medium and a beam of radiation focused and made visible 10 Mpc from the central galaxy).

The composite description discussed above is illustrated in Figure 10-8. The nature of the underlying object (referred to in the figure as a spinar) is not clear. In general, the rest of the figure represents our current view of the morphology of active galaxies.

Figure 10-8. Schematic diagram of a typical strong source (not to scale). Numbered regions are as follows: 0, Spin axis: spinar with angular momentum J, mass $\approx 10^9 M_\odot$; $Q \approx 10^{-7}$ sec^{-1}. 1, Surface of spinar: poloidal field enhancement. 2, Synchrotron emission of infrared. 3, Compton-recoil emission of optical and X-ray continua. 4, Critical surface $R_{cr} \approx c/Q$. 5, Radio-burst clouds, radiofrequency emission (expanding as they move out) (Van der Laan, 1966). 6, Emission-line optical source, excited gas filaments (Osterbrock and Parker, 1966). 7, Absorption-line optical source, fast-moving cooled gas filaments (Rees, 1970). 8, Weak-field synchrotron plasma, radio frequency mainly $\lambda > 1$ m. Particle acceleration continues throughout region 2 and perhaps beyond. The general B field falls about as $1/r^2$ or $1/r^3$ (or steeper for multipoles higher than dipole) until the critical surface is reached; outside that, it is mainly toroidal, and falls as $1/r$. The whole volume is traversed by a large output of relativistic protons. From Morrison and Cavaliere (1971).

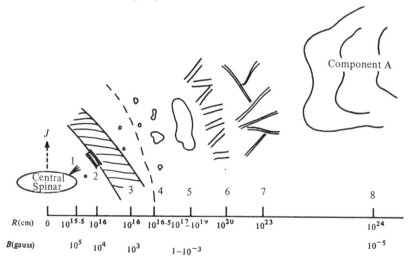

Relation of Active to Normal Galaxies

Before turning to the question of the ultimate energy sources in active galaxies, we should briefly mention their temporal evolution. Their spatial structure seems well defined, but their origin is a mystery. We must answer the following questions: Are active galaxies an evolutionary phase through which all galaxies pass? Or does activity arise because of a morphological peculiarity at birth? We do not know. It appears that the number density of early-epoch quasars at high redshift may be greater than the number density in the present epoch. This means either that activity is an *early* phase of galactic evolution, leading eventually to the formation of normal galaxies, or that those protogalaxies of the proper predisposition immediately become quasars. What predisposition? It could be, for example, that those systems which have high angular momentum per unit mass (J/M) are prevented from collapsing, and form spiral galaxies. Those of low J/M collapse more readily to become dense configurations. Such a scenario would leave unanswered the question of ongoing and *current* activity in, for example, nearby Seyfert galaxies. Perhaps Seyferts arise from some other cause. For example, if Seyfert galaxies have stellar populations which evolve fast enough at their centers, they may become dynamically unstable. Such sequential stellar coalescence could lead to recent activity.

The fact is, we are not at all certain whether activity is a phenomenon primarily associated with the birth or with death of galaxies. The answer may be a combination of the two: a kind of cosmic reincarnation. The Seyfert phenomenon probably occurs in many spiral galaxies with no other peculiar properties at their peripheries. BL Lac type objects are probably more closely related morphologically to elliptical galaxies. However, the nature of the quasar phenomenon is at present obscure. More than likely, they represent an alternative path for the evolution of galaxies, bypassing the formation of ellipticals and spirals. In this view, activity at the centers of ellipticals and spirals comes at the end point of galactic evolution and the quasar phenomenon at the beginning. In either case, similar objects are formed.

Theories of Active Galaxies

Dozens of models have been proposed for the source of activity at the centers of active galaxies and quasars. They range from the sublime to the ridiculous, and attempt to account for a variety of observed properties. A plausible theory for the phenomenon of activity should be able at least to account for the following properties:

1. Systems that radiate up to 10^{48} erg/sec containing as much as 10^{63} erg in nonthermal modes—fast particles and magnetic fields.
2. Ejection of matter, both in the form of gas and relativistic plasma. Relative to the central object, the ratio E/M of ejected energy to mass is perhaps as high as 1% or even more.
3. Small scale, indicated by variability, self-absorption, etc., of perhaps 10^{13}–10^{15} cm.
4. Also, the model should consider the origin and evolution of the source.

In the following, I briefly consider a sampling of these models. None explains everything observed; some explain nothing; some probably even contain a seed of the truth. I discuss these in a subjectively determined order of probable validity. Others in this field may not agree with my choices.

Spinars

The most developed and quantitative theory proposed to date is that of Morrison (1969) and of Ozernoi (1966). By analogy with pulsars, they propose the source of activity to be a giant (10^8–10^{10} $M_\odot$), rotating, magnetized superstar, formed by the collapse of a magnetized gas cloud. Conservation of angular momentum and magnetic flux increase the rotational energy W as $1/R^2$ until the rotational frequency Ω becomes of order $1/\sqrt{G\rho}$ for the object, where $\rho \simeq M/R^3$. At this point, the presence of magnetic fields causes the system to slowly radiate away energy and angular momentum. Whether this radiation is caused by acceleration of fast charges or by magnetic dipole emission, a huge amount of energy can be stored and then liberated in a nonthermal form. This can be seen as follows. Imagine a magnetic dipole, of scale size R_0, strength B_0, so that $B(R) = B_0(R_0/R)^3$. If one spins the dipole at angular frequency Ω, at a radius $R = c/\Omega$, field lines can no longer co-rotate with the dipole—they must be "sheared off" to rotate away. This radiation must then propagate with velocity c. Shearing off the lines at a cylinder of radius $\sim R$ then leads to the radiative flux

$$\frac{dE}{dt} \simeq \frac{B_0^2 R_0^6 \Omega^4}{c^3} .$$

For $L \simeq 10^{48}$ erg/sec, one requires a mass $M \simeq 10^9 \, M_\odot$, $B_0 \simeq 10^5 \, G$, $R \simeq 10^{17}$ cm, $T = 2\pi/\Omega \simeq 1$ year. Such a rotating system can store energy for 10^6 years, radiating very efficiently into nonthermal forms of energy. The spinar, then, is a gigantic machine converting gravitational binding energy into electromagnetic radiation by means of rotation. As the system radiates away energy and angular momentum, unlike pulsars of fixed size, it spins faster and radiates more. A good account of the model can be found in the paper by Morrison and Cavaliere (1971).

This model gained support from the observation of quasi-periodic outbursts from the QS0 3C 345 once every 320 days. These sharply peaked, periodic outbursts not only suggest a regular clock, but also suggest some coherent emission region from the object. No other object has displayed such regular activity. Nonetheless, the model is extremely attractive in accounting for the source of energy. Many questions remain. How are coherent objects or jets ejected from the central spinar? One answer might be that galactic nuclei contain several or many spinars, which are themselves occasionally ejected. Although their formation from diffuse matter is understandable, their eventual fate is unclear. Whether they eventually rotate so fast that they break up into many smaller components, or whether they eventually reach the stage of complete collapse into a Schwarzschild singularity, is unclear. Most crucial, the connection between the energy liberated and the observed spectra of radiation (i.e., power laws of the form $F(\nu) \propto \nu^{-\alpha}$) remains unclear. This, I believe, is not to be held against the model, since it is the same problem that theories of pulsars must explain. Yet the pulsar in the Crab Nebula manages, somehow, to produce a nonthermal spectrum of electrons, at a power level consistent with the energy output of the entire nebula.

Superstars

In 1963 Hoyle and Fowler suggested that the gravitational contraction of 10^6–10^8 $M_\odot$ would lead to the release of enormous energies. Collapse close to the Schwarzschild singularity can give rise to a very efficient energy release as well: 10^{60}–10^{62} erg. It is not at all clear what form the energy release can take. In particular, production of synchrotron radiating electrons and strong magnetic fields seems difficult. Thermodynamically, spinars concentrate a large amount of energy in one degree of freedom (rotation), thus offering a huge entropy source. General collapse, on the other hand, would appear to lead to a catastrophic but thermal release of energy. Furthermore, the time scale for collapse would be short compared with 10^6 years. A variant on this idea, in which nuclear burning occurs, has been discussed by Fowler (1966).

The superstar can then be stabilized temporarily against collapse, but at the expense of introducing the low efficiency of nuclear reactions. A good discussion of the entire topic appears in the article by Wagoner (1969).

Stellar Collisions

Spitzer (1971) has suggested that at the core of a galaxy is a dense cluster of stars. The collisions between stars of density 10^{11} pc^{-3} moving at $\sim 10^4$ km/sec will occur many times per year. Each such collision would release about 10^{51} erg. Besides again confronting the difficulty of accounting for the release of nonthermal forms of energy, the process will run its course in only about 10^2–10^3 years, implying a very short lifetime for the active phase of quasars. This appears to be inconsistent with the total numbers of such objects observed at present.

Multiple Supernovae

Shklovsky (1961) has suggested that the rate of supernova events in galactic nuclei is higher than elsewhere in a galaxy. With a total energy release per event of 10^{51} erg, this notion requires 10^4 supernovae per year to account for the strongest sources. Although this again leads to a short lifetime for the active phase of galactic nuclei, it has the advantage that we do see evidence for synchrotron radiation in supernovae. Thus, however it occurs, one expects the process to lead to nonthermal emission. However, for this model, as well as other random large-event models, one might expect evidence of great variability (by orders of magnitude in days) in active galaxies. Such large variations occur rarely, if at all.

Galactic Flares

Sturrock (1966) and others have suggested that, by analogy with the sun, flares on superstars can accelerate fast particles. We do not understand how flares on the sun accelerate charged particles, but we do know that relative to the entire luminosity of the sun, they are produced with very low efficiency. Thus unless flares contribute an enormously important part of superstar radiation, it would seem premature to examine them in detail; in any case, we understand essentially nothing about the solar flares on which the model is based. Nonetheless, acceleration of fast particles by magnetic-field line reconnection or destruction is an attractive possibility.

Potpourri of Other Theories

Most other theories attack the problem of the ultimate energy source, with little regard for the form the energy takes. Ambartsumian (1958) has suggested that quasars represent the creation of galaxies—white holes. This is just the time-reversed version of black holes, suggested by Lynden-Bell (1969) and by many others as an energy source. Accretion of matter onto a giant (10^8 $M_\odot$) black hole should lead to the release of gravitational binding energy. Matter-antimatter annihilation turns matter into energy with 100% efficiency. But what separated the matter to begin with? And where are the expected resulting γ-rays? If massive quarks exist ($M \gtrsim 5$ GeV), then binding them with less massive hadrons will yield energy with great efficiency. Again the origin of the quarks and their produced spectrum of radiation are both unsolved mysteries. Models involving galaxy formation, intergalactic accretion, clusters of pulsars, and gravitational lensing have all been proposed: the ideas go on and on. See Burbidge and Burbidge (1967) and O'Connell (1971) for a more detailed elaboration of these ideas and others.

Summary

To summarize this brief review of active galaxies, we should emphasize first the main observational conclusion. There exist galaxies showing gravitational disequilibrium (explosions, jets, and mass ejection) on time scales short compared with their dynamical time of 10^8 years. Strong evidence exists for nonthermal radiation and therefore, by inference, for cosmic-ray electrons and strong magnetic fields in these objects. Total energies of 10^{62} erg or more somehow produced from masses of 10^8–10^9 $M_\odot$ seem to be general features of active galaxies. How the energy is released, where it is stored during the normal lifetime of the galaxy, and how it is converted into the constituents which are observed, all are questions which remain to be answered.

Appendix

The minimum energy in magnetic fields and fast electrons necessary to produce an observed synchrotron radiation luminosity L at frequency ν may be estimated as follows (for optically thin incoherent emission). The total luminosity produced by a spectrum of cosmic ray electrons $N(E)$ is

$$L = \int N(E)(dE/dt)dE, \tag{10-1}$$

where

$$dE/dt = \alpha B^2 E^2 (\text{erg/sec}) \tag{10-2}$$

is the total power radiated by a single electron ($\alpha \simeq 6.08 \times 10^{-9}$ for B in gauss, E in GeV). The power is radiated predominantly at frequency

$$\nu \simeq \beta B_L E^2 (\text{GHz}) \tag{10-3}$$

($\beta \simeq 1.6 \times 10^4$). For simplicity, let us take a delta-function distribution of electrons, $N(E) = N_e \delta(E - E_e)$. Therefore, the total number of radiating electrons is $\int N(E)dE = N_e$, and their total energy is

$$W_e = \int N(E)EdE = NE_e. \tag{10-4}$$

The total radiated luminosity is therefore

$$L = \alpha B^2 N E^2 = \alpha \nu^{1/2} \beta^{-1/2} B^{3/2} W_e, \tag{10-5}$$

or, solving for W_e,

$$W_e = L\beta^{1/2}\alpha^{-1}\nu^{-1/2}B^{-3/2} \equiv C_e B^{-3/2} L. \tag{10-6}$$

The total magnetic energy W_B is

$$W_B = \int (B^2/8\pi)d(\text{vol}) = C_B B^2 V, \tag{10-7}$$

where V is the source volume. The minimum total energy which will give rise to a radiated luminosity L at frequency ν is found by finding the minimum of $W_{\text{total}} = W_e + W_B$ with respect to variations in B. From $dW/dB = 0$ one finds that

$$B_{\text{min}} = \left(\frac{3}{4}\frac{C_e}{C_B}\frac{L}{V}\right)^{2/7}, \tag{10-8}$$

so that $W_{e,\text{min}} = (4/3)W_{B,\text{min}}$. Thus the minimum total energy is also roughly given by the sum of the equipartition values for W_e and W_B. For a power law distribution of radiating particles, $N(E) \propto E^{-\alpha}$, the same result follows.

References

Alfvén, H., and Herlofson, N.   1950. Cosmic radiation and radio stars, *Phys. Rev. 78*, p. 66.

Ambartsumian, V. A.   1958. On the evolution of galaxies, in R. Stoops, ed., *La Structure et l'Evolution de l'Universe,* pp. 241–280, Institut Solvay, Bruxelles.

Baldwin, J. A., Burbidge, E. M., Robinson, L. B., and Wampler, E. J.   1975. The nature of BL Lacertae, *Astrophys. J. (Letters) 195*, L55–59.

Burbidge, E. M., Burbidge, G. R., and Prendergast, K. H.   1959. Mass distribution and physical conditions in the inner region of NGC 1068, *Astrophys. J. 130*, 26–37.

Burbidge, G. R.   1970. The nuclei of galaxies, *Annu. Rev. Astron. Astrophys. 8*, 369–460.

Burbidge, G. R.   1971. Theoretical considerations regarding non-thermal emission and ejection of matter from galactic nuclei, in D. J. K. O'Connell, ed., *Nuclei of Galaxies,* pp. 411–433, American Elsevier, New York.

Burbidge, G. R., and Burbidge, E. M.   1967. *Quasi Stellar Objects,* Freeman, San Francisco.

Burbidge, G. R., Burbidge, E. M., and Sandage, A. R.   1963. Evidence for the occurrence of violent events in the nuclei of galaxies, *Rev. Mod. Phys. 35*, 947–972.

Demoulin, M.-H., and Burbidge, G. R.   1968. Non-thermal optical radiation from galaxies, *Astrophys. J. 154*, 3–20.

Disney, M. J., Peterson, B. A., and Rodgers, A. W.   1974. The redshift and composite nature of AP Librae (PKS 1514–24), *Astrophys. J. (Letters) 194*, L79–82.

Elliot, J. L., and Shapiro, S. L.   1974. On the variability of the compact nonthermal sources, *Astrophys. J. (Letters) 192*, L3–6.

Evans, S. E., ed.   1972. *External Galaxies and Quasi-Stellar Objects.* D. Reidel, Dordrecht-Holland.

Field, G. B., Arp, H., and Bahcall, J. N.   1973. *The Redshift Controversy,* W. A. Benjamin, Reading, Mass.

Fowler, W. A.   1966. Supermassive stars, quasars, and extragalactic radio sources, in L. Gratton, ed., *High Energy Astrophysics,* pp. 316–366, Academic Press, New York.

Hoyle, F., and Fowler, W. A.   1963. On the nature of strong radio sources, *Monthly Not. Roy. Astron. Soc. 125*, 169–176.

Hoyle, F., Burbidge, G. R., and Sargent, W. L. W.   1966. On the nature of quasi-stellar sources, *Nature 209*, 751–753.

Lynden-Bell, D.   1969. Galactic nuclei as collapsed old quasars, *Nature 223*, 690–694.

Lynds, C. R., and Sandage, A. R.   1963. Evidence for an explosion in the center of the galaxy M 82, *Astrophys. J. 137*, 1005–1021.

Markarian, B. Y. 1972. On the nature of galaxies with ultraviolet continuum. 1. Principal spectral and colour characteristics. *Astrofizika 8,* 165–176.

Matthews, T. A., and Sandage, A. R. 1963. Optical identification of 3C 48, 3C 196, and 3C 286 with stellar objects, *Astrophys. J. 138,* 30–56.

Matthews, T. A., Morgan, W. W., and Schmidt, M. 1964. A discussion of galaxies identified with radio sources, *Astrophys. J. 140,* 35–49.

Morgan, W. W. 1968. A comparison of the optical forms of certain Seyfert galaxies with the N-type radio galaxies, *Astrophys. J. 153,* 27–30.

Morrison, P. 1969. Are quasi-stellar radio sources giant pulsars? *Astrophys. J. (Letters) 157,* L73–79.

Morrison, P., and Cavaliere, A. 1971. Spinars—a progress report, in D. J. K. O'Connell, ed., *Nuclei of Galaxies,* 485–509, American Elsevier, New York.

Morrison, P., Solinger, A. B., and Markert, T. 1976. M 82 without explosions, *Astrophys. J.,* submitted for publication.

O'Connell, D. J. K., ed. 1971. *Nuclei of Galaxies,* American Elsevier, New York.

Oke, J. B., and Gunn, J. E. 1974. The distance of BL Lacertae, *Astrophys. J. (Letters) 189,* L5–8.

Osterbrock, D. E., and Parker, R. A. R. 1966. Excitation of the optical emission lines in quasi-stellar radio sources, *Astrophys. J. 143,* 268–270.

Ozernoy, L. M. 1966. A theory for the formation and structure of quasistellar radio sources, *Soviet Astron.-AJ 10,* 241–249.

Pacholczyk, A. G., and Weymann, R. J. 1968. *Seyfert Galaxies and Related Objects,* in Proceedings of the Conference on Seyfert Galaxies and Related Objects, *Astron. J. 73,* 836–918.

Rees, M. J. 1970. On multiple absorption redshifts in quasi-stellar objects, *Astrophys. J. (Letters) 160,* L29–32.

Rees, M. J., and Sargent, W. L. W. 1972. The structure of Seyfert nuclei, *Comments Astrophys. Space Sci. 4,* 7–14.

Robinson, I., Schild, A., and Schucking, E. L., eds. 1965. *Quasi-Stellar Sources and Gravitational Collapse,* University of Chicago Press, Chicago.

Schmidt, M. 1963. 3C 273: A star-like object with a large red-shift. *Nature 197,* 1040.

Schmidt, M. 1964. Models of quasi-stellar sources, in K. N. Douglas, I. Robinson, A. Schild, E. L. Schucking, J. A. Wheeler, and N. J. Woolf, eds., *Quasars and High Energy Astronomy,* Proceedings of the Second Texas Symposium on Relativistic Astrophysics, pp. 55–59, Gordon and Breach, New York.

Seyfert, C. K. 1943. Nuclear emission in spiral nebulae, *Astrophys. J. 97,* 28–40.

Shklovsky, I. S. 1961. Radio galaxies, *Soviet Astron.-AJ 4,* 885–896.

Spitzer, L. 1971. Dynamical evolution of dense spherical star systems, in D. J. K. O'Connell, ed., *Nuclei of Galaxies,* pp. 443–471, American Elsevier, New York.

Sturrock, P. A. 1966. A model of quasi-stellar radio sources, *Nature 211,* 697–700.

Van den Bergh, S. 1975. The classification of active galaxies, *Roy. Astron. Soc. Canada 69,* 105–125.

Van der Laan, H.   1966. A model for variable extragalactic radio sources, *Nature 211*, 1131–1133.

Verschuur, C. L., and Kellerman, K. I., eds.   1974. *Galactic and Extragalactic Radio Astronomy*, Springer-Verlag, Berlin.

Wagoner, R. V.   1969. Physics of massive objects, *Annu. Rev. Astron. Astrophys. 7*, 553–576.

Weedman, D. W.   1972. Emission-line intensities and UBV magnitudes for twenty-three Markarian galaxies, *Astrophys. J. 171*, 5–12.

Weymann, R. J.   1969. Seyfert galaxies, *Sci. Amer. 220 (no. 1)*, 28–37.

Willis, A. G., Strom, R. G., and Wilson, A. S.   1974. 3C 236, DA 240: The largest radio sources known, *Nature 250*, 625–630.

Zwicky, F.   1964. Compact galaxies and compact parts of galaxies. I. *Astrophys. J. 140*, 1467–1471.

11

Galaxies and Cosmology

Marc Davis

Galaxies play a central role in the large-scale structure of the universe; any discussion of cosmology therefore should be prefaced by a discussion of the observed properties of galaxies. This chapter is divided into two parts: the first describes some characteristics of galaxies and galaxy clusters, and develops a statistical description of the clustering phenomenom. The second deals with cosmology. In our discussion of cosmology we describe the nature of the Robertson-Walker-Friedmann universe, which has become the standard cosmological model. Then we concentrate on whether the universe is closed or open, and discuss the classical tests of cosmology and some of the difficulties with these tests. Finally, we consider discussions of the primeval fireball radiation, the mass density of the universe and the stability of galaxy clusters, and the question of the deuterium abundance and its interpretation.

Cosmology has long been a field with an excess of speculation and a shortage of facts. Today we have more facts, but we are still a long way from answering the central question: Is the universe open, or is it closed!

Galaxies

Taxonomy, or classification, helps to organize a field of study. In extragalactic astronomy, a sizable lexicon has evolved to classify galaxies, which are the most apparent observable components of the extragalactic universe and which have a wide variety of shapes, sizes, and luminosities. Further details are given by de Vaucouleurs (1959) and by Sandage (1961).

To classify galaxies by shape (i.e., morphologically), Hubble proposed the terms: elliptical, spiral, and irregular, forming a sort of evolutionary sequence in which ellipticals are "early" type, and spirals and irregulars are "late" types, in a scheme similar to the Hertzsprung-Russell stellar diagram. Since it is now believed that all galaxies formed at the same time, "evolution" refers to the development of structure and to the reduction of symmetry. Figure 11-1 shows some galaxies considered to be representative of their type.

The elliptical galaxies, which account for about 10% of all galaxies, are elliptical in shape and show a total absence of spiral structure. They are predominantly red in color, usually contain no trace of dust, and show no evidence of ongoing star formation. The luminosity profile of an elliptical closely follows a relationship of the form

$$I(r) = I_0 \left/ \left(1 + \frac{r}{a} \right)^2 \right., \tag{11-1}$$

where I_0 is the central surface brightness, r is the radius, and a is a scale length, which is typically less than one kpc. The surface brightness profiles of several elliptical galaxies have been traced to radii in excess of 100 kpc, but at some point the brightness profile must become steeper than equation (11-1) predicts if the total luminosity of the galaxy is to remain finite.

The spiral galaxies account for more than 50% of the galaxies observed to a distance of 100 Mpc and brighter than 14.5 mag. The spirals are further subdivided into barred and normal spirals. The spiral arms of barred spiral galaxies originate at the ends of a luminous straight bar, which projects symmetrically through the nucleus. Spiral galaxies do have extensive dust lanes in the plane of rotation, and the spiral arms are blue from the light of the young supergiant stars that are always found near gas and dust clouds. The spiral nuclei are red, however. The size and luminosity of the nucleus of spirals vary widely, and there is evidence for nonthermal emission and recent explosive activity in some very bright nuclei. Galaxies with this property are termed Seyfert, or N-type galaxies, as discussed by Brecher in Chapter 10. The luminosity brightness profile of spirals follows a power law similar to equation (11-1) in the nuclear bulge, but behaves exponentially farther out, both along and normal to the disk. This behavior is also characteristic of SO galaxies.

Lenticular, or SO galaxies, were later added to the classification scheme by Hubble to bridge the gap between ellipticals and spirals, and approximately 20% of the galaxies we see are of this type. SO galaxies have sym-

Figure 11-1.   Representative nearby galaxies. (a) NGC 4278, an E 1 elliptical; (b) NGC 5866, an E7/SO galaxy; (c) NGC 1300, a barrel spiral (SBb); and (d) NGC 5457 (M 101), an Sc spiral galaxy. (From Sandage, 1961.)

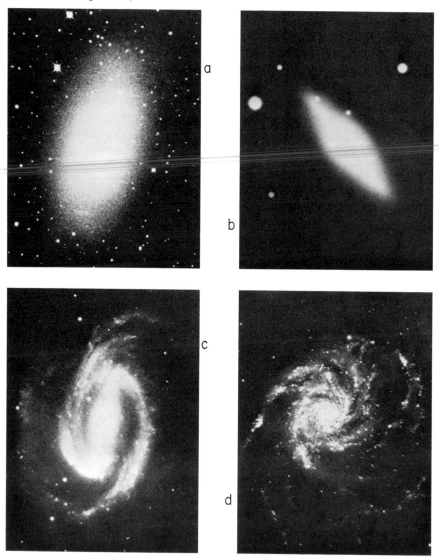

metrical forms more elongated than ellipticals, and they have no spiral structure or trace of bars. They show a bright nucleus, a central lens surrounded by a faint and sometimes extensive envelope, and occasionally circular absorption lanes. An interesting point is that if we removed the dust, spiral structure, and luminous young blue stars that are characteristic of spiral galaxies, the luminosity profile of SO galaxies would result.

Irregular galaxies show no recognizable structure, and have no rotational symmetry. They are relatively rare, and usually are intrinsically faint and small. This in turn implies that they cannot be seen to great distances.

The classifications given above are each further subdivided into numerous subclassifications. Of course, since galaxies are large macroscopic systems, it is unrealistic to expect to place them neatly into clearly distinct bins, when in fact the distribution function of shapes of galaxies is surely continuous. Morphological classification of a galaxy near a bin division is a matter of personal judgment. The numerous galaxies which have some characteristics of several different bins are often termed "peculiar."

Advancing through the morphological sequence, the galaxies tend to become bluer and flatter. The ratio of major to minor axis for ellipticals is always less than 3:1, while for spirals it is always greater than 3:1. The standard hypothesis is that the flatter systems possess more angular momentum per unit mass. This seems reasonable, but is unproved.

The distribution of masses of elliptical galaxies extends from 10^{12} to $10^5 \, M_\odot$, and thus includes objects the size of globular clusters. Many low-mass galaxies, often termed dwarf ellipticals or dwarf irregulars, can be seen in nearby groups of galaxies. For example, 7 out of 17 galaxies in the Local Group of galaxies are of dwarf elliptical type (de Vaucouleurs, 1975). Dwarf galaxies are frequently associated with larger galaxies.

In addition to a distribution in shape, galaxies have a broad distribution in luminosity, usually called $\Phi(M)$ and measured in units of number per Mpc^3 per magnitude interval. Figure 11-2 shows a recent measurement of $\Phi(M)$ for galaxies within 20 Mpc of the Milky Way. This distribution agrees quite well with that obtained from distant clusters of galaxies. There is no evidence that ellipticals have a luminosity distribution function different in shape from that of spirals, except possibly at the bright end of the distribution, where the statistics are poor because the objects are so rare. This could be a coincidence, but it may be a point of fundamental significance. The shape of the distribution function at the bright end is an important parameter for the cosmological tests which are discussed below. The faint end of the luminosity distribution function is also poorly known because the objects are difficult to observe at large distances.

Since the blue stars found in late-type galaxies are more efficient radiators than evolved red stars, the mass-to-light ratio should be distinctly larger in elliptical galaxies than in spirals. The measurement of galactic masses is difficult, but the observations described below do indicate mass-to-light ratios of at least 50 for elliptical galaxies, and 10 for spiral galaxies. Conceivably, however, spiral galaxies contain large amounts of nonluminous matter (e.g., meteoritic material, perhaps Jupiter-sized bodies, or small dwarf stars), which would defy detection and could bring the mass-to-light ratio to a value closer to the elliptical ratio.

Clusters of Galaxies

All self-gravitating systems have a tendency to form clumps, or density concentrations, and this clustering phenomenon is observed on all scale sizes. Witness the existence of planets, stars, star clusters and gas clouds, galaxies, clusters of galaxies, and clusters of clusters of galaxies (superclusters).

Galaxy clustering shows a continuous hierarchy of structure. Most gal-

Figure 11-2. The luminosity function of nearby galaxies. The upper curve represents the function for spiral and irregular galaxies, while the lower curve displays the results for elliptical galaxies. $\Phi(M)$ is given in units of number per cubic megaparsec per magnitude interval. Apart from a normalization factor the two curves are statistically indistinguishable. (From Shapiro, 1971.)

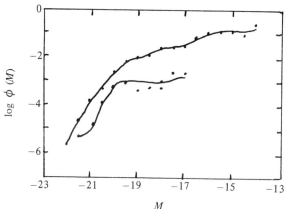

axies occur in pairs or in small multiple groups, which in turn are often clustered around larger groups. Larger clusters are themselves clustered with other large clusters. For example, most small groups of galaxies near our Local Group are in the direction of the large Virgo cluster, $10\,h^{-1}$ Mpc away (where $h \sim \frac{1}{2}$) (see equation 11-27), suggesting that our own Local Group could be on the edge of a supercluster of galaxies.

Most galaxies seem to be associated with clusters of some type, and several classification schemes have been developed to categorize these various associations, based on such parameters as the number of galaxies in the association and the density contrast with the surrounding region. Most galaxy associations are small and have a low density contrast. These are the "poor" clusters. Some clusters, however, are very rich and contain thousands of member galaxies at a much higher space density than average. This large aggregation of galaxies ensures that some members will be drawn from the bright end of the luminosity distribution function, and therefore will be visible to great distances. For this reason, rich clusters are important for cosmological measurements.

There is actually a continuum of cluster sizes, just as there is a continuum of galaxy sizes. There is no preferred length scale in a gravitational field, and therefore categorizing galaxy associations as either "poor" or "rich" is somewhat arbitrary. Similarly, deciding how to divide the galaxies in a given region into associations is a subjective judgment for all but the very rich, compact clusters. Decide for yourself how to associate the galaxies in Figure 11-3.

Galaxy Clustering and Correlation Analysis

An objective, statistical description of galaxy clustering, recently expounded by Peebles (1975), utilizes two-particle correlation functions. Instead of examining galaxy clusters on an individual basis, correlation analysis provides measures of the average properties of all clusters and is therefore complementary to the more traditional study of individual groups of galaxies. We will summarize some of the major results obtained in several important correlation studies carried out recently. First, however, we provide some mathematical background on correlation analysis.

Consider a region R of space and a statistical ensemble of galaxy distributions in R. Our universe is one realization from this ensemble. In another realization of the ensemble, the galaxies would be distributed differently, but presumably with the same statistical characteristics. If n is the mean number density, then the infinitesimal probability dP of finding a galaxy (considered as a point particle) in a volume element dV is

$$dP = ndV. \qquad (11\text{-}2)$$

The probability of finding more than one particle in the volume element dV is an infinitesimal of higher order, and we ignore it. If the ensemble is homogeneous, n is independent of position. Of course, n will not be independent of position within any realization of the ensemble. Our statistical statements

Figure 11-3. Distribution of galaxies brighter than 15 mag in the northern galactic hemisphere with declination $\delta > 0$. The concentric circles are at galactic latitudes $b_{\mathrm{II}} = 0, 30,$ and 60. The empty region in the lower right corresponds to $\delta < 0$. Note the decreasing density of points for low galactic latitudes, where galactic absorption becomes important. (From Peebles, unpublished.)

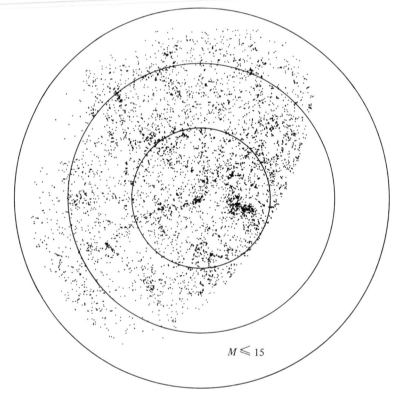

apply strictly to the ensemble of galaxy distributions, and only when averaged over a large region can we expect our universe to approximate the ensemble.

Consider, next, two distinct volume elements dV_1 and dV_2 at positions r_1 and r_2. We define the infinitesimal joint probability of finding a particle in each element as

$$dP \equiv n^2[1 + \xi(r_1,r_2)]dV_1 dV_2 \qquad (11\text{-}3)$$

where $\xi(r_1, r_2)$ is the two-particle correlation function. For a homogeneous, isotropic ensemble ξ is a function of $r = |\mathbf{r}_1 - \mathbf{r}_2|$; ξ is a dimensionless function in the range $-1 \leq \xi \leq \infty$. If $\xi = 0$, the two probabilities are independent of each other, and equation (11-3) describes a random Poisson distribution of points. A value $\xi(r)$ greater than zero implies clustering, while $\xi(r) < 0$ implies anticlustering on a scale r. Given one particle, $n\xi(r) \, dV$ is the probability in excess of the random probability ndV that a volume element dV located a distance r from the first particle will contain another particle. Thus an alternative but equivalent statement of equation (11-3) is to write the probability that a particle will be found in a volume element dV a distance r from a given particle:

$$dP = n[1 + \xi(r)]dV. \qquad (11\text{-}4)$$

For the correlation function to be useful in a statistical sense, $\xi(r)$ must approach 0 for a distance much smaller than the size of the universe. This condition is well satisfied since there is no evidence for galaxy clustering on sizes greater than 100 Mpc; as we discuss later, the size of the universe is of order 6000 Mpc.

The description of clustering given above applies to point particle distributions, but it can be very simply generalized for density distributions in a continuous fluid having a density $\rho(\mathbf{x})$. We can denote the average density as

$$n = \langle \rho(\mathbf{x}) \rangle$$

where $\langle \, \rangle$ denotes a spatial average value.

We can define a quantity $\xi_\rho(\tau)$ by

$$\begin{aligned} \xi_\rho(\tau) &= \langle [\rho(\mathbf{x} + \tau) - n][\rho(\mathbf{x}) - n] \rangle / n^2 \\ &= [\langle \rho(\mathbf{x} + \tau)\rho(\mathbf{x}) \rangle - n^2]/n^2 \end{aligned} \qquad (11\text{-}5)$$

and we have

$$\langle \rho(\mathbf{x} + \tau)\rho(\mathbf{x}) \rangle = n^2[1 + \xi_\rho(\tau)], \qquad (11\text{-}6)$$

in close analogy with equation (11-3). The function $\xi_\rho(r)$ is sometimes called the "normalized lagged product" and is completely analogous to the function $\xi(r)$ of the point particle description. We will drop the subscript ρ in the discussions below when using the continuous fluid description of clustering.

A few simple models of galaxy clustering will demonstrate possible shapes of the correlation function. If we assume that all galaxies are in spherical clusters, each of diameter D, and that the clusters are themselves distributed randomly in space, then $\xi(r)$ would have the functional form of Figure 11-4. To see this, consider choosing a random galaxy and then employ equation (11-4). The random galaxy is in a cluster, so for distances $r < D$, there is a greater than random chance of finding another galaxy in an infinitesimal volume element dV, and therefore $\xi(r) > 0$. For $r > D$, the second galaxy would be in a different cluster, which is randomly distributed with respect to the cluster of the first galaxy; therefore $\xi = 0$. The shape of the shoulder of ξ in the vicinity $r = D$ is dependent on the assumed density profile within the clusters.

Figure 11-4.   Expected correlation function if all objects are in clusters of size D. The shape of the edge at $r = D$ will depend on the density profile of objects within the cluster.

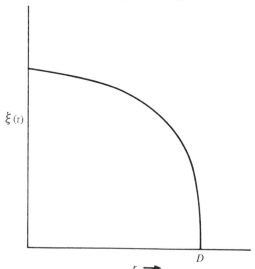

Another recipe for galaxy clustering is the following: Start with galaxies positioned randomly, unclustered throughout the universe, and then partition space into simply connected cells of size R_2. Draw a fraction f of the galaxies in each cell into a cluster of size R_1, and position the cluster near the center of the cell. The remaining fraction $(1 - f)$ of galaxies remains randomly distributed throughout the cell.

If there are N galaxies in a cell, the density remaining outside the cluster is

$$n_u \simeq (N/R_2{}^3)(1 - f)$$

while the density within the cluster will have average density $n_c = (N/R_1{}^3)f$. If the density in the center is constant, we can perform the integration of equation (11-6). The average density $n = N/R_2{}^3$. For $\tau = 0$ the integral on the left side of equation (11-6) consists of two parts:

$$n^2[1 + \xi(0)] \simeq (1/R_2{}^3)\int d^3r\rho^2(r)$$
$$\simeq (1/R_2{}^3)(n_c{}^2R_1{}^3 + n_u{}^2R_2{}^3)$$

and

$$\frac{N^2}{R_2{}^6}[1 + \xi(0)] \simeq \frac{N^2}{R_2{}^3}\left[\left(\frac{f}{R_1{}^3}\right)^2 R_1{}^3 + \left(\frac{1-f}{R_2{}^3}\right)^2 R_2{}^3\right].$$

Hence,

$$\xi(0) \simeq f^2[1 + (R_2/R_1)^3] - 2f.$$

For the region $R_1 <\cdot \tau < R_2$ equation (11-6) becomes approximately

$$n^2[1 + \xi(\tau)] \simeq (1/R_2{}^3)(2n_c n_u R_1{}^3 + n_u{}^2R_2{}^3),$$

so that

$$1 + \xi(\tau) \simeq 2f(1 - f) + (1 - f)^2,$$

and

$$\xi(\tau) \simeq -f^2. \tag{11-7}$$

For $\tau > R_2$, unless the cells are arranged in a periodic lattice, $\xi(\tau) = 0$. This recipe for galaxy clustering thus incoporates an anticorrelation in the region $R_1 < \tau < R_2$. However, as we shall see below, $\xi(\tau)$ is observed to be non-negative, which apparently implies that this simple picture does not describe how clusters were formed.

A final recipe is to compute $\xi(\tau)$ in a model where the density of galaxies within a cluster varies as $r^{-\epsilon}$, where r is the radius from the center of the cluster, and ϵ is a power law constant. Rich clusters of galaxies and globular star clusters are both observed to have $\epsilon \simeq 2$ at sufficient distance from the center (which is the prediction of an isothermal gas sphere). To find $\xi(\tau)$ we will again approximate the point particle distribution by a continuous fluid and use equation (11-6):

$$n^2[1 + \xi(\tau)] = \langle \rho(\mathbf{r})\rho(\mathbf{r} + \boldsymbol{\tau}) \rangle$$
$$\propto \int d^3r \rho(\mathbf{r})\rho(\mathbf{r} + \boldsymbol{\tau})$$
$$\propto \int_{-1}^{1} d \cos \theta \int_{0}^{\infty} dr\, r^{2-\epsilon}(r^2 + \tau^2 - 2r\tau \cos \theta)^{-\epsilon/2}.$$

How does this vary with τ? By defining a new variable $y = r/\tau$, we can convert the integral into a dimensionless number dependent only on ϵ, and factor out the τ dependence. For $\xi \gg 1$ we have

$$\xi(\tau) \propto \tau^{3-2\epsilon}I(\epsilon), \tag{11-8}$$

where

$$I(\epsilon) = \int_{-1}^{1} d \cos \theta \int_{0}^{\infty} dy\, y^{2-\epsilon}(y^2 + 1 - 2y \cos \theta)^{-\epsilon/2}.$$

Thus if all galaxies are in clusters with density varying as $r^{-\epsilon}$, $\xi(r)$ will vary as $r^{3-2\epsilon}$.

Recently a considerable effort has been made by Peebles and coworkers to measure $\xi(r)$ and to interpret the results in terms of models of galaxy and cluster formation; see Peebles (1975) and references therein. However, $\xi(r)$ cannot be measured directly by observing the distribution of galaxies on the sky, which provides only two-dimensional information.

What can be measured directly from a catalog of galaxy positions is $w(\theta)$, the two-particle, angular correlation function. If we know something about the distribution of luminosities of galaxies, we can relate $w(\theta)$ to $\xi(r)$. This analysis is detailed below.

If there are $\mathcal{N}$ galaxies per steradian in the sky, then the probability that an object will be found in an infinitesimal solid angle $d\Omega$, is

$$dP = \mathcal{N}d\Omega. \tag{11-9}$$

The joint probability that one object will be found in $d\Omega_1$, and another in

$d\Omega_2$, separated from the first by an angle θ, is defined as

$$dP \equiv \mathcal{N}^2 d\Omega_1 d\Omega_2[1 + w(\theta)]. \tag{11-10}$$

To derive $w(\theta)$ in terms of $\xi(r)$, consider also that the differential luminosity function $\Phi(M)$ is known (see Figure 11-2) and can be used to supply some distance information. The probability dP that a galaxy having an absolute magnitude between M and $M + dM$ will be found in a random volume dV_1 is

$$dP = \Phi(M)dM \, dV_1. \tag{11-11}$$

The joint probability of finding one galaxy in a volume element dV_1, absolute magnitude interval $M_1, M_1 + dM_1$, and another galaxy at a distance r_{12} from the first in a volume element dV_2 and magnitude interval $M_2, M_2 + dM_2$ is similarly defined as

$$dP = [\Phi(M_1)\Phi(M_2) + G(r_{12},M_1,M_2)]dV_1dV_2dM_1dM_2, \tag{11-12}$$

where $G(r_{12}, M_1, M_2)$ is the correlation function, analogous to $\xi(r)$ but now includes information about the luminosities of the correlated pairs of galaxies.

Of course, in the construction of a catalog of galaxies, there are numerous biases introduced by observer selection, the chief one being the decision that a particular galaxy is above or below a particular apparent-magnitude threshold. We can define an observer selection function $f(m - m_c)$ as the probability that a galaxy of magnitude m will be included in a particular catalog of galaxies brighter than the magnitude cutoff m_c.

The solid curve in Figure 11-5 shows the expected shape of f. Approximating f by the dotted line will be adequate for computations. Well below m_c, all galaxies will be included and $f = 1$, but near $m = m_c$, errors in magnitude estimates and the faintness of the image on the photographic plates will cause the observer to fail to count some galaxies below the threshold, and to include some from above the threshold.

Place the observer at the position $r = 0$, and consider the probability dP_1 that a galaxy of any magnitude is at radius r to $r + dr$ in a solid angle $d\Omega$, and will be counted by the observer at $r = 0$ (see Figure 11-6). This is given as

$$dP = d\Omega \, r^2 dr \int_{-\infty}^{\infty} \Phi(M)f(m - m_c)dM, \tag{11-13}$$

Figure 11-5. The observer selection function $f(X)$ has a shape similar to the solid line, but the dotted line is an approximation satisfactory for computation. The cutoff magnitude of the sample occurs at $X = 0$.

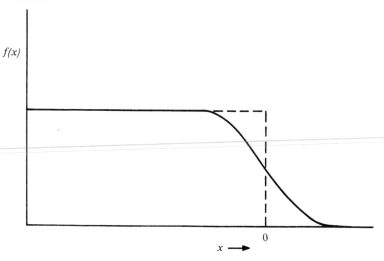

Figure 11-6. The volume element employed in equation (11-13).

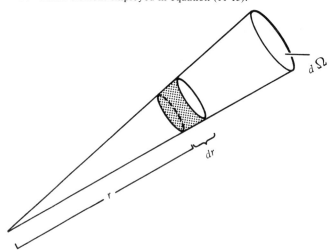

We have integrated over magnitudes weighted by the selection function, because we are interested in counting *every* observable galaxy. If we integrate equation (11-13) over r and compare the result with equation (11-9) we see that $\mathcal{N}$, the number of observed galaxies per steradian, is given as

$$\mathcal{N} = \int_0^\infty r^3 dr \int_{-\infty}^\infty \Phi(M) f(m - m_c)\, dM. \tag{11-14}$$

We now transform $m - m_c$ to absolute magnitudes. A characteristic magnitude M^* is defined, which corresponds to a conspicuous inflection point of the measured function $\Phi(M)$ (for example, $M^* = -19.5 + 5 \log h$; see Figure 11-2). If we define D^* as the distance at which an object of absolute magnitude $M = M^*$ would have an apparent magnitude $m = m_c$, then we have

$$m - m_c = M - M^* + 5 \log(r/D^*). \tag{11-15}$$

If we make the substitution $r = xD^*$ then

$$\mathcal{N} = (D^*)^3 E,$$

where

$$E = \int_0^\infty x^2 dx \int_{-\infty}^\infty \Phi(M) f(M - M^* + 5 \log x)dM. \tag{11-16}$$

Here E is a universal constant dependent on the luminosity function, and $\mathcal{N}$ scales as the cube of the distance, as expected; D^* could be considered to be the characteristic distance of a set of galaxies brighter than the limiting magnitude of m_c.

Note that throughout this discussion we have characterized the universe by Euclidean space and we have assumed that there is no obscuration of distant galaxies, either by intergalactic absorption or scattering, or by intervening galaxies. The first assumption is valid because the catalogs are limited to samples of galaxies within 400 Mpc of the Milky Way, where Euclidean space is an excellent approximation. The second approximation is valid because the intervening galaxies out to the distances sampled occupy a very small fraction of 4π steradians, and because no intergalactic absorption or scattering has been detected even to much larger distances.

Assuming that the observer selection function is independent of the densities of objects in a region of the photographic plate, we can consider in analogy to equation (11-13) the joint probability that two galaxies exist at radii r_1 and r_2 and that both are counted by the observer (see Figure 11-7).

Then

$$dP = d\Omega_1 d\Omega_2 r_1{}^2 dr_1 r_2{}^2 dr_2 \int\int_{-\infty}^{\infty} dM_1 dM_2$$

$$\times f_1 f_2 [\Phi(M_1)\Phi(M_2) + G(r_{12},M_1,M_2)], \quad (11\text{-}17)$$

where

$$r_{12}{}^2 = r_1{}^2 + r_2{}^2 - 2r_1 r_2 \cos \theta,$$

and where we have written $f_i = f(m_i - m_c)$, for compactness. If equation (11-17) is integrated over r_1 and r_2 we have, on comparison with equations (11-14) and (11-10), the expression

$$\mathcal{N}^2 w(\theta) = \int_0^{\infty} r_1{}^2 dr_1 \int_0^{\infty} r_2{}^2 dr_2 \int_{-\infty}^{\infty} dM_1 \int_{-\infty}^{\infty} dM_2$$

$$\times f_1 f_2 G(r_{12},M_1,M_2). \quad (11\text{-}18)$$

What information in the integral equation (11-18) can be gleaned with a minimum of assumptions about G? Without solving anything, we can determine how $w(\theta)$ should scale with D^*, the depth of the catalog-set of galaxies.

To do this we define new coordinates

$$u = (r_1 + r_2)/2,$$

$$v = r_1 - r_2,$$

Figure 11-7.   The volume elements employed in equation (11-17).

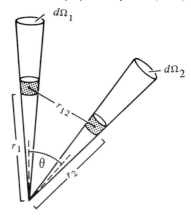

and make the small-angle approximation

$$\cos \theta \simeq 1 - \tfrac{1}{2}\theta^2.$$

Then

$$r_{12}{}^2 \simeq v^2 + (u^2 - \tfrac{1}{4}v^2)\theta^2.$$

Since we expect G to be a steeply falling function of r_{12}, the dominant contribution to the integral in equation (11-18) will occur for $r_1 \simeq r_2$. If $G(r)$ is negligible for $r \ll u$, we can use the approximation $v \ll u$ to obtain

$$r_{12}{}^2 \simeq v^2 + u^2\theta^2. \tag{11-19}$$

Substituting this approximation in equation (11-18) we have

$$w(\theta) = \mathcal{N}^{-2} \int_0^\infty u^4 du \int\int\int_{-\infty}^\infty dv dM_1 dM_2$$
$$\times f_1 f_2 G[(v^2 + u^2\theta^2)^{1/2}, M_1, M_2]. \tag{11-20}$$

Note that θ on the right side of equation (11-20) appears only in conjunction with u. If we substitute $u = xD^*$ we can find the scaling behavior of equation (11-20). Recall that $\mathcal{N}$ scales as $(D^*)^3$ (equation 11-16). We then have

$$w(\theta) = (D^*)^{-1} F(\theta D^*), \tag{11-21}$$

where F is a universal function of one variable given by

$$F(Y) = E^{-2} \int_0^\infty x^4 dx \int\int\int_{-\infty}^\infty dv dM_1 dM_2\, f_1 f_2 G[(v^2 + x^2 Y^2)^{1/2}, M_1, M_2].$$

Equation 11-21 is a prediction of how the observed $w(\theta)$ should scale with the depth D^* of the catalog of galaxies, and serves as an important check on the observations. It simply states that the observed angular scale size of the correlation function should vary inversely with the depth of the catalog, and that as D^* increases, chance superposition of unassociated background and foreground galaxies should decrease the over-all magnitude of the observed angular correlations. Both of these effects are expected.

Several catalogs of galaxies, which extend to various limiting magnitudes and therefore different values of D^*, have been analyzed by Peebles (Peebles and Hauser, 1974; Peebles, 1975); the results are shown in Figure 11-8. The analysis can be most easily performed simply by counting the excess number of pairs of galaxies having separations θ. Shown in the figure are the Zwicky catalog, which contains some 3700 galaxies with mag < 15, galactic latitude $b_{II} > 40°$, and declination $\delta > 0$; the Shane-Wirtanen catalog containing over 10^6 galaxies with mag < 19 and $\delta > -23°$, with the data published as counts of galaxies per degree square, and the Jagellonian Field, which is a catalog of 10^4 galaxies with mag < 21 in an area of the sky $6° \times 6°$ at high galactic latitude.

It is important that the analysis be confined to high galactic latitudes, so that obscuration by galactic dust does not produce any artificial gradients in the number density of galaxies. Gradients will also enter if the plates taken of one region of the sky are more sensitive than those taken of another region. There are many other systematic effects, including observer selec-

Figure 11-8. Correlation functions for the various galaxy catalogs: Jagellonian (squares), Zwicky (triangles), and Shane-Wirtanen (circles). (From Peebles, 1975.)

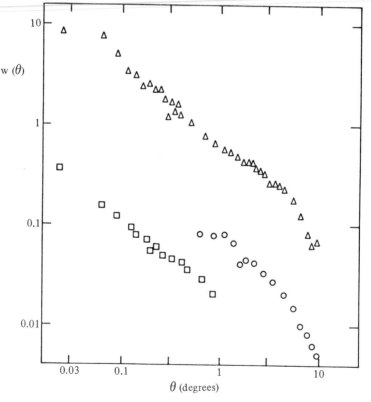

tion effects, which must be measured and controlled when producing such a catalog.

If we choose $M^* = -19.5 + 5 \log h$, then D^* for the Zwicky catalog is $50\ h^{-1}$ Mpc. After including a color correction for the redshifted spectra, the Jagellonian field samples galaxies five magnitudes fainter, or roughly ten times more distant, than the Zwicky catalog. The scaling of $\mathcal{N}$, the number density of sources per steradian, for the three surveys does vary as D^{*3} to good accuracy, as expected.

If the three catalogs are scaled according to equation (11-21), we find excellent agreement, as shown in Figure 11-9. The catalogs are affected by numerous systematic errors, and it is remarkable that, although each was produced independently and sampled different galaxies, they should agree so well. Figure 11-9 shows that apparently the universe is statistically homoge-

Figure 11-9. Test of the scaling relation. The data have been scaled to remove the expected effects of the different depths of the surveys. The abscissa is a measure of linear scale; the circles represent the Shane-Wirtanen catalog, the triangles display the Zwicky catalog, and the squares show the Jagellonian catalog. (From Peebles, 1975.)

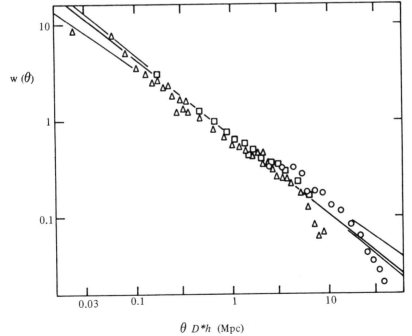

$\theta\ D^*h$ (Mpc)

neous, and that our neighborhood ($\sim 50\ h^{-1}$ Mpc) is a fair sample of the universe. The observed function $w(\theta)$ can be approximated by a power law of the form $w(\theta) = A\theta^{-\beta}$, where $\beta \simeq 0.77 \pm 0.06$. There is no a priori reason for choosing a power law to fit Figure 11-8, but a power law is sufficient in view of the uncertainties of the data.

Since $w(\theta)$ can be approximated as a power law, we would like to know whether $\xi(r)$ can also be approximated as a power law. To find the form of $\xi(r)$ we must assume that $G(r, M_1, M_2)$, introduced in equation (11-17), is a separable function of its arguments. Let

$$G(r_{12}, M_1, M_2) = \xi(r_{12})\Phi(M_1)\Phi(M_2). \tag{11-22}$$

Equation (11-22), known as the Limber hypothesis, states that the luminosity distribution function is the same for galaxies that are strongly clustered as for weakly clustered galaxies, or those not clustered at all. This is approximately what is observed, even though galaxies in rich clusters tend to be predominantly elliptical and lenticular, whereas most galaxies not in rich clusters are spirals.

Equation (11-22) allows us to simplify our expression for F (equation 11-21). If we define an integrated luminosity function $\varphi(x)$ by

$$\varphi(x) = \int_{-\infty}^{\infty} dM\Phi(M)f(M - M^* + 5\log x)$$

we can write equation (11-21) as

$$w(\theta) = (D^*)^{-1}E^{-2}\int_0^{\infty} dx\, x^4\varphi^2(x) \int_{-\infty}^{\infty} dv\xi\{[v^2 + (x\theta D^*)^2]^{1/2}\}. \tag{11-23}$$

Rather than invert the integral equation (11-23) to find $\xi(r)$, we can substitute a guess for $\xi(r)$ and fit parameters. If we substitute a power law

$$\xi(r) = Br^{-\gamma},$$

where B and γ are constants, and introduce the scaling substitution $v = x\theta D^*Y$, we can factor out the θ dependence of equation (11-23) and find

$$A/\theta^\beta = B(\theta D^*)^{1-\gamma}(D^*)^{-1}$$

$$\times [E^{-2}]\left[\int_0^{\infty} x^{5-\gamma}\varphi^2(x)dx\right]\left[\int_{-\infty}^{\infty}(Y^2 + 1)^{-\gamma/2}dY\right]. \tag{11-24}$$

The three bracketed quantities are universal integrals dependent on the shape of the luminosity distribution function; they can easily be tabulated.

Examination of equation (11-24) indicates that $\gamma = 1 + \beta$. Hence, the spatial correlation function $\xi(r)$ varies one power more steeply than the angular correlation $w(\theta)$.

After performing the integrals of equation (11-20), we have

$$\xi(r) \simeq 20/(hr)^\gamma,$$
$$\gamma \simeq 1.77 \pm .06, \qquad\qquad (11\text{-}25)$$

where r is in megaparsecs.

The factor h is defined below (equation 11-27) and is in the range of 0.5 to 1. The uncertainty in the numerical factor of equation (11-25) is about a factor of 2, arising primarily from uncertainty in $\Phi(M)$. Note that the normalization of $\Phi(M)$ does not affect B. Equation (11-25) quantifies what is obvious to the eye: galaxy clustering is a very pronounced effect. Given a galaxy at a point in space, the probability that a point 1 Mpc away from the first will also contain a galaxy is 21 times the expected random probability!

There are several remarkable features about the correlation function. A power law is a reasonable fit over a scale range of 400, from $50\,h^{-1}$ kpc $< r < 20\,h^{-1}$ Mpc; that is, from the size of individual galaxies to the size of the largest clusters of galaxies. This strongly suggests that there is a common physical mechanism which should explain the formation of clusters on all scales.

The term $\xi(r)$ is a smooth function showing no shoulders (i.e., no points of inflection) and defining no characteristic lengths. This is not unexpected for systems in which gravitation is the dominant interaction because gravity is a power-law force that defines no length scale itself.

If all clusters were of size R_c, then there would be a shoulder in the correlation function at R_c. But the clusters we observe do not have defined edges, and therefore would not produce a shoulder in the correlation function at large radii. However, the power-law density of all observed rich clusters flattens out at small radii (known as the core size, ~ 0.25 Mpc), as it must if the density at the center is to be finite. The correlation function for any one cluster would therefore have a shoulder at the core radius. If we postulate the existence both of rich and of poor clusters, then there must be a sufficient number of poor clusters to somehow cancel out the shoulder in the correlation function produced by the rich clusters.

The hierarchical picture of clustering provides a more plausible explanation of the smoothness of $\xi(r)$. If there exists a continuous hierarchy of clusters within larger clusters within still larger clusters, and if none of the component levels of the hierarchy have a preferred scale length, then the observed correlation function will be a featureless power law. In this picture there is no demarcation point between rich and poor clusters. The hierarchical picture of galaxy clustering is further supported by the analysis of three-particle correlation functions (Peebles and Groth, 1975).

It would be desirable to produce galaxy catalogs that were subject to less systematic error, so that one could determine whether the power-law slope is constant within 10% over a range in r of 10^3. In the Shane-Wirtanen and the Zwicky catalogs, the correlation function apparently falls below the power-law prediction at large angles. It is not clear whether this effect is real, or is an artifact of some systematic error in the catalogs. More data are required to clarify the situation.

The most interesting feature of the correlation analysis is the slope, $\gamma = 1.77 \pm 0.06$. What exactly does this slope tell us about the nature of galaxy clustering? Presumably the clustering we observe today has grown from a more uniform state at earlier times. It is known that self-gravitating matter is unstable against density perturbations. In an expanding universe, density perturbations will grow, but they grow so slowly that there is considerable doubt as to whether the large inhomogeneities so obvious today, galaxies and clusters of galaxies, could have grown from small density fluctuations in the early universe.

The slope of the correlation function today is probably due to the spectrum of initial density perturbations in the early universe. However, in different cosmological models, density fluctuations grow at different rates, and perhaps one can use the observed correlation function as a test of cosmological models. Unfortunately, the equations governing the time dependence of $\xi(r)$ are quite complicated and have not yet been solved.

Correlations of Different Morphological Types of Galaxies

In compact clusters of galaxies, spirals are found mainly in the outer regions. One possible explanation is that spirals that happen to pass through the central regions of the cluster are stripped of their gas and dust by intracluster gas. The orbits of the stars of the spiral galaxy are not affected by the wind of the intracluster medium as it blows by at high velocity. However, this wind is moving at supersonic speed with respect to the gas of the spiral galaxy and imparts a flow velocity to the gas relative to the stars. The gas simply remains behind in the cluster core.

Soon after passing through the cluster center, the blue stars burn out and the galaxy becomes a lenticular galaxy. A typical velocity dispersion for rich clusters is 10^3 km/sec, so that the time required for a spiral galaxy to move 6 Mpc across the cluster is 2×10^9 years. Thus, the crossing time is short enough for galaxies a few Mpc from the center of the cluster to have passed a few times through the cluster core and therefore to transform into lenticular galaxies.

Correlation analysis of spirals with spirals reveals less clustering than is found for lenticulars with lenticulars, which is in turn less than that shown by ellipticals with ellipticals. The above mechanism provides an explanation only of why spiral-spiral correlations would be smaller than lenticular-lenticular correlations in compact clusters, where we observe considerable intracluster gas (although not enough to bind the clusters).

However, even in poor clusters, which have lower velocity dispersions and longer crossing times, spirals still cluster less than lenticulars, which cluster less than ellipticals. Thus an old question is again raised, but again left unanswered. Did the diverse morphologies of galaxies arise because of different physical conditions at the time of galaxy formation, or were all galaxies formed in the same way, but later differentiated according to interactions with their environments? The latter hypothesis is supported by arguments like the transformation of spirals into lenticulars, described above, but the argument is insufficient to explain all lenticulars and all ellipticals.

Our understanding of galaxy formation is in a primitive state, and the questions raised above are likely to remain with us for some time to come.

Cosmology

In 1929 Hubble published his discovery that the apparent velocity of recession v of a galaxy, as determined by the Doppler shift of the spectral lines and corrected for the rotation of the solar system about the center of our Galaxy, is directly proportional to the estimate of the distance l to the galaxy;

$$v = H_0 l. \qquad (11\text{-}26)$$

Today the constant of proportionality H_0 is known as Hubble's constant, and the best estimate for H_0 is approximately 55 km sec^{-1} Mpc^{-1}. In view of the possible uncertainties of H_0, we will write

$$H_0 = 100\, h \text{ km sec}^{-1} \text{ Mpc}^{-1}, \qquad (11\text{-}27)$$

where h is probably about $\frac{1}{2}$. Note that H_0^{-1} has units of time, and the value $H_0^{-1} = 10^{10}/h$ years is an upper limit to the age of the universe in the models discussed below. Equation (11-26) is the fundamental phenomenon predicted by all models of an expanding universe, in the limit $v \ll c$.

The simplest cosmological models are obtained by assuming the universe to be homogeneous and isotropic. This assumption is known as the cosmological principle; it should be approximately valid if we average over a sufficiently large region in space.

In an expanding universe, it is convenient to use the concept of the co-moving observer, who is an observer at rest with respect to matter in his vicinity. By the homogeneity principle we can synchronize a cosmic time t by instructing all comoving observers to use the density of matter in their vicinity as a measure of time. All comoving observers therefore share the cosmic time t.

The symmetries implied by the cosmological principle require that the most general form of the line element is

$$ds^2 = c^2 dt^2 - R^2(t) du^2 \qquad (11\text{-}28)$$

where du is a distance element in the comoving space coordinates. If we consider a spherical coordinate system centered on a random comoving point, then the only form of du^2 which is homogeneous and isotropic is

$$du^2 = \frac{dr^2}{1 - kr^2} + r^2(d\theta^2 + \sin^2 \theta d\phi^2), \qquad (11\text{-}29)$$

where r, θ, and ϕ are comoving coordinates. The curvature index k has values ± 1, 0. The value $k = -1$ describes a universe of negative curvature for which the surface of a saddle is a two-dimensional analogue. Positive curvature universes are described by $k = 1$, for which the surface of a balloon is a two-dimensional analogue. $k = 0$ is known as the Einstein-de Sitter, or cosmologically flat universe. Equations (11-28) and (11-29) define the Robertson-Walker line element, and apply to any metric theory of gravity in a homogeneous, isotropic universe. A Euclidean universe, for example, is prescribed by $k = 0$ and $R(t) = $ constant.

The function $R(t)$ in equation (11-28) is the size of the universe as a function of cosmic time. The equations that govern this function must be supplied by a theory of gravity—Newton's theory, Einstein's theory, or any other theory.

The increment of proper distance dl of two observers separated by coordinate distance dr when $d\theta = d\phi = dt = 0$ is

$$dl = R(t)(1 - kr^2)^{-1/2}dr.$$

If $r \ll 1$, we can integrate this expression to obtain

$$l \simeq R(t)r.$$

Comoving observers are fixed to their coordinates, so that the velocity of separation of two observers is

$$v = dl/dt = \dot{R}(t)r = [\dot{R}(t)/R(t)]l = Hl. \tag{11-30}$$

We identify $H_0 = \dot{R}(t)/R(t)|_{t=t_0}$ (t_0 is the present cosmological time) and thus obtain Hubble's law (equation 11-26) for small r. Again recall that we have assumed only the cosmological principle to derive this result, and have assumed only that $R(t)$ is a differentiable function.

The information we receive from distant galaxies consists of photons which travel on null geodesics defined by $ds^2 = 0$. Consider a photon moving in the $-r$ direction with the Galaxy at the center of the coordinate system. The equation of motion is

$$ds^2 = 0 = dt^2 - R^2(t)\frac{dr^2}{1 - kr^2}.$$

If a photon leaves a distant galaxy at position r_e, time t_e, it reaches us at a time t_0 given by the expression

$$\int_{t_e}^{t_0} \frac{dt}{R(t)} = -\int_0^{r_e} \frac{dr}{(1 - kr^2)^{1/2}} = \begin{cases} \sin^{-1} r_e \ , k = 1 \\ r_e \quad , k = 0 \\ \sinh^{-1} r_e, k = -1 \end{cases} \tag{11-31}$$

Suppose the next wave crest of the emitted light leaves at time $t_e + \Delta t_e$ and arrives at time $t_0 + \Delta t_0$.
Then

$$\int_{t_e}^{t_0} \frac{dt}{R(t)} = \int_{t_e + \Delta t_e}^{t_0 + \Delta t_0} \frac{dt}{R(t)},$$

which implies that

$$\int_t^{t + \Delta t} \frac{dt}{R(t)} = \int_{t_0}^{t_0 + \Delta t_0} \frac{dt}{R(t)}.$$

Assuming $R(t)$ is constant over the short time intervals Δt_e and Δt_o, we have

$$\frac{\Delta t_0}{\Delta t_e} = \frac{\nu_e}{\nu_0} = \frac{R(t_0)}{R(t_e)}, \qquad (11\text{-}32)$$

where ν_e is the frequency of the emitted radiation as measured at r_e and ν_0 is the frequency of the radiation observed by the observer at the origin. We define a redshift parameter z as

$$1 + z \equiv \frac{\lambda_0}{\lambda_e} = \frac{\nu_e}{\nu_0} = \frac{R(t_0)}{R(t_e)}. \qquad (11\text{-}33)$$

When $z > 0$, spectral lines are redshifted, indicating an expanding universe, and when $z < 0$ they are blueshifted, implying a contracting universe. This relation compares to the first-order Doppler shift $z = v/c$ of a source moving at velocity v. However, it is unphysical to think of the redshift entirely as a special-relativistic Doppler effect. Rather, it is a natural consequence of the expanding universe.

Cosmological Field Equations

It is at the point of deriving the differential equation for $R(t)$ that alternative gravitational theories differ. The standard cosmology model (known as the Friedmann model) assumes the correctness of Einstein's theory of gravity, with cosmological constant $\Lambda = 0$. This model is not in disagreement with any observations, but neither are several other theories.

The steady-state universe, on the other hand, is no longer regarded as a viable theory by most cosmologists because, among other features, it fails to provide a reasonable explanation of the observed cosmic microwave background radiation, which will be discussed below. We shall limit ourselves to considering the standard Friedmann cosmologies. Remarkable features of the field equation are its simplicity and its classical interpretation.

Assume that the universe is filled with a homogeneous gas of density ρ and negligible pressure. Consider a small spherical volume of radius l, and remove the mass from inside. Newtonian gravitational theory tells us that an object inside a hollow spherical mass shell feels no force from the shell. In general relativity, the same is true in this situation: a particle in a hollow sphere feels no force from matter outside the sphere. Now return the matter into the sphere and realize that, if the gravitational potential of the mass M inside the sphere is small, $(GM/lc^2) \ll 1$, Newtonian gravity is adequate to describe the dynamics. Thus we have an isolated self-gravitating sphere, and the differential equation for l is simply

$$\frac{d^2 l}{dt^2} = - \frac{GM}{l^2}.$$ (11-34)

If we multiply equation (11-34) by dl/dt we see that the first integral of this equation is

$$\left(\frac{dl}{dt}\right)^2 = \frac{2GM}{l} - K = \frac{8\pi}{3} G\rho(t)l^2 - K,$$ (11-35)

where K is a constant. If each point is comoving we have

$$l(t) = R(t)l_0$$

and

$$\dot{R}^2 = \frac{8\pi}{3} G\rho R^2 - k \qquad (k = Kl_0^2).$$ (11-36)

Equation (11-35) simply states that the sum of the gravitational and kinetic energies is a constant, and this formula is exactly what would result from a full analysis of Einstein's field equations, although the constant would be derived more naturally. The closed universe models (positive curvature or $k = 1$) have a negative total energy, which implies that the universe will eventually collapse. The open universe models (negative curvature or $k = -1$) have positive energy, which implies that the universe will expand forever. The Einstein-de Sitter universe is the intermediate case having *zero* total energy.

In a matter-dominated universe where pressure is unimportant, the mass density ρ is dominated by the rest masses of the particles. Conservation of particles then implies that

$$\rho(t)R^3(t) = \rho_0 R_0^3 = \text{constant},$$ (11-37)

where the subscript 0 pertains to the present time. Hence, equation (11-36) can be written as

$$\left(\frac{dR}{dt}\right)^2 = \left(\frac{8\pi}{3} G\rho_0 R_0^3\right) \frac{1}{R} - k.$$ (11-38)

If we multiply equation (11-38) by R and differentiate with respect to t we have

$$2\dot{R}\ddot{R}R + \dot{R}^3 + k\dot{R} = 0$$

which implies that

$$k = -\dot{R}^2[1 + 2(R\ddot{R}/\dot{R}^2)].$$

Evaluated at the present time, we have

$$k = H_0^2 R_0^2 (2q_0 - 1). \tag{11-39}$$

The dimensionless number $q_0 = -R_0\ddot{R}_0/\dot{R}_0^2$ is known as the deceleration parameter and is an observable quantity; $q_0 = \frac{1}{2}$ in the Einstein-de Sitter model, $q_0 > \frac{1}{2}$ for a closed universe, and $q_0 < \frac{1}{2}$ for an open universe. Note that q_0 is time independent only when $k = 0$.

Equation (11-38) evaluated at the present epoch implies that

$$\rho_0 = \frac{3}{4\pi G} H_0^2 q_0. \tag{11-40}$$

The universe is closed if the present density of the universe exceeds a critical density ρ_c given by

$$\rho_c = 1.9 \times 10^{-29} h^2 \text{ gm/cm}^3 \tag{11-41a}$$

or

$$n_c = 1.1 \times 10^{-5} h^2 \text{ protons/cm}^3. \tag{11-41b}$$

We can use equations (11-39) and (11-40) to express equation (11-38) in terms of q_0 and H_0, the two observable parameters:

$$[\dot{R}(t)/R_0]^2 = H_0^2[1 - 2q_0 + 2q_0R_0/R(t)]. \tag{11-42}$$

For $q_0 = \frac{1}{2}$ the solution to this equation is

$$R(t)/R_0 = \left(\frac{3}{2} H_0 t\right)^{2/3}. \tag{11-43}$$

For $q_0 < \frac{1}{2}$, $R(t)$ is a monotonically increasing function larger than that given by equation (11-43). For $q_0 > \frac{1}{2}$, $R(t)$ describes a cycloid (see Figure 11-10).

The standard cosmology thus reduces to the determination of two numbers, H_0 and q_0, which together specify R_0 and ρ_0 through equations (11-39) and (11-40).

The cosmic scale parameter, H_0 has been the object of intensive investigation. Sandage and Tammann (1975) have recently completed a long study using several independent methods, and conclude that $H_0 = 55 \pm 7$ km sec^{-1} Mpc^{-1}. For $q_0 = \frac{1}{2}$, the age of the universe is

$$t_0 = \tfrac{2}{3}H_0^{-1} \simeq 13 \times 10^9 \text{ years},$$

an age in comfortable agreement with those derived by nucleocosmochronology and with models of globular clusters and galaxies.

Whereas H_0 is presumably known within a 15% error, the present estimate of q_0 is 1 ± 1 (the error brackets include $q_0 < 0$, but this value does not have physical meaning in Friedmann cosmologies, although it is permissible in cosmologies having a nonzero cosmological constant.). For the past twenty-

Figure 11-10. The time development of the expansion parameter $R(t)$ for the various Friedmann cosmologies. Note that H_0^{-1} is larger than the present age of the universe.

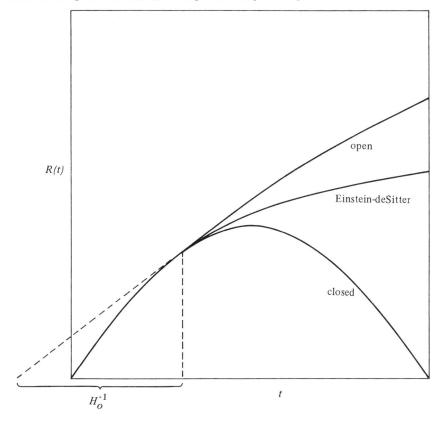

Frontiers of Astrophysics

five years astronomers have spent a large amount of time observing with the 200-inch telescope at Mount Palomar, trying to determine whether the universe is closed or open. Yet today the situation remains unresolved. Several tests of cosmology are theoretically capable of determining the curvature of space, but each test is subject to systematic errors dependent on poorly understood astrophysical processes. We will discuss the more critical tests, and their uncertainties, below. Further details of cosmology and cosmological tests are given in Peebles (1971), Weinberg (1972), Rose (1973), Sciama, (1973), Gott et al. (1974), and Sandage (1975).

The Magnitude-Redshift Test

The magnitude-redshift test is the classical test for the effect of the curvature of space on light traveling from distant objects, and it has been the most studied of the cosmological tests. To look at distant objects is to look back in time when the universe was smaller and objects were closer to us. Thus distant galaxies subtend a larger solid angle than they would in a static universe (see Figure 11-11).

Suppose a galaxy at a coordinate distance r from us emits light which we detect at time t_0. According to the Robertson-Walker line element, the proper area of a spherical surface of coordinate radius r_1, is

$$A = 4\pi r_1^2 R_0^2.$$

Figure 11-11. Galaxy of constant proper size D was closer to us at time t_e when it emitted photons, which we receive at time t_0. Therefore it occupies a larger apparent angular extent than it would in a static universe.

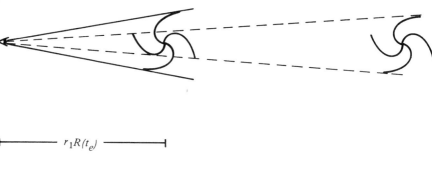

If L is the bolometric luminosity, the observed bolometric luminosity per unit area f is given by

$$f = \frac{L}{4\pi r_1^2 R_0^2} \left(\frac{1}{1+z}\right)^2. \tag{11-44}$$

One power of $(1 + z)$ in equation (11-44) arises because of the redshift of photons traveling to us. The second power of $(1 + z)$ is due to the reduced photon arrival rate from the source, which is moving away from us.

If we define a luminosity distance d_L by the value it would have in a Euclidean universe,

$$d_L = \left(\frac{L}{4\pi f}\right)^{1/2},$$

we see that in a Friedmann universe

$$d_L = R_0 r_1 (1 + z). \tag{11-45}$$

With equation (11-42), $r_1 R_0$ can be shown to be expressible in terms of z, H_0 and q_0, yielding

$$f = \frac{LH_0^2}{4\pi c^2} \left[\frac{q_0^2}{q_0 z + (q_0 - 1)(1 + 2q_0 z)^{1/2}}\right]^2. \tag{11-46}$$

For small z this can be reduced to

$$f \simeq \frac{L}{4\pi} \left(\frac{H_0^2}{c^2 z^2}\right) [1 + (q_0 - 1)z]. \tag{11-47}$$

Equation (11-47) is a close approximation to equation (11-46) for $z < 1$, and is exact for $q_0 = 0$ and $q_0 = 1$. Note that the first term in parenthesis in equation (11-47) is the Euclidean prediction for the flux, and the second term in parenthesis represents departures from the Hubble law, equation (11-26), caused by the curvature of space. Thus the effects of q_0 on f are of second order. The first-order effects on f (namely the Doppler shift of the arriving photons, and the fact that the galaxy was closer to us when it emitted the photons we now detect) have balanced each other. For larger values of q_0, the universe has decelerated more since the time the light left the galaxy. Therefore the galaxies were closer to us, subtended a larger solid angle, and should be more luminous, in agreement with equation (11-47) Unfortunately, equation (11-47) predicts only a small effect out to the distance of the farthest observed galaxies ($z \simeq 0.5$). At this redshift the apparent luminosity changes by only 0.4 mag as q_0 varies from 0.1 to 1.0. Thus a 10% uncertainty of q_0 requires a photometric accuracy of 0.05 mag in observing faint, distant objects, an accuracy very difficult to obtain.

To apply the magnitude-redshift test, one would like to observe a "standard candle" at different redshifts and compare the observed luminosities with the values given by equation (11-46). The first task, of course, is to decide what to use for a standard candle. The candle must be very luminous in order to be seen at large redshift ($z > 0.2$), and it should maintain a constant luminosity in time, since the light-travel time to $z = 0.4$ is several billion years.

Bright elliptical galaxies in rich clusters of galaxies have been used as the standard candle for all the observations to date because they are bright and can be seen at large redshifts.

The "first brightest" (i.e., most luminous) galaxy in rich clusters has a reasonably "standard" luminosity, with a dispersion in absolute magnitude ranging from 0.20 to 0.40 mag, depending on the selection procedure. Quasars, on the other hand, have a much brighter intrinsic luminosity and can be seen to much greater distances, but the dispersion in their absolute luminosity is several magnitudes, rendering them useless for the magnitude-redshift test.

Unfortunately, bright ellipticals probably have not maintained constant luminosity over the past several billion years, and it is not clear whether their luminosity increases or decreases in time. Most models of the evolution of galaxies predict a decrease in luminosity with time. It is assumed that young galaxies had high initial star-formation rates, and so were much more luminous at early epochs. The rate of change of a galaxy's luminosity is dependent on the assumed distribution function of stellar masses.

Many of the first brightest galaxies in rich clusters are known as cD galaxies, and they tend to have lower central surface brightnesses and flatter luminosity profiles than most elliptical galaxies. Furthermore, they often have multiple nuclei. Recently it has been suggested (Ostriker and Tremaine, 1975) that the largest galaxies in a cluster, which are found near the center of the cluster, may accrete some of the smaller galaxies by dynamical friction processes. If the largest and brightest galaxies are formed in this way, then all first brightest galaxies in clusters will have a mass and a luminosity that increase with time. This effect counteracts the expected evolutionary correction described above.

To describe the effect of evolution on the measurement of q_0, we can write the luminosity of the galaxy in a power series expansion in time, measured from the present time t_0:

$$L(t) = L_0[1 + E_0(t - t_0)] \simeq L_0[1 - E_0(z/H_0)]$$

where E_0 is the first-order time evolution rate of the luminosity of the galaxy. Including this effect in equation (11-47) then gives

$$f = \frac{L_0 H_0^2}{4\pi z^2 c^2}\left[1 + (q_0 - 1)z - \frac{E_0 z}{H_0}\right]. \tag{11-48}$$

Therefore, the q_0 measured by the magnitude-redshift test is

$$q_0^{\text{measured}} = q_0 - E_0/H_0. \tag{11-49}$$

Thus E_0 must be known before the true q_0 can be derived from the test. Progress has been made towards understanding galaxy evolution, but at the present time estimates of E_0 are quite uncertain.

Formidable systematic errors must be overcome even when galactic evolution is better understood. One well-known systematic bias, known as the Scott effect, is the preference toward selecting intrinsically brighter sources as the objects move further away. Unless the luminosity function of galaxies has a cutoff at the bright end, the Scott effect will operate in selecting sources for the magnitude-redshift test, and the effect will be to overestimate the average luminosity of distant sources and thereby to overestimate q_0. There is absolutely no evidence to show that the luminosity function does have a cutoff at the bright end; the rarity of the galaxies with extremely large luminosity has precluded a definitive statistical test.

Two studies of the magnitude-redshift test have recently been published, using different observing and sampling techniques. After considerable analysis, Sandage and Hardy (1973) report a measured value of $q_0 = 1 \pm 1$, whereas Gunn and Oke (1975) report as a maximum $q_0 = 0.33 \pm 0.68$. (The estimates are made before an evolutionary correction is applied.) Gunn and Oke selected their sample in a manner intended to minimize the influence of the Scott effect, and their lower estimate of q_0 may reflect this selection.

To improve the limits on q_0 with this test, many more clusters at large redshift must be studied. Unfortunately, the distant clusters are very faint ($m_V > 20$) and to measure their redshift requires long integration using the most modern detectors on a large telescope. Even with improved statistics, it is essential to fully understand the time evolution of elliptical galaxies in order to correct for its effect on the measurements. Within the foreseeable future, this test is unlikely to permit us to conclude definitely whether the universe is closed or open.

The Angular Diameter-Redshift Test

Another test of q_0 is to measure the angular diameter of a source as a function of its redshift. Consider a source of fixed proper size D. In Euclidean space it would subtend an angle

$$\theta = D/d_A,$$

where d_A is the distance. In an expanding universe, consider light that leaves the source at coordinate distance r_1 and at time t_e, arriving at the origin at time t_0. The light from the edges of the object travels to us along fixed angles (see Figure 11-11), and for small angle θ, we have

$$D = R(t_e)r_1\theta$$

so that

$$d_A = R(t_e)r_1.$$

Note by comparison with equation (11-45) that

$$\frac{d_A}{d_L} = \frac{1}{(1 + z)^2}. \tag{11-50}$$

The angular diameter-redshift test is measuring essentially the same effect as the magnitude-redshift test, but it may prove easier to perform. To employ this test, it is not necessary to fully understand the luminosity evolution of galaxies. The test can be used on any distant source that has a well-defined edge. Unfortunately, galaxies have kinks in their luminosity profiles only at very small radii, generally smaller than the seeing disk (1 arcsec). Possibly a diffraction-limited large space telescope could observe distant galaxies and perform this test.

Clusters of galaxies are easier objects to employ in this test. The surface density of galaxies in a cluster can be measured from photographs and fitted to a profile, such as an isothermal sphere, which has a measurable core radius (the radius where the density is one-half the central surface density). From dimensional analysis,

$$R_c \propto \left(\frac{\langle v^2 \rangle}{GN_0}\right)^{1/2}$$

where N_0 is the central surface density and $\langle v^2 \rangle$ is the velocity dispersion of galaxies in the cluster. Figure 11-12 shows a plot of this measured core radius for clusters of small redshift. The scatter in the core radius is about 20%, and if approximately 100 clusters are observed at large redshift ($z \sim 0.3$), this test could measure q_0 with perhaps 10% accuracy.

The major theoretical uncertainty, of course, concerns the time evolution of the "core radius" in the large clusters. In addition, if larger, richer clusters have larger core radii, then an analogue of the Scott effect can produce a selection bias which causes q_0 to be overestimated because richer clusters will be preferentially observed at larger redshifts. Rich clusters do appear as relaxed, isothermal spheres, but it is not known for how long they have existed in such a state. They could have condensed within the past 10^9 years, or they could be as old as 10^{10} years. The dynamics of rich clusters must be further investigated before the angular-diameter test can be used to measure q_0.

For both of these classical cosmological tests, it has been assumed that the mass of the universe is homogeneously distributed, so that the curvature of the universe is constant and the Robertson-Walker metric applies. It is obvious, however, that the universe is lumpy on a small scale. If we assume that all the mass of the universe is contained within galaxies, then, when we observe a distant object, we preferentially observe along a line of sight that contains a smaller than average mass density. This occurs because foreground galaxies obscure more distant sources. The light bundle traveling to us from distant sources therefore encloses a cone containing a smaller than average mass density.

Figure 11-12. Core radii of clusters of galaxies displayed as a function of redshift. The curves show the observed angular sizes as a function of cosmological model. (From Bahcall, 1975.)

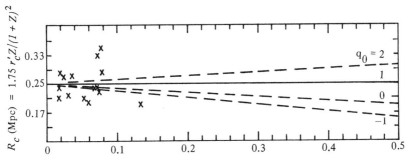

In a Friedmann universe distant sources subtend a larger angle than in a static universe for two reasons: (1) the sources were closer to us when the light was emitted than when it is detected, and (2) the intervening matter between the source and the observer gravitationally deflects and focuses the light bundle, increasing the apparent angular size of the source (see Figure 11-13). Without the focusing caused by the intervening matter, the angular size of the distant sources will be smaller than predicted from the Robertson-Walker metric. In other words, the angular-diameter distance will increase, as will the luminosity distance (by equation 11-50), and the net effect is to underestimate q_0. Depending on the fraction α of the mass of the universe that is smoothly distributed in space, this effect can produce up to a 30% underestimate of q_0 for sources at $z = 0.5$, and a much greater error for larger z (Roeder, 1975). Measurements of q_0 derived from the magnitude-redshift or angular diameter-redshift relations must be corrected by an amount dependent on the unknown parameter α.

This correction further reduces the usefulness of observing distant sources to determine the curvature of the universe. The two classical cosmological tests are beset by numerous serious difficulties, which are likely to remain unresolved in the near future. Therefore, more attention has recently been devoted to alternative methods of determining q_0. One method is to estimate the average mass density ρ_0 and to compare it with the critical density (given by equation 11-41) that divides the closed from the open models. Most estimates of ρ_0 at present are well below ρ_c, and those who are philosophically inclined toward closed cosmological models have undertaken a great quest to find the "missing mass" of the universe. Some of the methods of estimating ρ_0 will be discussed below.

The Mean Mass Density of the Universe

The most direct method of determining the curvature of the universe is to somehow measure the average mass density and to compare it with the critical density ρ_c (equation 11-40). The observed mass density is subject to dispute, but almost all estimates of the observed density are less than ρ_c. This is not conclusive evidence for an open universe, because not all significant contributions to the mass density need reside in forms readily detectable by us. In effect, measurement of the local mass density is most useful for setting a lower bound on q_0.

Since galaxies are the most conspicuous objects in the universe, we

Figure 11-13. (a) Homogeneous universe; focusing of light bundle by curvature of space between the Milky Way and a distant galaxy. (b) Absence of focusing in an inhomogeneous universe when the light bundle encloses a tube devoid of matter.

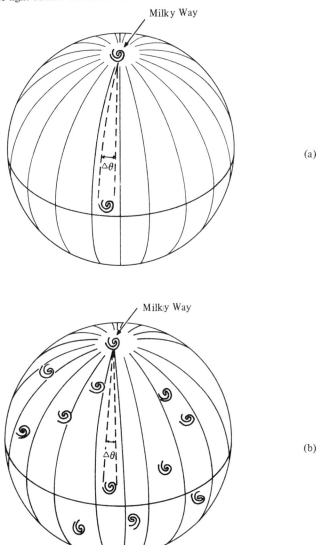

(a)

(b)

address first the question of the mean mass density due to galaxies. If we can assume that all galaxies of one type have a constant mass-to-light ratio, then we can estimate the mass density if we measure the volume emissivity (luminosity/volume) of galaxies:

$$\rho_{gal} = \mathcal{L}_{sp}(M/L)_{sp} + \mathcal{L}_{el}(M/L)_{el}. \tag{11-51}$$

Here, all galaxies are classified as either spirals or ellipticals. By adding the light contributed by all galaxies with magnitude $m_V < 12$, Shapiro (1971) has determined that

$$\mathcal{L}_{sp} = 3 \times 10^8 hL_\odot \mathrm{Mpc}^{-3},$$
$$\mathcal{L}_{el} = 2 \times 10^7 hL_\odot \mathrm{Mpc}^{-3}. \tag{11-52}$$

The effect of the local supercluster of galaxies could have caused an overestimate of the mean volume emissivity, but probably by not more than a factor of 1.5. Uncertainties in the faint end of the luminosity distribution function $\Phi(M)$ (see Figure 11-2) introduce an uncertainty of at least a factor of two in the volume emissivity.

The mass-to-light ratios must be determined before the density can be estimated. Several methods are used to measure the masses of galaxies (see Chapter 12); most of them utilize the virial theorem, which should be applicable for condensed systems that are in approximate equilibrium. For self-gravitating condensations, the virial theorem states that

$$2K + U = 0. \tag{11-53}$$

Given particles of mass m and velocity v, K is the kinetic energy

$$K = \frac{1}{2} \sum_i m_i v_i^2, \tag{11-54}$$

where the sum is over all the particles in the system, and U is the total potential energy

$$U = -\frac{1}{2} \sum_i \sum_{j \neq i} \frac{Gm_i m_j}{|r_i - r_j|}. \tag{11-55}$$

The factor $\frac{1}{2}$ enters in this equation because each pair is summed twice. Dimensionally, the virial theorem says that the mass of the system M is given as

$$M \propto \langle v^2 \rangle R/G$$

where R is the characteristic size and $\langle v^2 \rangle$ is the measured velocity dispersion.

In nuclei of galaxies it is possible to determine $\langle v^2 \rangle$ of the stars in the center by measuring the shapes of the Doppler-broadened stellar lines. These measurements generally give $M/L \sim 7\text{--}10\ M_\odot/L_\odot$ for *all* types of galaxies.

In spiral galaxies, the bright stars and gas move in approximately circular orbits in the disk. Knowing the circular velocity $\theta(r)$ from the Doppler shift in the spectral lines across the disk, one obtains the gravitational acceleration θ^2/r. If reasonable assumptions are made for the thickness of the mass distribution, one can solve for the approximate mass of the galaxy, with the result that

$$M/L \sim 10\,h\ M_\odot/L_\odot$$

for radii of about 20 kpc. Recently, however, rotational velocities determined by 21-cm measurements of neutral hydrogen have shown that the mass of large spiral galaxies continues to increase beyond 20 kpc and that possibly the mass-to-light ratio increases with r, approximately as

$$M/L \propto r.$$

Page (1962) has analyzed binary systems of galaxies, pairs of galaxies apparently in tight orbit about each other (with projected separations $< 50\ H^{-1}$ kpc). For each pair one knows only the relative radial velocity of the two galaxies and their projected radial separation. Assuming that the galaxies move in circular orbits with random orientations toward us, one can determine the mean mass-to-light ratio for a collection of binary pairs. For elliptical galaxies, Page finds $M/L \sim 50\,h$, and for spirals he finds $M/L \sim 1\,h$. This last number is based on a very small statistical sample and should not be considered too seriously. A recent reexamination of this problem by Turner (1976), who uses a larger set of data, suggests that M/L for spirals is approximately $130\,h$.

It has been commonly assumed that $(M/L)_{\mathrm{sp}} = 7\,h$ and $(M/L)_{\mathrm{el}} = 50\,h$. With these values of M/L, the mass density of galaxies is only 1.2% of the critical density, so that the mass of galaxies apparently is far below the mass required for a closed universe.

Using these mass-to-light ratios, one can compute T and U for clusters of galaxies, and check to see whether the virial theorem is satisfied. In every observed cluster, the virial theorem apparently is *not* satisfied, so that the clusters have positive energy, with the kinetic energy typically 3 to 100 times larger than the potential energy. If this is indeed true, then all galaxy clusters must be unbound and will disperse in a time scale of the order of the galaxy crossing time, which for compact clusters is less than 10^9 years! Yet many of these compact clusters, like Coma, appear to be relaxed systems, and have density profiles that fit an isothermal sphere model, for which the virial theorem is satisfied. One strongly suspects, therefore, that there must be hidden mass somewhere in these clusters to satisfy the virial theorem.

There are only two possible explanations of the mass discrepancy in clusters: (1) clusters form out of "white holes," which are singularities in spacetime, and which disperse in one crossing time, or (2) we have underestimated the masses of galaxies by factors of 5 to 50, and all clusters having a crossing time short compared with the Hubble time (10^{10} years) are bound. The second explanation is considerably easier to accept, especially in view of the difficulty of measuring galactic masses. Recall also that measures of rotational velocity or velocity dispersions at a radius R can determine only the mass interior to R, and they say nothing about the mass exterior to R. Applying the virial theorem to galaxy clusters determines the mass of the component galaxies inside a much larger R.

It is important to realize that the clusters cannot be bound by the addition of a uniform background of mass at any density, since the dynamics of galaxies depends only on the fluctuations from the average densities. The density profile of galaxies in rich clusters like Coma fits an isothermal shape; this fact implies that the missing mass in the cluster must also fit the isothermal density profile, although the mass need not be "attached" to the individual galaxies.

Recently Spinrad (personal communication) has measured the velocity dispersion of small lenticular galaxies that are apparently satellites of a giant elliptical galaxy in the Coma cluster, and from them he determined $M/L \sim 360\,h$ for the elliptical galaxy. This is in good agreement with determinations of $M/L \sim 300\,h$ for the Coma cluster itself, derived from the virial theorem and the assumption that the cluster is bound. Since Coma is composed mostly of lenticular and elliptical galaxies, this estimate of the mass is 6 times greater than the previous estimate.

In poorer clusters, most of the galaxies are spirals, and here the mass discrepancies determined from the virial theorem range from factors of 20–200

(Rood, Rothman, and Turnrose, 1970). Yet those clusters with crossing times $\Delta t < 0.25\, H_0^{-1}$ are surely bound, and, in an average over all poor clusters, the virial theorem should be approximately satisfied. A conservative estimate of the mass discrepancy in the spiral galaxies in clusters with this crossing time (or less) is approximately a factor of 25, and if the same factor applies to all spirals, then $(M/L)_{sp} \sim 200\, h$, which is close to the estimate of $(M/L)_{el}$.

Inserting these values in equation (11-51) gives $\rho_{gal}/\rho_c \simeq 0.25$, so that galaxies apparently do not close the universe, even if the clusters are bound. This estimate has an uncertainty of a factor of at least two. Here we have actually measured the mass necessary to bind clusters and have assumed that all this mass is contributed by "galaxies." Whether one considers the nonluminous matter as part of a galaxy or as something distinct is a question of semantics.

How can galaxies have such large M/L ratios, and why do the outer regions of galaxies have a different M/L ratio than inner regions? In what form of condensation is the mass in the halo regions of the galaxy? It is not in the form of neutral hydrogen, and it probably is not all ionized gas, as will be discussed in Chapter 12. It would be easy to hide a substantial fraction of the mass of the galaxy in M dwarf stars, and even easier in Jupiter-sized bodies, which would be far too faint to be observed. We simply do not understand the evolution of galaxies well enough to predict what type of star formation, if any, should occur in halos of galaxies.

Is intergalactic space filled with low-mass objects having an M/L ratio similar to that of a galaxy halo? Perhaps galaxy and cluster formation is an inefficient process and only a small fraction of the matter in the universe has condensed into galaxies. The remainder probably exists in a gaseous form, and current limits on this density do not preclude the possibility of a density comparable to ρ_c. This issue is discussed further in the following chapter.

The dynamics of irregularly spaced lumps of matter can provide only lower limits to the mean mass density of the universe, and the observations do not rule out a closed universe, although one has to imagine exotic ways to hide the missing mass. The missing mass does not reside in any electromagnetic background radiation, because most of the electromagnetic energy is stored in the 2.7-K microwave background radiation, which has an equivalent mass density of

$$\rho_{rad} \simeq 2.3 \times 10^{-5} h^{-2} \rho_c;$$

hence its contribution to the mean mass density of the universe today is neg-ligible. Neutrino background radiation and gravitational radiation, generated in the very early universe when all these components may have been in thermal equilibrium, probably have thermal-type spectra with temperatures less than 2.7 K and therefore make a negligible contribution to the mean mass density. The universe could be heavily populated with small black holes created by fluctuations in density at the time of the big bang; it is diffi-cult to rule them out because they would be virtually undetectable today.

Thus, the sum of all known components is considerably less than the criti-cal density needed to close the universe, and implies only that q_0 is probably greater than 0.12.

With recent advances in instrumentation, it has become much easier to measure the redshifts of faint galaxies. Thus, velocity determinations of many nearby galaxies are becoming available. When a sufficient number of velocities have been determined, it may be possible to reconstruct the dynamics of galaxies in various regions and then to resolve the problem of the virial mass discrepancy of clusters.

Cosmological Nucleosynthesis

In the past several years there has been much activity concerned with the measurement of the interstellar abundance of deuterium. The recent Coper-nicus satellite has measured a ratio of D/H $\sim 2 \times 10^{-5}$ by observing the rela-tive intensity of ultraviolet Lyman absorption lines of interstellar atomic hy-drogen and deuterium in front of bright stars having an ultraviolet continum (Rogerson and York, 1973). Unlike the heavy elements, the present deu-terium abundance is unlikely to have originated by the normal route of stellar nucleosynthesis. The proton-proton reaction

$$p + p \longrightarrow D + e^+ + \nu_e \qquad (11\text{-}56)$$

produces deuterium in the sun, but the deuterium is rapidly burned by the much faster reaction

$$p + D \longrightarrow He^3 + \gamma \qquad (11\text{-}57)$$

which results in a very small equilibrium abundance of D. (Here e^+ = posi-tron, ν_e = neutrino, and γ = photon.) Indeed the deuterium abundance of the sun is much less than that observed in interstellar space. Stellar nucleo-synthesis destroys deuterium, so the observed interstellar deuterium must have been made by some other means. To produce deuterium requires an environment of high energy, and to preserve deuterium requires an environment of low density. The early stages of a hot big-bang universe provide an environment that meets both of these conditions.

The standard model of element production in the early Friedmann universe relies on a number of assumptions, all of which seem reasonable. First of all, it is essential to assume that the observed microwave background radiation is indeed the remnant of the primeval fireball radiation, which has adiabatically cooled as the universe has expanded. We must make the bold extrapolation that at some time in the early universe the temperature was in excess of 10^{11} K, indeed much hotter than the 3 K temperature of the microwave radiation observed. It is necessary also to assume that no significant quantities of antimatter existed at this epoch, so that baryon particle numbers are conserved. Furthermore, one needs to assume that the universe is not filled with a large density of neutrinos, which could be virtually undetectable but could have impeded the formation of free neutrons in the early universe. Finally, it is necessary to assume that the Friedmann equations (cf. equation 11-36) govern the expansion rate of the universe at this early epoch.

Under these conditions, at temperatures in excess of 10^{10} K, all heavy nuclei will disintegrate to form a sea of neutrons and protons, which will be held in approximate equilibrium by the weak interactions

$$n \longrightarrow p + e^- + \bar{\nu}_e$$
$$e^+ + n \rightleftarrows p + \bar{\nu}_e$$
$$\nu_e + n \rightleftarrows p + e^- \qquad (11\text{-}58)$$

(here e^- = electron and $\bar{\nu}_e$ = antineutrino). The presence of free neutrons allows the reaction

$$n + p \rightleftarrows D + \gamma \qquad (11\text{-}59)$$

to occur, which is much faster than reaction (11-56) because it does not involve weak interactions and because there is no Coulomb barrier to overcome. At high temperatures reaction (11-59) is favored toward the left (photodissociation), but at 10^9 K and lower temperatures, reaction (11-59) favors the right (deuterium production). At this temperature, from reaction (11-58) the neutron-proton equilibrium abundance is ~ 0.14 and most of the free neutrons are absorbed in the formation of deuterium. The buildup of the deuterium in turn allows the formation of the heavier elements, notably He³ and He⁴. Figure 11-14 shows the results of detailed calculations of the abundances of the various elements as a function of the cosmic time. Nucleosynthesis quickly ceases after the free neutrons are used up, and as the densities and average energies decrease.

Most of the deuterium and He³ burns to form He⁴, and the fraction remaining depends inversely on the baryon density. The results of detailed calculations show the expected abundances as a function of ρ_0 (see Figure 11-15). Thus deuterium, especially, is an excellent probe of the early universe. If we assume that the observed interstellar abundance of deuterium is equal to the primordial abundance, the recent Copernicus data suggests that $\rho_0/\rho_c \simeq 0.20$, but a more realistic estimate is that half the primordial deuterium has been consumed in stars, and that $\rho_0/\rho_c < 0.10$. This is in embarrassing disagreement with the mean mass density required to bind clusters, which suggests that $\rho_0/\rho_c > 0.25$.

The deuterium abundance is the strongest piece of evidence suggesting an open universe, and for such an important problem it is essen-

Figure 11-14. Abundances of various elements as a function of temperature and age of the universe. (From Schramm and Wagoner, 1974.)

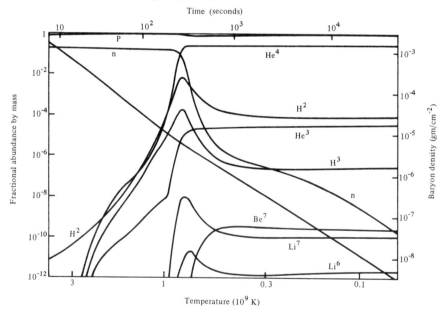

tial to consider all alternative explanations for the high observed abundance. This is a problem now under intensive investigation, and some possibilities are mentioned below. If there are no sources of deuterium other than primordial nucleosynthesis, then interstellar matter that has been processed through a star will be underabundant in deuterium, causing ρ_0 to be overestimated.

It has been shown that cosmic rays spallating heavy elements in the interstellar medium can produce deuterium, but the cosmic ray density is only high enough to produce 1 % of the observed deuterium abundance. Supernova shock waves could produce energies high enough to spallate He^4 via the reaction $p + He^4 \rightarrow D + He^3$, which has a center-of-mass threshold energy of 18.35 MeV. It is at present uncertain whether sufficient material is accelerated to high enough energies in supernova shocks to explain the deuterium abundance. Furthermore, detailed calculations indicate that under most conditions, spallation in shockwaves produced by supernovae is more

Figure 11-15. The dependence of the final cosmic mass fractions of various nuclides on the postulated value of the present baryon density, based on the present photon temperature of 2.7 K, is shown here for standard big-bang models. (From Schramm and Wagoner, 1974.)

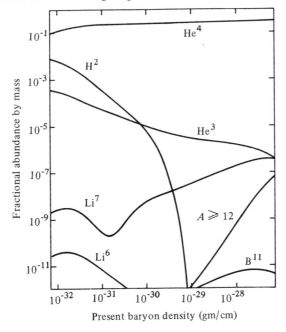

likely to produce Li, Be, and B than D, in disagreement with the observed relative abundances of these elements. This suggests that most deuterium did not originate in supernova shockwaves, although this issue is not fully resolved.

The question of primordial versus nonprimordial formation of deuterium could possibly be resolved with more observational data. At present, we have interstellar absorption data for only a half dozen bright stars in our neighborhood. The next generation of satellites for ultraviolet observations will have improved sensitivity, enabling observations of fainter, more distant stars. If deuterium is formed in some type of stellar nucleosyntheses, then the interstellar medium closer to the galactic center should have a higher relative abundance of D, analogous to the higher observed metal abundances of stars closer to the galactic center. If deuterium is produced only primordially, then its relative abundance should decrease closer to the galactic center. These crucially important questions concerning deuterium production may be resolved within the next decade.

The Primeval Fireball Radiation

In 1965, Penzias and Wilson discovered an anomalous excess instrumental noise, amounting to about 3 K, in a radio telescope used for early communication-satellite experiments. No local source for the radiation could be found, and it was soon realized that this universal microwave background radiation was probably the remnant of the primeval fireball.

A primeval fireball is thermal black-body radiation, presumably from an earlier epoch when the universe was much hotter and denser than it is now. If a perfectly absorbing cavity is in thermal equilibrium at a temperature T, then general arguments of statistical equilibrium imply that in that cavity there will be present a photon gas having a black-body Planck radiation flux

$$u(\nu) = \frac{8\pi h\nu^3}{c^2} \left[\exp(h\nu/kT) - 1\right]^{-1} \quad (\text{erg sec}^{-1}\,\text{cm}^{-2}\,\text{Hz}^{-1}). \quad (11\text{-}60)$$

Assume that at some time in the past the universe was in thermal equilibrium. What happens as the universe expands? Assume that the photons have no interaction with matter, and that the photons are conserved.

Conservation of energy per comoving volume in an expanding universe implies that

$$\frac{d}{dt}\{[R(t)]^3 u[\nu(t)]\} = 0. \qquad (11\text{-}61)$$

Recall that as the universe expands, each individual photon will be red-shifted according to the relation

$$\nu(t) \propto R(t)^{-1}. \qquad (11\text{-}62)$$

If the spectral density at epoch t_e is $u[\nu(t_e)]$, then equation (11-61) says that at any later time t_0, the radiation density is

$$u[\nu(t_o)] = \left[\frac{R(t_e)}{R(t_o)}\right]^3 u\left[\nu\,\frac{R(t_o)}{R(t_e)}\right]. \qquad (11\text{-}63)$$

Thus, if the initial spectrum was a black-body spectrum (equation 11-60) with temperature T_e, the spectrum observed at epoch t_o is

$$u(\nu) = \frac{8\pi h\nu^3}{c^2}\,[\exp(h\nu/kT_o) - 1]^{-1} \qquad (11\text{-}64)$$

where

$$T_o = T_e\left[\frac{R(t_e)}{R(t_o)}\right]. \qquad (11\text{-}65)$$

That is, the spectrum preserves its black-body form, and only the effective temperature of the radiation changes.

Figure 11-16 shows the experimentally measured microwave spectrum; the solid line is a 2.7-K black-body spectrum. The observations at the low-frequency, or Rayleigh-Jeans region, have been accurately known for ten years. The high-frequency measurements are more difficult to obtain; positive detection of the background beyond the characteristic turnover of the spectrum at 2 mm was reported recently by Robson et al. (1974) and by Woody et al. (1975). There is now little doubt that the microwave background does have a black-body shape.

The cosmic microwave radiation is the strongest piece of evidence supporting hot big-bang cosmologies. At early epochs in these models, all matter was ionized, and Thomson scattering provided an efficient means of maintaining thermal contact between radiation and matter. When the universe cooled to a temperature of 3000 K, the hydrogen plasma rapidly recombined, removing the scattering electrons, thus decoupling the radiation from the matter.

Unless the matter has been somehow subsequently reionized since that time, the microwave photons observed today last interacted with matter at a redshift

$$1 + z = \frac{3000}{2.7} \simeq 1000.$$

Beyond this redshift the matter is opaque to the radiation, and therefore $z \simeq 1000$ is the greatest distance (or time) to which we can ever "look back."

What does the microwave radiation tell us about the universe at this epoch? In addition to measuring the spectrum, it is possible to measure the isotropy of the radiation. The results to date indicate that the radiation is indeed quite isotropic on both large and small angular scales. Small-scale isotropy measurements use large radio telescopes to search for fluctuations in the microwave radiation on angular scales of 10–100 arc min. No anisotropy has been observed, and upper limits to possible anisotropy are as small as $\Delta T/T < 10^{-3}$. These limits are important in models of galaxy formation, since large density fluctuations at the recombination epoch should produce detectable small-scale fluctuations in the observed microwave background.

Figure 11-16. The observed spectrum of the cosmic microwave background, with the solid line representing the 2.7 K black-body curve. The box centered at 1 mm is the result of recent experiments which had a wide bandwidth. (Adapted in part from Peebles, 1971.)

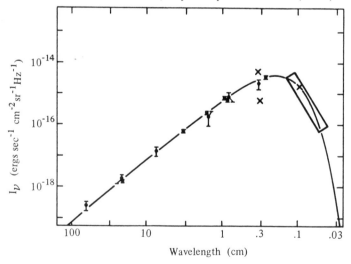

The absence of detectable fluctuations argues either that density fluctuations at the recombination epoch were small, or that there is a substantial density of ionized gas in the universe (see Chapter 12). This problem is under active investigation from both an observational and a theoretical viewpoint.

The large-scale isotropy experiments employ pairs of small antennae, looking in opposite directions of the sky. The microwave radiometers used are sensitive to the difference in intensity received by the two antennae. Best results have been obtained from balloon-borne receivers, where contamination from nonuniform atmospheric radiation is reduced. Again, no anisotropy has been detected to a limit of approximately $\Delta T/T < 10^{-3}$.

This result implies that since the epoch of the last scattering of the microwave photons, the universe has undergone an isotropic expansion. That is, the Hubble constant is independent of direction, and the simple Robertson-Walker metric is indeed a good large-scale approximation of the universe today. Another point about the large-scale anisotropy is of great interest to cosmologists. Even in a Friedmann universe, the microwave background is isotropic only to comoving observers. Observers moving relative to this preferred frame will detect the microwave photons arriving from their direction of motion to be blueshifted, and photons traveling in their direction to be redshifted. This is merely the familiar Doppler effect, and it can be expressed as a variation in observed black-body temperature, given to first order ($v/c \ll 1$) by

$$T(\theta) = T_o \left(1 + \frac{v}{c} \cos \theta \right),$$

where $T_0 = 2.7$ K and θ is the angle between the velocity vector and the direction of observation. The microwave background thus serves as a method to determine the motion of the earth with respect to the preferred frame of the universe. It is reasonable to conclude from experiments thus far that $|v_{\oplus}| \leq 300$ km/sec.

We know that the earth revolves around the sun at 30 km/sec, and the solar system revolves around the Galaxy at about 250 km/sec. Corrected for galactic rotation, the Milky Way and Andromeda (M 31) are moving toward each other at a velocity of about 90 km/sec. The Virgo cluster of galaxies at a distance of 10 h^{-1} Mpc is a relatively large mass fluctuation which may have deflected the Local Group of galaxies (i.e., primarily M 31 and the Milky Way) from the comoving reference frame. Observations are now in progress that should either improve the upper limit of $v_\oplus$, or determine its value. Perhaps in the near future this technique will measure the motion of the earth with respect to the "aether," and it might even provide a new measurement of the mass of the Virgo cluster of galaxies, which would certainly be of interest for cosmology.

Summary

The detection of microwave background radiation is the most significant cosmological discovery since Hubble showed that the universe is expanding. The microwave background strengthens our faith in the naive, hot big-bang cosmology, and the remarkable isotropy of the radiation implies that the Robertson-Walker line element is also approximately correct. The large-scale structure of the universe is apparently adequately described by the simplest cosmology consistent with general relativity.

To determine further which of the simple Friedmann models describes our universe has been a frustrating task. The data today are contradictory, as described above; and a resolution of the problems will take time and further careful observations. The correlation function is a new description of the clustering phenomenon, one which may yield new clues for cosmology. Future analysis is likely to show that different cosmologies produce different correlation functions but, if the effect is like most in cosmology, the predictions will not be free of ambiguity.

References

Bahcall, N. A. 1975. Core radii and central densities of 15 rich clusters of galaxies, *Astrophys. J. 198*, 249–254.

de Vaucouleurs, G. 1959. Classification and morphology of external galaxies, *Handb. Phys. 53*, 275–372.

de Vaucouleurs, G. 1976. Nearby groups of galaxies, in A. Sandage, M. Sandage, and J. Kristian, eds., *Stars and Stellar Systems*, vol. 9, *Galaxies and the Universe*, pp. 557–600, University of Chicago Press, Chicago.

Gott, J. R., Gunn, J. E., Schramm, D. N., and Tinsley, B. M. 1974. An unbound universe? *Astrophys. J. 194*, 543–553.

Gunn, J. E., and Oke, J. B. 1975. Spectrophotometry of gaint cluster galaxies and the Hubble diagram: An approach to cosmology, *Astrophys. J. 195*, 255–268.

Hubble, E. 1929. Distance and radial velocity among extra-galactic nebulae, *Proc. Nat. Acad. Sci. 15*, 168–173.

Ostriker, J. P., and Tremaine, S. D. 1975. Another evolutionary correction to the luminosity of giant galaxies, *Astrophys. J. (Letters) 202*, L113–117.

Page, T. 1962. M/L for double galaxies: A correction, *Astrophys. J. 136*, 685–686.

Peebles, P. J. E. 1971. *Physical Cosmology*, Princeton University Press, Princeton, N. J.

Peebles, P. J. E. 1975. Statistical analysis of catalogs of extragalactic objects. VI, The galaxy distribution in the Jagellonian field, *Astrophys. J. 196*, 647–652.

Peebles, P. J. E., and Groth, E. J. 1975. Statistical analysis of catalogs of extragalactic objects. V. Three point correlation function for the galaxy distribution in the Zwicky Catalog, *Astrophys. J. 196*, 1–12.

Peebles, P. J. E., and Hauser, M. G. 1974. Statistical analysis of catalogs of extragalactic objects. III. The Shane-Wirtanen and Zwicky Catalogs, *Astrophys. J. Suppl. 28*, 19–36.

Penzias, A. A., and Wilson, R. W. 1965. A measurement of excess antenna temperature, *Astrophys. J. 142*, 419–421.

Robson, E. I., Vickers, D. G., Huizinga, J. S., Beckman, J. E., and Clegg, P. E. 1974. Spectrum of the cosmic background radiation between 3 mm and 800 μm, *Nature 251*, 591–592.

Roeder, R. C. 1975. Apparent magnitudes, redshifts, and inhomogeneities in the universe, *Astrophys. J. 196*, 671–674.

Rogerson, J. B., and York, D. G. 1973. Interstellar deuterium abundance in the direction of Beta Centauri, *Astrophys. J. (Letters) 186*, L95–98.

Rood, H. J., Rothman, V. C. A., and Turnrose, B. 1970. Emprical properties of the mass discrepancy in groups and clusters of galaxies, *Astrophys. J. 162*, 411–423.

Rose, W. K. 1973. *Astrophysics*, Holt, Rinehart, and Winston, New York.

Sandage, A. 1961. *The Hubble Atlas of Galaxies,* Carnegie Institution of Washington.

Sandage, A., and Hardy, E. 1973. The redshift-distance relation. VII. Absolute magnitudes of the first three ranked cluster galaxies as functions of cluster richness and Bautz-Morgan cluster type: The effect on q_0, *Astrophys. J. 183,* 743–758.

Sandage, A., and Tammann, G. A. 1975. Steps toward the Hubble constant. V. The Hubble constant from nearby galaxies and the regularity of the local velocity field, *Astrophys. J. 196,* 313–328.

Sandage, A., Sandage, M., and Kristian, J., eds. 1976. *Stars and Stellar Systems,* vol. 9, *Galaxies and the Universe,* University of Chicago Press, Chicago.

Schramm, D. N., and Wagoner, R. V. 1974. What can deuterium tell us? *Physics Today 27,* 12, 40.

Sciama, D. W. 1973. *Modern Cosmology,* Cambridge University Press, Cambridge, England.

Shapiro, S. 1971. The density of matter in the form of galaxies, *Astron. J. 76,* 291–295.

Turner, E. 1976. Binary galaxies. I. A well-defined statistical sample. II. Dynamics and mass-to-light ratios, *Astrophys. J.,* in press.

Weinberg, S. 1972. *Gravitation and Cosmology,* Wiley and Sons, New York.

Woody, D. P., Mather, J. C., Nishioka, N. S., and Richards, P. L. 1975. Measurement of the spectrum of the submillimeter cosmic background, *Phys. Rev. Letters, 34,* 1036–1040.

12

The Mass of the Universe: Intergalactic Matter

George B. Field

As we saw in the previous chapter, application of the virial theorem to clusters of galaxies shows that they contain more mass than one would calculate from the numbers of galaxies observed, and from the masses of individual galaxies. Part of the "missing mass" may be in faint outlying parts of galaxies; also, various types of nonluminous objects such as faint stars, collapsed stars, and planets, may lie between the galaxies and contribute to the hidden mass.

By analogy with our own Galaxy, the Milky Way, where interstellar gas and dust contribute significantly to the mass density, it is also possible that diffuse matter between the galaxies in a cluster is responsible for some of the hidden mass. In this chapter we consider such intergalactic matter, or IGM. We also saw that the masses of galaxies are lower than those required to provide a mean cosmic density equal to the critical value, $\rho_c = 4.7 \times 10^{-30}$ g/cm^3, which divides open from closed models of the universe, as explained in equation (11-41a) of the previous chapter. In this chapter we take $H_0 = 50$ km sec^{-1} Mpc^{-1}, corresponding to $h = \frac{1}{2}$ in the notation of the previous chapter. If one adopts the conservative mass-to-light ratios $M/L = 25$ for ellipticals and 3.5 for spirals, and defines ρ_{gal} to be the mean density in galaxies, then $\rho_{\text{gal}}/\rho_c = 0.012$.

Because we use the ratio of mean densities to ρ_c so often in this chapter, we shall follow convention in defining

$$\Omega = \frac{\rho}{\rho_c}, \tag{12-1}$$

where ρ is the mean density of any component of the universe. Thus, $\Omega_{\text{gal}} = 0.012$, compared to the critical value $\Omega = 1$. For conventional Friedmann models in which the effects of pressure are negligible, we saw in the last chapter that

$$\Omega = 2q_0 \qquad (12\text{-}2)$$

so that the critical value of q_0 which separates open and closed models is $\frac{1}{2}$.

Even if the much larger values of M/L inferred from the binding of clusters of galaxies are adopted for galaxies in general, $\Omega_{gal} \simeq 0.25$. Thus galaxies appear to be unable to close the universe. Again we are led to consider diffuse intergalactic matter, this time lying outside of clusters of galaxies and far from any individual galaxies. Such matter would contribute to Ω, but would not affect the dynamics of clusters, so it would not be counted in the effective value of $\Omega_{gal} \simeq 0.25$ derived earlier from the requirement that clusters be bound.

In summary, then, intergalactic matter within clusters may help in binding them, while that between clusters may affect the dynamics of the universe as a whole. We are thus led to search for astrophysical evidence for intergalactic matter by the realization that it might have important gravitational effects. As recently as 1970, astrophysicists reviewing the evidence (Brecher and Burbidge, 1970) found no compelling astrophysical evidence for the existence of intergalactic matter, but recent data from radio and X-ray astronomy suggest that it does exist (Field, 1972, 1974a). In this chapter we will examine the evidence, and consider the effects which intergalactic matter would have.

Let us first consider the much better understood phenomenon of interstellar matter, that is, gas and dust lying between the stars in our own Galaxy. The earliest studies of the Milky Way showed that stars in it are affected by patchy obscuration, now known to be caused by interstellar clouds of small ($\sim 10^{-5}$ cm) particles of dust. Interstellar extinction, as this effect is called, becomes significant for all stars beyond about 1 kpc, from which it has been deduced that the mean density of dust in the Galaxy is roughly 10^{-26} g/cm^3. Since the dust is very likely composed of heavy elements such as silicon, oxygen, and iron, which comprise only $\sim 1\%$ by mass of normal galactic material, this led to the suspicion that the density of interstellar matter in all forms is at least 10^{-24} g/cm^3; there could be more if not all the heavy elements are bound up in dust particles. Most of this would be hydrogen and helium, by far the most abundant elements. This line of argument was interesting because it had already been determined from the virial theorem applied to stars that the mean density of gravitating matter in the plane of the Milky Way near the sun is 10×10^{-24} g/cm^3 (Oort, 1965), of which only 4×10^{-24} g/cm^3 can be accounted for by observed stars (Allen, 1973). Just as in the case of clusters of galaxies, it is certainly possible that much of the missing mass in the Milky Way, about 6×10^{-24} g/cm^3, is attributable to invisible objects like very faint stars. On the other hand, the inference from the amount of interstellar dust suggested that a search for interstellar hydrogen would be worthwhile (helium being much harder to detect because all of its ground-state atomic transitions lie in the inaccessible ultraviolet).

In 1951, radio astronomers first detected the hyperfine line emission of interstellar hydrogen atoms (H I) at 21-cm wavelength. The Galaxy has now been mapped at this wavelength, permitting one to infer that H I with its accompanying helium atoms (presumed to be 10% of H by number from stellar measurements) contributes 1.2×10^{-24} g/cm^3 to the mass density (Kerr and Westerhout, 1965). This is only 20% of the missing mass, but the story does not end there.

Hydrogen can also occur as a hydrogen molecule or H_2, which can be detected directly only in the ultraviolet region accessible with space telescopes. Although H_2 has therefore been detected only in a few regions, it is believed to be the dominant form of H whenever the interstellar cloud in which the gas is located is thick enough to shield the H_2 from stellar ultraviolet light, which dissociates H_2 into H I atoms. In 1971, Hollenbach, Werner, and Salpeter made careful counts of the numbers of thick, or dark, interstellar clouds, and estimated their masses. Adding up the results, they concluded that there is at least as much H_2 as H I, or an additional 1.2×10^{-24} g/cm^3. Thus, H I and H_2 together account for at least 2.4×10^{-24} g/cm^3, or 40% of the missing mass near the sun. As we shall see, it is likely that, in a similar way, intergalactic gas accounts for a substantial fraction of the missing mass in clusters, and may contribute significantly to the entire mass of the universe.

We start by considering intergalactic dust. Unlike interstellar space, which becomes opaque beyond distances of 1 kpc because of the presence of interstellar dust, intergalactic space is transparent out to huge distances. Recently, galaxies have been found with redshifts greater than 0.5, implying that they lie at distances greater than 3000 Mpc. These galaxies show no sign of intergalactic extinction, from which we conclude that the mean free path for light is greater than about 3000 Mpc. This is more than 3×10^6 times the corresponding distance for interstellar space in the Galaxy. Since the mean free path is inversely proportional to the density of absorbing material, one concludes that the density of intergalactic dust is less than 3×10^{-33} g/cm³, showing right away that intergalactic dust itself does not contribute significantly to $\rho_c = 4.7 \times 10^{-30}$ g/cm³. If we postulated that normal chemical abundances apply to intergalactic matter, and that the fraction of heavy elements locked up in dust is the same as in interstellar matter, we would infer a total amount of intergalactic matter which is 200 times the amount of intergalactic dust. This gives an upper limit, $\rho_{IGM} < 6 \times 10^{-31}$ g/cm³, from which one would be tempted to conclude that $\Omega_{IGM} < \frac{1}{8}$. However, this conclusion would be unreliable, because the ratio of gas to dust is probably much greater than 200 in intergalactic space. The reason is that heavy elements are produced in stars and, therefore, in general they remain within galaxies rather than mixing with intergalactic matter. This statement is supported by the fact that the oldest stars in the Galaxy, which were made of material closest in composition to the intergalactic medium at the time the Galaxy formed, are in fact deficient in heavy elements, sometimes by factors of 100 or more. We conclude that if one is interested in intergalactic matter, one must try to observe the main constituents, hydrogen and helium. Helium is presumed to be present as the result of nucleosynthesis in the big bang, as explained in the previous chapter, but it is difficult to observe, as explained above.

It seems intelligent to look for intergalactic hydrogen first within clusters of galaxies, for the expected densities are highest there. To get an idea of how much might be present, we consider the Coma cluster of galaxies, a cluster with a redshift of 7000 km/sec, which contains about 1000 E and SO galaxies; its radius on the sky is at least 2°. The dynamics of clusters is determined by a balance between the tendency to expand (associated with the kinetic energy K) and the tendency to contract (associated with the gravitational potential energy U), as explained in the last chapter. There we saw that the mathematical expression of this balance is the virial theorem,

$$2K + U = 0. \tag{12-3}$$

Since K is proportional to the mass M_{cl} of the cluster and U is proportional to M_{cl}^2, equation (12-3) permits one to solve for M_{cl}, using the observed dispersion of the galaxies in space and in velocity. The mass M_{cl} so calculated is called the "virial theorem mass," M_{VT}. Any one component i of the cluster, say the galaxies, has a mass $M_i \leq M_{VT}$. We can define

$$\omega_i = \frac{M_i}{M_{VT}}, \qquad (12\text{-}4)$$

in analogy to Ω for the universe. The fact that a cluster is gravitationally bound implies that

$$\omega \equiv \sum_i \omega_i = \frac{1}{M_{VT}} \sum_i M_i = 1, \qquad (12\text{-}5)$$

but any one ω_i can be less than unity. The problem is then to identify all the components of a cluster and to calculate their corresponding ω_i.

Rood et al. (1972) calculate $M_{VT} = 5 \times 10^{15} \, M_\odot$ for the Coma cluster. Using the counts of galaxies and their masses deduced from Page's (1962) M/L ratios, Rood et al. calculate that $M_{cl} = 6 \times 10^{14} \, M_\odot$, so that

$$\omega_{cl} = 0.12. \qquad (12\text{-}6)$$

Where is the missing mass? As already pointed out in the previous chapter, it may lie in the outer parts of galaxies, as indicated by recent work of H. Spinrad (personal communication). Alternatively, it may consist of intergalactic gas. According to Rood et al., the missing mass must be distributed in space in the same way as are the galaxies, otherwise the gravitational potential well would not have the right shape. The corresponding virial theorem density at the center of the cluster is $\rho_{VT} = 2 \times 10^{-25} \, \text{g/cm}^3$ (King, 1972), about 10^5 times the cosmological density. Surely one ought to be able to observe gas of such density if it is there. In the next section we describe attempts to do so.

Evidence Concerning Intergalactic Matter in Clusters

Zwicky (1962) found that the number of clusters of galaxies observed behind the Coma cluster is smaller than in neighboring fields, and interpreted this in terms of extinction caused by IGM within the Coma cluster, amounting to about 0.3 mag at visual wavelengths at about $1°$ from the center of the cluster.

His method is this: suppose one is observing a homogeneous population of objects (as clusters of galaxies are thought to be) at different distances r. By

the inverse-square law of brightness and the exponential nature of extinction,

$$F = \frac{L}{4\pi r^2} e^{-\tau} \tag{12-7}$$

is the total flux at the earth in terms of the luminosity L and optical depth τ.
The distance can be eliminated by noting that the number of objects within a
distance r is

$$N = \frac{4\pi}{3} nr^3 \tag{12-8}$$

in Euclidean space (which is valid for redshifts $\ll 1$). Here n is the number
density of objects. Then

$$F = \frac{L}{4\pi} \left(\frac{4\pi n}{3N} \right)^{2/3} e^{-\tau}, \tag{12-9}$$

which, for objects of fixed n and L implies that

$$N = \text{const } F^{-3/2} e^{-(3/2)\tau}. \tag{12-10}$$

The first term expresses the increase in number of objects observed with decreasing F, just due to geometrical effects, while the second expresses the
reduction in the number due to extinction. If one counts distant clusters in
the absence of a nearby cluster, one evaluates the first term. Counting clusters behind the nearby cluster, one gets both terms. Dividing the two expressions yields τ.

Karachentsev and Lipovetskii (1968) observed the same effect for
Coma, deriving 0.4 mag for wavelengths in the blue part of the spectrum.
As these wavelengths are shorter than the visual wavelengths studied by
Zwicky, this is consistent with the fact that extinction by dust should be
larger at shorter wavelengths, as is observed for interstellar extinction. They
went on to study 14 other nearby clusters, finding an average extinction of
0.2 mag. They showed that the effect is much smaller in the red, as expected
for wavelength-dependent extinction. Finally, they showed that the effect
decreases smoothly as one goes away from the center of the cluster,
reaching about half of the central value at a radius equal to the "cluster radius" quoted by Zwicky.

A number of authors have questioned the reality of this effect. Reaves
(1974) points out that it will be more difficult to identify distant clusters be-

hind nearby ones because of confusion effects, and refers to observations in another part of the sky to support his position. There he finds that distant clusters *are* deficient in number behind nearby ones, as Zwicky suggests, but that the biggest deficiency is for those clusters immediately beyond the nearby clusters, suggesting that the confusion effect is present. However, he also finds that the effect persists for the most distant clusters observed (where confusion effects should be minimal), and that the magnitude of the effect agrees well with that found by Karachentsev and Lipovetskii.

In reviewing this earlier (Field, 1972) I pointed out a problem in believing that the observed effect, if real, has anything to do with the missing mass in the Coma cluster or other clusters. It has been demonstrated by Rood et al. (1972) that whatever is providing the gravitating mass in the Coma cluster must be distributed in the same way as the galaxies. Since the projected density of galaxies at 1° from the center of the cluster is only 2% of the value at the center (King, 1972), the same has to be true of intergalactic matter, if it is responsible for binding the cluster. This would mean that the extinction through the center of the cluster would be 50 times greater than at 1°, or 15 mag in the visual. The galaxies in the center would exhibit an extinction half of that, or 7.5 mag, which would be a huge effect. In fact, Karachentsev and Lipovetskii (1968) find that the effect at the center of various clusters is no larger than that in the outer parts—a few tenths of a magnitude in the blue. Although this argument seems to vitiate the proposal that the dust is related to the missing mass, the argument does not take into account that conceivably the heavy elements in the outer regions of the cluster are concentrated in dust grains, while those at the center may have been released from the dust into the gas, by bombardment of the dust by atoms of the hot gas now known to be present (see below). One finds that the hot gas is dense enough to destroy the grains over the lifetime of the cluster ($\sim 10^{10}$ years) everywhere within about 1.5° (or 2.4 Mpc) of the center of the Coma cluster (Smart, 1973). Perhaps the extinction observed by Zwicky and by Karachentsev and Lipovetskii originates in a shell of dust which is so far out in the cluster that it has not been destroyed by the dense hot gas near the center of the cluster. Hence the observations of extinction still stand as an interesting, albeit ambiguous, indication of intergalactic matter within the Coma cluster, observations that should be repeated and evaluated further.

Another indication that intergalactic matter exists in clusters of galaxies comes indirectly, from studies of radio galaxies within such clusters. Powerful radio galaxies appear in the telescope as luminous elliptical galaxies which, like most ellipticals, have little interstellar dust and gas, or as bright young stars. What they do have is two radio-emitting clouds, often disposed like giant wings stretching up to 10^5 pc or more on either side of the optical galaxy. These wings are composed of clouds of relativistic electrons which radiate by the synchrotron radiation process.

Synchrotron radiation results from circular or spiral motion of relativistic electrons in a magnetic field, as in a synchrotron. The radiation they emit is directed into a narrow "searchlight" along the instantaneous direction of motion, owing to the large aberration effect of the relativistic velocity. The observer therefore sees a pulse of radiation each time the searchlight sweeps past, and, because the pulse is short, it contains frequencies much higher than the basic gyrofrequency. For magnetic fields of the order of 10^{-6} gauss and Lorentz factors of the order of 10^4, the gyrofrequency is of order 10^{-4} Hz, and the radiated frequency is of order 10^8 Hz. Further discussion of synchrotron radiation is given in Chapter 7.

The amounts of energy involved, 10^{60} erg or more, required explosive acceleration of 10^6 solar masses of particles to relativistic speeds in a short time; the mechanism operating here is still not understood, but that is another story. What interests us here is that, in several clusters, evidence has been found that both radio clouds ejected by radio galaxies are displaced in the same direction from the parent galaxy, strongly suggesting that they have been swept back by a wind of some kind. A natural interpretation is that the wind is due to the orbital motion of the parent galaxy which is responsible for generating the radio clouds, through intergalactic gas which is either at rest or moving radially in the cluster. The effect is very similar to that of the solar wind blowing on the magnetic field of the earth, forming a "magnetospheric tail." The theory of this phenomenon has been worked out for NGC 1265 in the Perseus cluster of galaxies, a cluster of 500 members whose distance is about 100 Mpc. Figure 12-1 shows a radio map of this object, made with the Westerbork Aperture Synthesis Array in the Netherlands at a wavelength of 6 cm. The theory (Miley, Wellington, and van der Laan, 1975) indicates that the tail can be explained if the intergalactic density is of order 10^{-3} particles per cm³, some 1000 times larger than the cosmological density. Of course this applies within a cluster, where the density of intergalactic gas, like that of galaxies, is presumably greatly enhanced.

The most impressive evidence of intergalactic matter in clusters comes

Figure 12-1.   The radio tail of the radio galaxy NGC 1265 in the Perseus cluster of galaxies. A 5000-MHz map of the region made at Westerbork is superposed on the Palomar Sky Survey blueprint of the optical galaxy NGC 1265 (Wellington, Miley, and van der Laan, 1973). Reproduced from *Nature,* vol. 244, p. 502, with the permission of Macmillan Journals, Ltd.

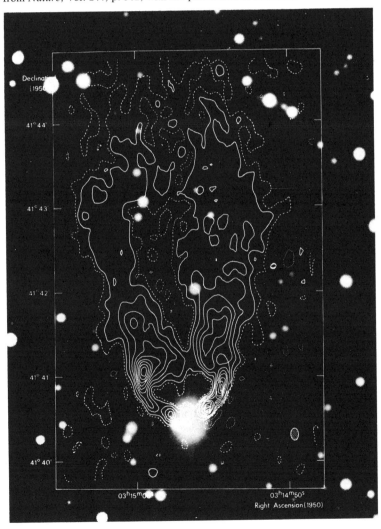

from X-ray astronomy. Starting with a few early reports of coincidences between X-ray sources and clusters of galaxies, there is now evidence that 21 sources detected by the first X-ray satellite, Uhuru, are clusters of galaxies (Kellogg, 1974).

There are several different X-ray emission mechanisms, some of which could be operating in the galaxies in the cluster, and some of which could be due to relativistic particles ejected from galaxies. One mechanism may operate with hot intergalactic gas in the cluster. This is the "thermal bremsstrahlung" process, in which pulses of radiation are emitted when free electrons in the gas are accelerated by passing ions. One can work out the formula for this using electromagnetic theory and the Maxwell-Boltzmann distribution of velocities; quantum theory has the effect of multiplying the formulae obtained classically by the Gaunt factor, g. The formula for the energy radiated per unit energy interval, per unit solid angle, per unit time, and per unit volume of hot gas (the so-called emissivity j_E) is

$$j_E = 8.2 \times 10^{-13} g T^{-1/2} e^{-E/kT} n_e \sum_i n_i Z_i^2 \qquad (12\text{-}11)$$

(in units of keV cm^{-3} sec^{-1} ster^{-1} keV^{-1}), where T is the temperature, E is the photon energy, n_e is the electron density, and n_i is the density of ions of charge Z_i.

One can understand the equation crudely as follows. The energy emitted in an encounter between an electron and an ion is proportional to (acceleration)^2 × (passage time), or to $Z_i^2 r^{-4} \Delta t$. The energy per unit frequency (or energy) interval is the total divided by $\Delta \nu$, and for a pulse of length Δt, $\Delta \nu \simeq (\Delta t)^{-1}$. Hence we multiply again by Δt, getting $Z_i^2 r^{-4} \Delta t^2$. Since $\Delta t \simeq r/v$, this is $Z_i^2 r^{-2} v^{-2}$. The number of collisions between r and $r + dr$ is proportional to $n_e n_i v r\, dr$, so the spectral power per unit volume is proportional to $Z_i^2 n_e n_i v^{-1} r^{-1}\, dr$. Integration over r yields a logarithmic factor, and averaging over a Maxwellian converts v^{-1} to $T^{-1/2}$, accounting for all the factors in equation (12-11) but $e^{-E/kT}$. This factor enters because in averaging over v for a given photon energy E, we should include only those encounters with $\frac{1}{2} mv^2 > E$. There are exponentially few electrons available in a Maxwellian distribution meeting this criterion when $E > kT$.

Charge neutrality demands that the ions and electrons go together, so n_i is proportional to n_e, but both ions and electrons may be clumped locally, so that the spatial mean of the product of the densities over a large volume is greater than the product of the mean densities. We introduce a "clumping factor" $C = \langle n_e^2 \rangle / \langle n_e \rangle^2$ to take this into account, so that

$$n_e \sum_i n_i Z_i^2 = \langle n_e \rangle \, C \sum_i \langle n_i \rangle \, Z_i^2$$

$$= 1.68 \, C \, \langle n_H \rangle^2, \qquad (12\text{-}12)$$

where we have calculated the sum for a 10:1 mixture of H (density n_H) and He. The Gaunt factor g varies with T and E, but is generally between $\frac{1}{2}$ and 2 for the values of interest (Kellogg, Baldwin, and Koch, 1975). Although the original papers include it in the calculations, we take it equal to 1 in what follows.

Equation (12-7) tells a physical story and it has astronomical implications. The main characteristic of the emitted energy spectrum is its exponential behavior with E, which, when plotted on log-log coordinates, shows a sharp cutoff, not at all like the power-law behavior characteristic of other mechanism. This behavior, seen in Figure 12-2, is due physically to the fact that energetic photons come from energetic electrons, of which there are exponentially small numbers in a Maxwellian distribution. Normally, a Maxwellian distribution is set up by mutual elastic collisions between particles in the gas. According to Spitzer (1968, p. 94) the time required for an electron moving, say, at three times the thermal speed in a gas at $T = 10^8$ K to slow down to thermal speeds is about $4000/n_e$ years, which for $n_e > 10^{-3}$ cm^{-3} (as is the case in clusters) is $<4 \times 10^6$ years, a very short time on the cosmic scale. Hence we expect the electron gas to be Maxwellian, and the bremsstrahlung spectrum to behave like $e^{-E/kT}$; this is a major test of the hot gas hypothesis for the original of the X-rays.

We also see from equation (12-12) that the emissivity is proportional to the square of the local density, because two particles, an electron and an ion, are involved in the emission process. This means that the observed intensity is not proportional to the amount of gas present, but to its square, so that any density fluctuations result in a greater emission than would be the case for the same amount of uniformly distributed gas.

In order to assess this effect, we note that

$$\langle n_e \rangle = \frac{\langle n_e^2 \rangle^{1/2}}{C^{1/2}} . \qquad (12\text{-}13)$$

The observed emission determines the numerator via equation (12-11), so the total amount of gas, which is proportional to $\langle n_e \rangle$, is proportional to $C^{-1/2}$ for a fixed amount of emission. Any observation really determines only an upper limit on the mass of gas, derived by taking $C = 1$, since one could always account for the emission with less gas if $C > 1$. Since $\omega_{gas} = M_{gas}/M_{VT}$, we see that ω_{gas} is proportional to $C^{-1/2}$; alternatively, the observations determine the value of $C^{1/2} \omega_{gas}$.

The first question concerning X-rays from the Coma cluster is whether they fit the spectrum predicted by equation (12-11). Kellogg, Baldwin, and Koch (1975) have shown that they do, with $T = 1.0 \times 10^8$ K. Since the one-dimensional r.m.s. velocity of a particle at this temperature is 1160 km/sec,

Figure 12-2. The observed diffuse X-ray background is represented by the points, taken from Schwartz and Gursky (1973). The curve is a thermal bremsstrahlung spectrum calculated from the temperature variation in Figure 12-3, and adjusted for best fit to the data (Field, in preparation). The resulting present temperature is $T_0 = 2.6 \times 10^8$ K and the density parameter is $C^{1/2} \Omega_{gas} = 0.5$.

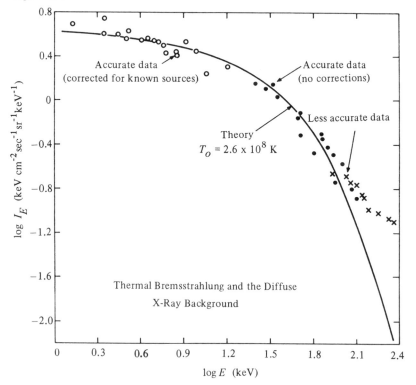

and the observed one-dimensional r.m.s. velocity of the galaxies in the center of the cluster (where most of the X-rays are generated) is observed to be 1060 km/sec, the particles should be distributed in the cluster gravitational field nearly like the galaxies. Knowing this, one can predict from equation (12-11) what distribution of brightness an X-ray telescope should see. Lea and colleagues (1973) showed that the observed distribution of X-rays on the sky fits the predicted values well. The mass of hot gas needed if $C = 1$ comes out to be $5 \times 10^{14} \, M_\odot$, which is 10% of the virial mass, so that $C^{1/2} \, \omega_{gas} = 0.10$. Therefore, hot gas is present in very significant amounts—about equal to the mass in galaxies if $C = 1$—but it does not suffice to bind the cluster. Note that if $C > 1$, this conclusion is strengthened. Examination of other clusters leads to similar conclusions. The central density of hot gas, 0.003 particles per cm³, is of the right order to explain the radio tails described previously.

This result that $\omega_{gas} < 1$ has stimulated the search for other components of gas which would not be visible at X-ray wavelengths. This is reasonable, in view of the fact that our own Galaxy contains components of gas with temperatures ranging from 20 K to 10^6 K living comfortably side by side (Field, 1975). This is possible because heat transfer is relatively ineffective in diffuse gases, and the sources of heat are irregularly distributed in the Galaxy. Radio astronomers have observed the Coma cluster at 21-cm wavelength to pick up any H I present, with negative results. Since H should be neutral for $T < 20,000$ K, this means there is not a significant amount of gas at those temperatures. Rocket experiments have sought Lyman-α line emission, which is expected if the temperature is between 10,000 K and 60,000 K, also with negative results. The soft X-ray data rule out significant missing mass with $T > 300,000$ K. This leaves a range from 60,000 to 400,000 K to be examined.

Both radio waves and visible light have actually been detected from among the galaxies in Coma. The radio waves may well be synchrotron radiation from electrons generated by the radio galaxies in the cluster, and the optical emission may well be from the outlying regions of galaxies. Hence this emission should be interpreted only as an upper limit for the radiation emitted by any hot gas in the cluster. Field (1974a) derived $C^{1/2} \, \omega_{gas} < 0.5$ from the optical emission. Field (1974b) considered the scattering of the light from galaxies in the cluster by free electrons in the intracluster gas. The observed optical light, interpreted in this way, shows that $\omega_{gas} < 0.2$; about half of the observed effect is due to the 10^8 K gas detected at X-ray wavelengths if $C = 1$. The conclusion is that the missing mass cannot be in the form of gas at any temperature. Hot gas, relativistic electrons, and perhaps dust, are all present. But the amounts are, alas, too small to account for the mass needed to bind the cluster.

Where does this lead? On the one hand, it leaves open the vexing question of the missing mass, for the observed galaxies and the hot gas together account for at most $12\% + 10\% = 22\%$ of the virial mass. Even if cooler gas is present in the maximum amount permitted by electron scattering, $\omega < 0.32$. On the other hand, it raises the question of the origin and evolution of the intracluster gas and its relation to galactic evolution. While the amount of gas present is insufficient to bind the cluster, it is not insignificant compared to the mass in galaxies. Where did it come from? Out of the galaxies? From intergalactic space? The answers to these questions should be enlightening.

Implications of Intergalactic Matter

How did galaxies form, and from what material? These questions are at the forefront of extragalactic astronomy and cosmology. One immediately thinks of the formation of stars in spiral galaxies like our own, where interstellar gas provides the required material, and where gravitational forces cause gas to collapse inward upon itself wherever the density exceeds a critical value. This picture has explained at least some of the features of the Galaxy.

The universe is different. There is not much gas evident at present; furthermore, the whole system is expanding, so that any gas present would tend to expand rather than to collapse. That these arguments cannot be taken at face value follows from evolutionary considerations. Our Galaxy is believed to be about 15×10^9 years old, as judged from the ages of the oldest stars, found in globular clusters. But the young stars which formed in spiral arms as recently as 1–100×10^6 years ago provide much of the light of the Galaxy, because they have not yet run out of nuclear fuel, as have the massive stars in globular clusters. Hence other spiral galaxies look young, in that their light is dominated by young stars, although they may be as old as the Galaxy. However, elliptical galaxies have no gas for current star formation and they do indeed look about as old as our globular clusters. Perhaps spirals as well as ellipticals formed about 15×10^9 years ago, and their different appearance at present is due to factors other than age. This is important because the universe is believed to be of the order of $1/H_0 = 20 \times 10^9$ years old; therefore, the universe would have been relatively young when the Milky Way and perhaps most other galaxies were formed. Hence conditions could have been much different at that time; in particular, there may have been much more intergalactic gas then, from which the galaxies could have formed.

What about the effects of expansion? Detailed calculations carried out by Lifshitz (1946) and reviewed by Peebles (1971) showed that this is not the problem that it first appeared to be. An expanding gas experiences the same gravitational instability as a static one, the only difference being that a clump of gas which becomes unstable at first continues to expand with the universe, but then reaches a maximum size, and finally collapses. In fact, the evolution of the clump mimics that of a closed Friedmann model (Figure 11-10 of the previous chapter). The theory demands only that there be fluctuations in the gas density amounting to a few percent for the instability to work upon. The overdense regions evolved to form galaxies, while the underdense regions continued expanding to form the intergalactic regions we see today.

When did the galaxies form? One can estimate a lower limit on the time they took to form by calculating their mean densities from observation. According to the picture sketched above, their present mean densities are somewhat larger than the densities of the clumps of gas which formed the galaxies at the time they reached their maximum size. The mean density of the Galaxy is about 3×10^{-24} g/cm³, so its density at maximum extension was less than this. One can calculate the time to reach maximum size, starting from a big-bang compressed state, by solving equation (11-36) of the previous chapter for the case of negative energy. We have

$$\dot{R}^2 = \frac{8\pi}{3} G\rho R^2 + \text{const} = \frac{2GM}{R} - \frac{2GM}{R_M}, \tag{12-14}$$

where we have chosen the constant so that $\dot{R} = 0$ when the clump of gas attains its maximum radius R_M. We let

$$R = R_M \sin^2 \theta, \tag{12-15}$$

where θ varies from 0 at the big bang to $\pi/2$ at the maximum extent. Then in terms of θ, equation (12-14) can be written as

$$\left(2R_M \sin \theta \cos \theta \, \frac{d\theta}{dt}\right)^2 = \frac{2GM}{R_M} \frac{\cos^2 \theta}{\sin^2 \theta}, \tag{12-16}$$

which has the solution

$$t(\theta) = \left(\frac{R_M^3}{2GM}\right)^{1/2} (\theta - \tfrac{1}{2} \cos 2\theta). \tag{12-17}$$

In particular, the time taken to reach maximum extent is

$$t_M = t(\pi/2) = \left(\frac{\pi^2 R_M^3}{8GM}\right)^{1/2} = \left(\frac{3\pi}{32G\rho_M}\right)^{1/2}. \tag{12-18}$$

Since $\rho_M < 3 \times 10^{-24}$ g/cm^3, it follows that $t_M > 4 \times 10^7$ years. It is reassuring that this time comes out greater than the time of decoupling from the blackbody background radiation, because theoretical work shows that galaxies would have been prevented from forming before decoupling by the disruptive effect of radiation pressure (which was not included in the above calculation).

On the other hand, we know that the Galaxy cannot have formed very late in the evolution of the universe, or it would be younger than it is observed to be. If we adopt an Einstein-de Sitter model, the present age of the universe is $\tfrac{2}{3}H_0^{-1} = 13.3 \times 10^9$ years. The globular clusters are believed to be at least 12×10^9 years old, so that one can surmise that the Galaxy formed within 10^9 years after the big bang. (This argument is not very accurate because it depends on the difference of two uncertain numbers.) We conclude that the Galaxy formed between 4×10^7 and 1×10^9 years after the big bang. A reasonable estimate would be 2×10^8 years after the big bang.

One can in principle test this by looking out so far that one sees objects whose age when the light left them was of this order, and seeing whether there are young galaxies among them. According to equation (11-42) of the previous chapter, the scale factor of the universe at this early time is given by

$$\frac{R}{R_0} = (\tfrac{3}{2}H_0 t)^{2/3} = 0.061, \tag{12-19}$$

and, from equation (11-33), this corresponds to a redshift

$$z = \frac{R_0}{R} - 1 = 15.4. \tag{12-20}$$

That is, the wavelengths of light from objects seen 2×10^8 years after the big bang would be 16.4 times the rest wavelengths. This would put the radiation into the infrared, where very sensitive measurements have not yet been carried out, so it is an open question whether galaxies exist at such distances. Peebles (1971) describes attempts to find young galaxies, in the context of a discussion of galaxy formation.

So far, the most distant galaxy found is 3C 123, with a redshift of 0.637 (Spinrad, 1975). But quasars are seen with high redshifts, the record being that of OQ 172, with $z = 3.53$ (Wampler et al., 1973). Most astronomers believe that the quasar redshifts are cosmological, and that quasars are an evolutionary phase of galaxies; hence this would argue that galaxies exist at least out to $z = 3.5$.

What about clusters of galaxies? The smoothed-out density within 3.9 Mpc of the center of the Coma cluster is 1.3×10^{-27} g/cm^3 (if the mass is 5×10^{15} $M_\odot$, as indicated by the virial theorem). Applying the same argument as above, one finds that the cluster must have formed more than 1.8×10^9 years after the big bang. It seems quite possible that the clusters formed after its constituent galaxies did, as a gravitational condensation in the number of galaxies in that region of space. This point of view is consistent with the results of the last chapter, which showed that the density fluctuations in the universe are strongest on the small scales. This would imply that galaxies form first, and later, clusters of galaxies form.

A prime question concerns the efficiency of galaxy formation. Let us define ϵ as the mass going into galaxies, divided by the total mass available, so that $\epsilon = \Omega_{gal}/\Omega$. Then $1 - \epsilon$ is the fraction of material left behind as intergalactic matter. In the case of star formation, one observes massive dense interstellar gas clouds which seem destined to collapse, and one also observes young clusters of stars like the Pleiades, which appear to have been formed in the recent past by the collapse of just such clouds. Both the dense clouds and the resulting clusters are estimated to have masses of the order of $300\ M_\odot$, so that ϵ appears to be of order unity for star formation in the Galaxy on observational grounds. However, this estimate of ϵ is very uncertain.

Unfortunately, theory is not much help on this question, as the processes involved are so complex. In the simplest case, a uniform gas cloud at rest collapsing to form a single object, one can convince oneself that $\epsilon = 1$. But usually gas clouds are not uniform: the inner, denser regions will collapse first, leaving the low-density outer regions behind. The gas in these regions would also fall in ultimately ($\epsilon = 1$) if the cloud were not rotating, nor affected by magnetic stresses, nor pushed away by the radiation of the newly formed object at the center. All of these effects are suspected to occur, with the result that probably $\epsilon < 1$, but one cannot estimate how much smaller it is.

The same problem exists with galaxy formation. Oort (1969) attempted to estimate ϵ from the requirement that in any region collapsing to form a galaxy the only gas which will make it into the galaxy is that endowed with a favorable initial velocity, the rest remaining behind in intergalactic space. His result, $\epsilon \sim 0.06$, cannot be completely trusted, as his arguments have not yet been confirmed by detailed hydrodynamic calculations.

In addition to the matter left behind, there will be some matter ejected into intergalactic space by galaxies. How might this happen? Any gas lost by stars in the disk of the galaxy (including novae, supernovae, planetary nebulae, red giants, and T Tauri stars, all of which are losing mass to the interstellar medium) quickly runs into fairly dense interstellar gas already in the disk, and shocks it to high temperature, slowing down in the process to become part of the interstellar medium. A possible exception to this is supernova shells, which move faster than 100 km/sec and hence shock the gas to temperatures exceeding 10^6 K, where radiative cooling is very inefficient. It seems possible (Cox and Smith, 1974; Field, 1975; Shapiro and Field, 1975) that such gas will percolate through the interstellar medium until it finds a channel to the surface of the disk, where its high temperature will cause it to flow into the halo of the galaxy. Conceivably, if some of it is hot enough, it will escape. Spitzer (1956) showed that gas hotter than 4×10^6 K cannot be permanently bound in the galaxy, as its thermal velocity exceeds the velocity of escape.

The situation in elliptical galaxies is different in two respects. There, the stellar orbits are quite unlike those in a spiral, where the orbital planes are closely aligned in a plane. Rather, the orbital planes are oriented at random, so that any one star may be moving several hundred km/sec with respect to other stars in its neighborhood. Hence any gas streams from neighboring stars will crash together at high speeds, even if the gas does not leave the star at high speed. This will heat the gas to high temperatures, such that the r.m.s. speed of particles in the gas will approach that of the stars in the galaxy. Because the gas density in ellipticals is observed to be low, the time required for the heat to be radiated away (which depends inversely upon the density) will be long. Hence large masses of hot gas will accumulate. (Note that the temperature imparted to the gas is not sufficient to cause escape from the galaxy, because the velocity of the stars is less than escape velocity.)

The hot gas will be heated further by any supernovae which go off in the galaxy. In elliptical galaxies there is no cool gas to impede its progress, and the hot gas may well escape. This possibility was explored by Mathews and Baker (1971), who found that the predicted behavior depends upon the relative amount of mass injected (primarily from planetary nebulae) and energy supplied by supernovae. If the energy/mass ratio is low, the hot gas is held in by the gravity of the galaxy; it slowly increases in density and, when it reaches a high enough density, it cools off and falls toward the center of the galaxy, forming a cloud of $\sim 10^7 \, M_\odot$ at $T \sim 10^4$ K. In fact, such clouds are seen at the centers of some ellipticals, but not all of them.

Alternatively, if the energy/mass ratio is high, the temperature of the gas increases before radiative cooling can set in, and finally the particles reach escape velocity. The result is a galactic wind, with the gas emitted by planetaries and heated by supernovae escaping from the galaxy altogether.

It has been estimated by Larson and Dinerstein (1975) that the mass loss in elliptical galaxies amounts to about 0.1 to 0.3 of the remaining galactic mass. For the Coma cluster, with $6 \times 10^{14} \, M_\odot$ in the galaxies, this would result in the loss of 6×10^{13}–$2 \times 10^{14} \, M_\odot$ of gas. What would happen to this gas? The gas stream from each galaxy will meet those from other galaxies at collision velocities comparable to the random velocities in the cluster, ~ 1000 km sec^{-1}, and thus be heated to temperatures approaching 10^8 K. At these enormous temperatures, radiative cooling is ineffective even over times as long as 10^{10} years, so we would still expect to see the gas at 10^8 K. This is just what is indicated by the X-ray observations. The predicted amount of gas, 0.6–$2 \times 10^{14} \, M_\odot$ can be compared with the mass of hot gas indicated by the X-ray observations, which is $5 \times 10^{14} \, C^{-1/2} \, M_\odot$. The gas ejected from the galaxies can account for that seen by the X-ray astronomers only if $C > 1$.

Once the gas has left the galaxies and is broadly distributed in the cluster, it is subject to further heating by a number of processes. One of the more effective ones is a process called dynamical friction, in which the more rapidly moving galaxies gravitationally attract the neighboring gas, causing it to converge, compress, and heat up as the galaxy passes by. After the galaxy is gone, the heat imparted causes the gas to expand. According to Yahil and

Ostriker (1973), this can supply enough energy to lift a considerable mass of material out of the gravitational potential well of the cluster. In short, such heating may cause a cluster wind, just as supernova heating can drive a galactic wind. The result would be that, over time, the amount of gas in intergalactic (as opposed to intracluster) space would increase.

Quite a different point of view as to the origin of the cluster gas has been proposed by Gunn and Gott (1972). They start with the proposition that a certain amount of gas is still left in intergalactic space between clusters ($\epsilon < 1$), and inquire into its history, given the gravitational effect of clusters that happen to be nearby. If one takes the gas to be cold, so there are no pressure forces to speak of, and if one also ignores magnetic forces, the gas moves solely under the influence of gravity, which, in the neighborhood of a cluster of galaxies, Gunn and Gott take to be contributed by the cluster alone. Hence one has a problem in classical ballistics. The gas originally is moving away from the cluster in its expansion with the universe. Gradually the gravitation of the cluster slows it down, reverses the flow, and accelerates it into the cluster itself. Solving this dynamical problem, Gunn and Gott conclude that if the amount of gas in intergalactic space were given by $\Omega_{gas} = 1$, more than $10^5 \ M_\odot$/year would flow into the cluster; in the $\sim 10^{10}$ year life of the cluster, this would result in more gas than is now observed there. They conclude from this that Ω_{gas} must be less than unity; in particular, $\Omega_{gas} = 0.05$ if this type of inflow accounts for the hot gas observed, and it is even less if the gas actually originates elsewhere (for example, in the galaxies). This argument cannot be accepted at face value, because it is possible that a cluster wind of the type discussed above would hold any intergalactic gas at bay.

How is one to decide whether the observed gas in Coma fell into the cluster from outside, or was ejected from galaxies inside the cluster? A key desideratum would be the abundance of heavy elements in the gas. Intercluster gas would be lacking in elements heavier than hydrogen or helium, because such elements are not made in the big bang, while if the gas originates in galaxies, it should exhibit more normal abundances of heavy elements as a result of stellar evolution. Larson and Dinerstein (1975) calculated a heavy element abundance equal to 0.02–0.03 by mass, much like the value for interstellar gas in the Galaxy. Unfortunately, in a gas at 10^8 K all of the abundant heavy elements will be almost completely stripped of their electrons, and so most characteristic spectral lines will be absent. One possibility for observation is the resonance line of the single-electron atom O VIII at 652 eV. One calculates that about 10^{-3} of the oxygen would be in this ionization state at 10^8 K, and that if the O/H ratio is normal, the optical depth of the line in absorption would be about unity.

The presence of heavy elements in the Coma cluster gas is very tentatively suggested by the observations referred to earlier, which indicate that dust may be present between the galaxies. I estimate from the magnitude of the effect (if it is indeed real), if the abundances of heavy elements are solar (2% by mass), and if all the heavy elements are in grains, that at least $1.5 \times 10^{14} \, M_\odot$ of gas are required to account for the effect.

Gas between Clusters

We have seen in the previous chapter that the abundance of deuterium requires that $\Omega < 0.2$. On the other hand, Ω_{gas} might actually be unity, with the result that the universe is closed, although this would be possible only if the deuterium we observe was actually created by some process other than the big bang. As explained in the previous chapter, this could be the case if various proposals concerning the origin of deuterium in massive stars prove to be correct.

Just as in the case of clusters, various attempts have been made to detect neutral and low-temperature ionized gas of very low density between the galaxies. (For reviews see Field, 1972, 1974a). All of these attempts have failed. One of the most significant observations in this area was the attempt to discover broadly distributed neutral hydrogen atoms by searching for their absorption effect at the Lyman-α resonance line in the spectrum of quasars of high redshift.

Any atoms near the quasar would be receding from us with nearly the same velocity as the quasar, and if the redshift of the quasar exceeds 2, the Lyman-α line is shifted from 1216 Å in the extreme ultraviolet to more than 3648 Å in the blue part of the spectrum. Hence the relevant observations can be made with ground-based telescopes. The observations show no absorption, permitting one to place the extraordinarily strong limit,

$$\Omega(\text{H I}) < 3 \times 10^{-7}, \tag{12-21}$$

at these great distances in the universe. It would be extremely surprising if galaxy formation were actually as efficient as this would imply, so one interprets this rather to mean that any gas present is ionized and hot, above 10^6 K. This is at least consistent with what we know of the gas in clusters.

More recent observations with high spectral resolution have disclosed large numbers of sharp absorption lines in quasar spectra, many of them identified as Lyman-α. Presumably these lines are caused by clouds of cool, relatively neutral gas, lying between us and the quasars, but their origin and location is controversial at the present time. Some investigators believe that the clouds have been ejected at high speed from the quasar itself, while others believe that at least some of the clouds are in objects, such as protogalaxies or galaxies, which happen to lie between us and the quasar, but which are too distant to be photographed with present equipment. Whatever the case, the amounts of neutral gas are small, and unless they are accompanied by large amounts of ionized gas, they are insignificant in terms of Ω.

On the other hand, there does exist a diffuse background of X-ray intensity covering the entire energy range from <1 keV to ~ 1000 keV (Schwartz and Gursky, 1973), which may be associated with true intergalactic gas having $T = 10^6$–10^9 K (Field and Henry, 1964). This background has been

Figure 12-3. The temperature of the intergalactic medium calculated by Field (in preparation), and plotted in units of the present temperature T_0. The abscissa is redshift z, defined by $z = (R_0/R) - 1$, where R is the scale factor of the universe. It was assumed that quasars turn on suddenly at $z = 3$ ($R = R_0/4$) and that they subsequently decay like $(R/R_0)^{-6}$, as indicated by counts of quasars. At first the quasars heat the gas, but ultimately the gas cools as a result of the cosmological expansion.

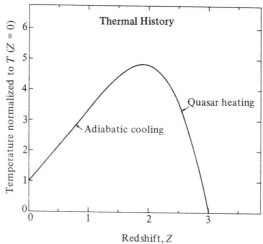

considered recently by Field (in preparation), who shows that the observed spectrum can be fitted with a gas having a present temperature of 2.6×10^8 K and a density corresponding to $C^{1/2} \, \Omega_{gas} = 0.2$–$2$ (Figure 12-2). To account for the high temperature, it is proposed that energy is injected into space by quasars in the form of fast, nonrelativistic particles. Consequently, each galaxy would have to emit $2 \times 10^{64} \, \Omega_{gas}$ erg during its quasar phases in order to account for the high temperature. The huge energy input is required in part because the cosmological expansion tends to cool the gas, just as would any expansion of a gas (Figure 12-3). This amount of energy is $2000 \, \Omega_{gas}$ times the amount of energy known to be ejected from quasars as relativistic electrons, as deduced from their radio and optical properties. If $\Omega_{gas} = 1$, it corresponds to 1% of the rest mass of the galaxy. In view of these facts, it seems unlikely—but not out of the question—that the X-rays originate in a gas with $C = 1$ and $\Omega_{gas} = 0.5$. If $C > 1$, they could be emitted by a gas with much smaller Ω. For example, one could account for the X-rays in this manner if $\Omega_{gas} = 0.05$ and $C = 100$. The energy requirements would then be 10 times less, and perhaps acceptable.

Summary

About 10% of the mass of the Coma cluster is hot gas at $\sim 10^8$ K. This gas may have originated in the cluster galaxies, or it may have fallen in from between the clusters. In the former case, the galaxies must be quite efficient at losing mass; in the latter case, one can estimate that intergalactic gas makes a significant contribution to the density of the universe and that $\Omega_{gas} \simeq 0.05$. Choosing between these alternatives may ultimately require high-resolution searches for X-ray absorption lines.

There may be hot gas between the clusters as well. If one accepts the constraint that $\Omega < 0.2$ from the deuterium abundance, one can account for the observed X-ray background between 1 and 100 keV with a gas having a present temperature of $T_0 = 2.6 \times 10^8$ K, $\Omega_{gas} = 0.05$, and $C = 100$, but enough intergalactic gas to close the universe cannot be ruled out. This gas may have been heated by quasar explosions.

References

Allen, C. W. 1973. *Astrophysical Quantities,* 3rd edition, The Athlone Press, London.

Brecher, K., and Burbidge, G. R. 1970. The diffuse cosmic X-ray background radiation, *Comments Astrophys. Space Phys. 2,* 75–83.

Cox, D. P., and Smith, B. W. 1974. Large-scale effects of supernova remnants on the Galaxy: Generation and maintenance of a hot network of tunnels, *Astrophys. J. (Letters) 189,* L105–108.

Field, G. B. 1972. Intergalactic matter, *Annu. Rev. Astron. Astrophys. 10,* 227–260.

Field, G. B. 1974a. Intergalactic gas, in M. S. Longair, ed., *Confrontation of Cosmological Theories with Observational Data, IAU Symp. 63,* pp. 13–28, D. Reidel, Boston, Mass.

Field, G. B. 1974b. Hot gas in and between galaxies, presented at IAU Colloquium No. 27, *Ultraviolet and X-Ray Spectroscopy of Astrophysical and Laboratory Plasmas,* Cambridge, Mass., Sept. 9, 1974.

Field, G. B. 1975. Heating and ionization of the interstellar medium; Star formation, in J. LeQuex and R. Balian, eds., *Atomic and Molecular Physics and the Interstellar Matter,* pp. 467–532, American Elsevier, New York.

Field, G. B., and Henry, R. C. 1964. Free-free emission by intergalactic hydrogen, *Astrophys. J. 140,* 1002–1012.

Gunn, J. E., and Gott, J. R. III 1972. On the infall of matter into clusters of galaxies and some effects on their evolution, *Astrophys. J. 176,* 1–20.

Hollenbach, D. J., Werner, M. W., and Salpeter, E. E. 1971. Molecular hydrogen in H I regions, *Astrophys. J. 163,* 165–180.

Karachentsev, I. D., and Lipovetskii, V. A. 1968. Absorbing material in clusters of galaxies, *Astron. Zh. 45,* 1148–1152. (English translation in *Sov. Astron. A-J 12,* 909–912, 1969.)

Kellogg, E. 1974. Extragalactic X-ray sources, in R. Giacconi and H. Gursky, eds. *X-Ray Astronomy,* pp. 321–357, Reidel, Dordrecht-Holland, Boston, Mass.

Kellogg, E., Baldwin, J. R., and Koch, D. 1975. Studies of cluster X-ray sources: Energy spectra for the Perseus, Virgo, and Coma clusters, *Astrophys. J. 199,* 299–306.

Kerr, F. J., and Westerhout, G. 1965. Distribution of interstellar hydrogen, in A. Blaauw and M. Schmidt, eds., *Stars and Stellar Systems,* vol. 5, *Galactic Structure,* pp. 167–202, University of Chicago Press, Chicago.

King, I. R. 1972. Density data and emission measure for a model of the Coma cluster, *Astrophys. J. (Letters) 174,* L123–124.

Larson, R. B., and Dinerstein, H. L. 1975. Gas loss in groups of galaxies, *Publ. Astron. Soc. Pacific 87,* 911–915.

Lea, S. M., Silk, J., Kellogg, E., and Murray, W.   1973. Thermal bremsstrahlung interpretation of cluster X-ray sources, *Astrophys. J. (Letters) 184,* L105–111.

Lifshitz, E. M.   1946. *Sov. J. Phys. 10,* 116–122.

Mathews, W. G., and Baker, J. C.   1971. Galactic winds, *Astrophys. J. 170,* 241–259.

Miley, G. K., Wellington, K. J., and van der Laan, H.   1975. The structure of the radio galaxy NGC 1265, *Astron. Astrophys. 38,* 381–390.

Oort, J. H.   1965. Stellar dynamics, in A. Blaauw and M. Schmidt, eds., *Stars and Stellar Systems,* vol. 5, *Galactic Structure,* pp. 445–511, University of Chicago Press, Chicago.

Oort, J. H.   1969. Infall of gas from intergalactic space, *Nature 225,* 1158–1163.

Page, T.   1962. M/L for double galaxies; A correction, *Astrophys. J. (Letters) 136,* L685–686.

Peebles, P. J. E.   1971. *Physical Cosmology,* Princeton University Press, Princeton, N.J.

Reaves, G.   1974. Distribution of clusters of galaxies, *Astron. Zh. 51,* 520–525 (English translation in *Sov. Astron. 18,* 307–310).

Rood, H. J., Page, T. L., Kintner, E. C., and King, I. R.   1972. The structure of the Coma cluster of galaxies, *Astrophys. J. 175,* 627–647.

Schwartz, D., and Gursky, H.   1973. The X-ray emissivity of the universe, in F. W. Stecker and J. I. Trombka, eds., *Gamma-Ray Astrophysics,* pp. 15–36, NASA SP-339, Washington, D.C.

Shapiro, P. R., and Field, G. B.   1975. Consequences of a new hot component of the interstellar medium, submitted to *Astrophys. J.*

Smart, N. C.   1973. The origin of the absorption in the centre of the Coma cluster, *Astrophys. Letters 14,* 233–235.

Spinrad, H.   1975. 3C 123: A distant first-ranked cluster galaxy at $z = 0.637$, *Astrophys. J. (Letters) 199,* L3–4.

Spitzer, L.   1956. On a possible interstellar galactic corona, *Astrophys. J. 124,* 20–34.

Spitzer, L.   1968. *Diffuse Matter in Space,* Interscience, New York.

Wampler, E. J., Robinson, L. B., Baldwin, J. A., and Burbidge, E. M.   1973. Redshift of OQ 172, *Nature 243,* 336–337.

Wellington, K. J., Miley, G. K., and van der Laan, H.   1974. High-resolution map of NGC 1265, *Nature 244,* 502–504.

Yahil, A., and Ostriker, J. P.   1973. Winds and X-rays from clusters of galaxies, *Astrophys. J. 185,* 787–795.

Zwicky, F.   1962. New observations of importance to cosmology, in G. McVittle, ed., *Problems in Extragalactic Research,* pp. 347–358, Macmillan, New York.

Contributors
Index

Contributors

Unless otherwise indicated, each author's academic position is with Harvard University; the research positions of Physicist, Astrophysicist, and Radio Astronomer are with the Smithsonian Astrophysical Observatory. The Smithsonian Astrophysical Observatory and the Harvard College Observatory together comprise the Center for Astrophysics, Cambridge, Massachusetts.

William R. Ward — Research Associate, Harvard College Observatory

Robert W. Noyes — Professor of Astronomy; Physicist; Associate Director for Solar and Stellar Physics, Center for Astrophysics

Stephen E. Strom — Astronomer, Kitt Peak National Observatory, Tucson, Arizona

Alastair G. W. Cameron — Professor of Astronomy; Associate Director for Planetary Sciences, Center for Astrophysics

Herbert Gursky — Professor of the Practice of X-Ray Astronomy; Astrophysicist

Giovanni G. Fazio — Lecturer on Astronomy; Physicist

Eric J. Chaisson — Assistant Professor of Astronomy

Alexander Dalgarno — Chairman of the Department of Astronomy; Professor of Astronomy; Physicist; Associate Director for Theoretical Astrophysics, Center for Astrophysics

James M. Moran — Lecturer on Astronomy; Radio Astronomer

Kenneth Brecher — Assistant Professor of Physics, Massachusetts Institute of Technology

Marc Davis — Assistant Professor of Astronomy

George B. Field — Paine Professor of Practical Astronomy; Physicist; Director of the Harvard College Observatory and the Smithsonian Astrophysical Observatory

Index